Reliability
Handbook

INTERNATIONAL SERIES IN DECISION PROCESSES

INGRAM OLKIN, Consulting Editor

A Basic Course in Statistics, 2d ed., T. R. Anderson and M. Zelditch, Jr.
Introduction to Statistics, R. A. Hultquist
Applied Probability, W. A. Thompson, Jr.
Elementary Statistical Methods, 3d ed., H. M. Walker and J. Lev
Reliability Handbook, B. A. Kozlov and I. A. Ushakov (edited by J. T. Rosenblatt and L. H. Koopmans)
Fundamental Research Statistics for the Behavioral Sciences, J. T. Roscoe
Statistics: Probability, Inference, and Decision, Volumes I and II, W. L. Hays and R. L. Winkler
An Introduction to Probability, Decision, and Inference, I. LaValle

B. A. Kozlov and I. A. Ushakov

Reliability Handbook

Edited by
L. H. Koopmans
University of New Mexico

Judah Rosenblatt
Case Western Reserve University

Translated by
Lisa Rosenblatt

HOLT, RINEHART AND WINSTON, INC.
New York · Chicago · San Francisco · Atlanta · Dallas
Montreal · Toronto · London · Sydney

Copyright © 1970 by Holt, Rinehart and Winston, Inc.
All Rights Reserved
Library of Congress Catalog Card Number: 79-85752

SBN: 03-081417-0

Printed in the United States of America

0   1   2   3   22   1   2   3   4   5   6   7   8   9

72⸍ 14040

PREFACE

The interest of a broad range of engineers in reliability theory and in computational methods for reliability requires the publication of handbooks as well as educational texts and special monographs. Handbooks, of course, may provide very different kinds of information, such as physical and technical characteristics of standard devices, computational formulas, and descriptions of products on the market with an indication of their possible uses.

This handbook contains both computational formulas and many illustrative examples of reliability computations. These examples introduce the reader to typical situations in which the formulas are used.

In the preparation of a handbook such as this, one is confronted with the problem of choosing from the literally boundless literature that which is actually necessary without omitting topics that are important but not strictly essential. One may or may not agree with the choices made by the authors, which are inevitably determined by their own areas of interest. In my opinion, for example, the authors are too concerned with stand-by problems of repairable systems, and the number of formulas given by them throughout the book could be reduced. This reduction should be accomplished by eliminating many special cases. At the same time, it is often very important in practical applications to have specialized formulas readily available. Thus, the restriction of the material to formulas of only the most general nature could give rise to unnecessary complications.

I think that the authors did well by leaving out of the handbook many results that are interesting and illuminating but are still not well developed. Thus, in problems of stand-by for repairable systems the authors restrict consideration to the case of elementary flow of failures and the exponential distribution law for the duration of restoration. The inclusion of isolated, disconnected results would have been valuable, but would at the same time have called forth feelings of frustration that there are still no formulas in other interesting cases. Moreover, there is the risk that results obtained for only a particular problem might unjustifiably be extended to cover a broader situation.

It is certain that this handbook is not free of defects. It should be considered, however, that the authors are pioneers in this difficult endeavor, and those in the forefront have the right to isolated deficiencies, to imperfection in their work. It is self-evident that constructive criticism by the readers will make it possible to improve the contents of this handbook and to find a more suitable design for it. I hope that even in this form the handbook will be of use.

B. Gnedenko

FROM THE AUTHORS

At the present time there are many scholarly books on the problem of reliability, of which first mention is due to the one by B. V. Gnedenko, Yu. K. Belyaev and A. D. Solov'ev, *Mathematical Methods in Reliability Theory;* also Ya. B. Shor, *Statistical Methods of Analysis and Control of Quality and Reliability;* the translated book of D. Lloyd and M. Lipow, *Reliability: Organization of Research, Methods, Mathematical Equipment;* and the book by R. Barlow and F. Proschan (with contributions by L. Hunter), *Mathematical Theory of Reliability*, in production by the Soviet Radio publishing house. However, all these books, which have many merits that this handbook is lacking, are designed for a thorough study of the subject as preparation for qualified professional activity in the field of reliability, and not for immediate application of formulas.

This handbook is intended for specialists working in the field of reliability, and is designed as a practical manual for the everyday work of ensuring reliability of radioelectronic equipment and various complicated systems. Although all the material presented in this handbook is of the cookbook type, the authors want to warn the reader against erroneous, simple-minded, and thoughtless use of these results. This is not a magical book that opens itself to the desired page and presents the reader with whatever solution he needs for any problem. The reader is assumed to have full knowledge of reliability theory. Most of all, the reader must be able to formulate his own problem properly and then find an adequate mathematical model for it in this handbook. The handbook basically deals with methods for a priori computation of the measure of reliability, and also with information on processing experimental data.

In Chapter 1 the basic concepts of reliability are presented and the most important measures of reliability of repairable and unrepairable structures are explained. The authors use generally accepted terminology wherever possible, drawing from the work of the Committee on Terminology of the USSR Academy of Sciences and the Leningrad Section of Reliability of the Scientific and Technical Society of Radio Engineering and Telecommunications named after A. S. Popov. However, in some cases they were forced to use different terminology that, in their opinion, was more suitable for this work.

In Chapter 2 basic measures of reliability for a unit, in the broad sense of the word, are examined.

The next three chapters treat methods for computing for stand-by systems. In Chapter 3 methods are given for computing for standard unrepairable systems, and in Chapter 4, methods for computing for standard repairable systems.

In Chapter 5 certain special stand-by problems are investigated that are most frequently encountered in analyzing reliability of complex systems. Undoubtedly the scope of problems here is far from being restricted to the models given, and the authors hope to obtain from the readers new formulations of problems that are dealt with in practical computations.

In Chapter 6 methods are given for computing for series systems whose separate units may or may not have stand-by units.

At the basis of the majority of the mathematical models considered in this handbook for repairable systems is the exponential law of distribution of the time of reliable operation and repair time. As practical work with complex radioelectronic equipment shows, the distribution of time of reliable operation is indeed well described by the exponential law. At this time no basic hypotheses for the distribution of repair time have been seriously proposed, although it is almost obvious that the exponential law a fortiori corresponds poorly to reality. However, as is known from analogous results from queueing theory, if the mean repair time is much less than the mean operating time, and the operating time has an exponential distribution, then the nature of the distribution of repair time does not play an essential role. Since this situation is most typical in practice, the results obtained in Chapters 4, 5, and 6 can be used for a broader class of systems under the assumption above.

In Chapter 7 general methods are given for estimating the efficiency of operation (operational efficiency) of complex systems where failure of separate units leads not to complete failure of the system as a whole but only to a certain deterioration of the quality of operation.

In Chapter 8, which deals with problems of optimal utilization, two types of problems are investigated: optimal stand-by and optimal search for faults. For the first problem, we give an exact method based on one of the modifications of the method of dynamic programming, and one of the possible approximate methods of solution (steepest descent) is given; this is the one considered by the authors to be the most intuitive and simple and at the same time sufficiently accurate for practical computations. The authors had no intention of touching on the problems of methods of computation and of optimal design of preventive maintenance since, from their viewpoint, there are at present no clear-cut methods for solving such problems that could be recommended for practical work.

In Chapter 9 some elementary methods are presented for processing statistical data obtained from special performance tests of radioelectronic equipment.

Some tables that are of general use for the entire handbook, and a series of mathematical formulas, are given in the Appendixes together with a brief exposition of basic derivations of probability theory and mathematical statistics.

The bibliography at the end of each chapter merely indicates appropriate books touching on the problems in that chapter. A much more extensive bibliography on reliability, probability theory, and mathematical statistics is given at the end of the handbook.

The monographs and collections of papers in the end-of-chapter and general bibliographies are all in Russian. The authors did not consider it possible to refer the reader of the handbook to all of the numerous domestic and foreign periodicals containing articles on reliability. Those interested in a more detailed bibliography for reliability can read the long survey paper by B. R. Levin and I. A. Ushakov, "Some aspects of the present state of reliability," *Radiotechnika*, **20**:4 (1965).

In writing this handbook, the authors aimed at making it possible to use the separate chapters and sections independently. However, the reader should first, if only cursorily, look it over as a whole.

The authors believed it necessary to present all the material with a unified attitude and design, and therefore much of it is original with them. Moreover, approximate methods for computing measures of reliability and error estimates for the computations are of practical interest, and in the majority of cases these too are presented for the first time.

In this book the authors did not succeed in making essential use of any "plug-in" results, since the majority of these pertained only to isolated special problems treated from different points of view, basically in periodicals and scattered through numerous domestic and foreign publications. The authors made direct use of material only from the proofs of the book by B. V. Gnedenko, Yu. K. Belyaev, and A. D. Solov'ev,

Mathematical Methods in Reliability Theory, which the authors kindly made available. Certain results and tables were also taken from the book by Ya. B. Shor, *Statistical Methods of Analysis and Control of Quality and Reliability*.

The authors would also like to mention the influence on the preparation of this handbook of the lectures and seminars on reliability held at Moscow State University that have been going on for several years. These lectures and seminars accomplished what probably would have been impossible to achieve otherwise: the turning out of scores of books on reliability.

The authors are very much indebted to B. V. Gnedenko, Yu. K. Belyaev, and A. D. Solov'ev for the attention they gave the manuscript and particularly their suggestions for its basic construction, which led to the reduction of numerous and sometimes obscure results into compact tables.

During the entire effort, encouragement and help were also given by B. R. Levin and Ya. B. Shor, who offered much valuable advice both on the general construction of the book and the writing of the individual parts, and then undertook the difficult task of reviewing the manuscript.

It is difficult to enumerate all to whom the authors are obliged for advice, support, and encouragement during the writing of the handbook. The latter was especially needed at a certain period of time. However, the authors would like to mention the interest shown by I. N. Kovalenko, who discussed some of the results with us, and I. I. Morozov and Ya. M. Sorin, who made some useful remarks.

As far as the authors know, this type of reference manual does not exist in the domestic or foreign literature. This made their task even more difficult, and as a result there may be mistakes and methodological errors, which are almost inevitable in such a situation. The authors will be happy to receive all suggestions and remarks on all the deficiencies of the book.

B. Kozlov, I. Ushakov

CONTENTS

Notation

$a_k(s)$	Laplace transform of the function $P_k(t)$
$f(t)$	density of the distribution function $F(t)$ or frequency of failures of the structure
$f_r(t)$	density of the distribution function $F_r(t)$
$F(t)$	distribution function of the random variable ϑ
$F_r(t)$	distribution function of the random variable ξ
H_k	state of a system, characterized by the fact that k of its units are in a failed state
$H_{i,j,\ldots,k}$	state of a system characterized by the fact that the units $i, j, \ldots, k$ are in a failed state
$H(t)$	state of a system at time t
$k(t)$	idleness coefficient of a simple structure
k	nonstationary idleness coefficient of a simple structure
K	coefficient of readiness of a structure
$K(t)$	nonstationary coefficient of readiness of a structure
p_k	stationary probability that a system is in state H_k
$p_k(t)$	probability that a system is in state H_k at time t
$P(t, t + t_0)$	conditional probability of proper operation of a structure in the time interval $[t, t + t_0]$ given proper operation at time t
$P(t_0)$	abbreviation for $P(0, t_0)$
$Q(t, t + t_0)$	conditional probability of failure of a structure in the time interval $[t, t + t_0]$ given proper operation at time t
$P\{A\}$	probability of the event A
$Q(t_0)$	abbreviation for $Q(0, t_0)$
$R(t, t_0)$	nonstationary coefficient of reliability of a structure in the time interval $[t, t + t_0]$ (also called the interval reliability)
s	the argument in the Laplace transform
$S(t, t_0) = 1 - R(t, t_0)$	the probability of failure-free operation in $[t, t + t_0]$
$R(t_0)$	coefficient of reliability of a structure
t_0	required time for proper operation of a structure
T_i	mathematical expectation of the random variable ϑ_i or mean time of proper operation of a structure from $(i - 1)$st repair up to the ith failure
T_1	mathematical expectation of the random variable ϑ or mean time of operation of a structure up to (the first) failure
T_2	mathematical expectation of the variable ϑ_2 or mean time of proper operation of a structure between the first and second failures; most often simply the mean operating time of a structure between failures

T	the general notation for the quantities T_1 and T_2 when $T_1 = T_2 = T$
$\tilde{Z}$	an approximate expression for Z
$\hat{Z}$	random variable or statistical estimator corresponding to the probability characteristic or parameter Z
$\gamma = \lambda/\mu$	or in the more general case $\gamma = \tau/T$
δ	absolute error of an approximate formula
δ_+	absolute error of an approximate formula yielding an underestimate compared to the true value
δ_-	absolute error of an approximate formula yielding an overestimate compared to the true value
θ_i	random duration of the time interval of failure-free operation between the $(i - 1)$st and ith failures

$$\Theta_k = \frac{\Lambda_0 \Lambda_1 \cdots \Lambda_{k-1}}{M_1 M_2 \cdots M_k}$$

$\lambda(t)$	conditional probability density of failures of a structure at time t, or the intensity of failures
Λ_j	intensity of passage of a system from state H_j to state H_{j+1}
$\mu(t)$	conditional probability density of repair of a structure at time t, or intensity of repair
M_j	intensity of passage of a system from state H_j to state H_{j-1}
ξ	random duration of repair time
τ	mathematical expectation of the random variable ξ (mean repair time)

Reliability
Handbook

1

DEFINITIONS AND CONCEPTS OF RELIABILITY MEASURES OF STRUCTURES

1.1 BASIC TERMS AND CONCEPTS OF RELIABILITY

In this section we present the basic terms and concepts used in this handbook. Not all of the terms are the most widely used ones. Some of them have been devised by the authors especially for this book.

All of these terms and concepts are arranged alphabetically for convenience.

The basic terms and concepts of probability theory and mathematical statistics used in this handbook are given in Appendix 1.

Component stand-by a stand-by method wherein separate parts of a structure are put on stand-by

Discrete failure the failure of the structure resulting from a discrete change of the values of one or more basic parameters of the structure

Duration of repair the duration of interruption of operation of a repairable structure due to the detection and elimination of failure

Efficiency a measure of the quality of proper functioning of a structure; the degree of usability of the structure for fulfilling given functions. The probability that the structure fulfills its functions under given conditions is generally chosen as the basic quantitative measure of efficiency.

Failure an event involving the total or partial loss of the operating ability of a structure

Failure density the probability density of operating time of the structure up to failure

Failure intensity the conditional density of failure of the structure at some moment of time, given no failure up to that time

Fitness that property of a structure of remaining operational during a given time interval under certain operating conditions

Idleness coefficient   the probability that a repairable structure in a stationary state of functioning is inoperable at any arbitrarily chosen moment of time

Inoperability   the state of a structure in which it fails to meet at least one of the demands made on it

Mean operating time between failures[1]   the mathematical expectation of the random operating time between failures

Mean operating time to failure   the mathematical expectation of the random operating time until the first failure

Mean repair time   the mathematical expectation of the time in the operation of a repairable structure spent in eliminating failure

Mean time of failure-free operation   an abbreviation for the mathematical expectation of the random time between successive failures, when the concepts of mean operating time up to first failure and mean operating time between failures are fully equivalent

Nonoperating stand-by   a stand-by method where the stand-by structure cannot fail before inclusion in the system

Nonstationary idleness coefficient   the probability that a structure is inoperable at some given moment of time

Nonstationary readiness coefficient   the probability that a structure is operable at some given moment of time

Nonstationary reliability coefficient   the probability that a structure operates during a time interval of given length, starting at some given moment of time (also called interval reliability)

Operability   the state of a structure whereby at a given moment of time it fulfills all requirements established for the working parameters

Operating stand-by   a stand-by method wherein the stand-by structure is in an operating state

Operating time   a time during which a structure, fulfilling its function, must operate without failure

Operating time between failures   the random time interval during which a structure operates from the moment of termination of the last repair to the moment of the following failure

Operating time until failure   the random time interval from the beginning of operation of a structure up to the first failure

Operationality   the state of a structure wherein it fulfills all requirements established relative to both its basic parameters and the second-stage parameters (characterizing ease of use, external physical shape required, and so on)

Output (yield)   the duration or amount of work completed by a structure under certain conditions

Partial failure   a failure without whose elimination at least partial use of the structure, as intended, is still possible

Partially operating stand-by   a method of stand-by where the stand-by unit is in a partially operating state

---

[1] In the general case one should distinguish mean operating time after the $(i - 1)$st repair up to the ith failure.

Permanent stand-by a stand-by method wherein the stand-by structures are connected to the basic ones during the entire operating time

Probability of failure the probability that there is at least one failure of a structure during a given time interval of operating under certain use conditions

Readiness coefficient the probability that a repairable structure in a stationary operating state is operable at any arbitrarily chosen moment of time

Redundancy a method for increasing the reliability of structures by using additional (stand-by) structures

Reliability that property of a structure's guaranteeing fulfillment of the required task in the required time under given use conditions

Reliability coefficient the probability that a repairable structure in a stationary operating state operates during a time interval of given length starting at an arbitrary moment of time

Repair the process of detecting and eliminating failure with the aim of restoring operability of a structure

Repairability that property of a structure consisting of the possibility of using it after repair or any other measures to eliminate failures have been carried out

Repairable structure a structure whose operation can be restored after failure by making the necessary repairs

Repair intensity the conditional density of repair time at some moment of time, given that the structure has not been repaired up to that moment of time

Replacement redundancy a stand-by method wherein the stand-by structures replace the basic ones after their failure

Self-maintenance the capability of a structure to detect and eliminate failures, and also to prevent them

Series connection a collection of structures for which an equivalent condition for failure is the failure of at least one structure in the collection

Structure in this handbook, a generalized concept equivalent to a unit or a system, depending on the context

Sudden failure a failure arising from an abrupt change in values of one or more basic parameters of the structure

System a collection of jointly operating objects that is designed for the fulfillment of an established task

Total failure a failure without whose elimination use of the structure as intended is impossible

Total output the total output of one or more structures over a given period of time

Total stand-by a stand-by method where a structure as a whole is put on stand-by

Unit of a system a part of a system designed to fulfill certain functions

Universal component stand-by a method of putting a group of reserve units on stand-by for a group of basic structures, where any of the stand-by structures can replace any failed basic structure

Unrepairable structure a structure whose operation after failure is considered impossible (at least under the conditions of the given mathematical model)

Table 1.1   Determination of Functions

GIVEN FUNCTION	FORMULAS FOR DETERMINING THE THREE REMAINING FUNCTIONS			
	$P(t)$	$Q(t)$	$f(t)$	$\lambda(t)$
$P(t)$	—	$1 - P(t)$	$-\dfrac{d}{dt} P(t)$	$-\dfrac{1}{P(t)} \dfrac{d}{dt} P(t)$
$Q(t)$	$1 - Q(t)$	—	$\dfrac{d}{dt} Q(t)$	$\dfrac{1}{1 - Q(t)} \times \dfrac{d}{dt} Q(t)$
$f(t)$	$\displaystyle\int_t^\infty f(x)\, dx$	$\displaystyle\int_0^t f(x)\, dx$	—	$\dfrac{f(t)}{\displaystyle\int_t^\infty f(x)\, dx}$
$\lambda(t)$	$e^{-\int_0^t \lambda(x)\, dx}$	$1 - e^{-\int_0^t \lambda(x)\, dx}$	$\lambda(t) e^{-\int_0^t \lambda(x)\, dx}$	—

1.2 MATHEMATICAL DEFINITION OF THE BASIC MEASURES OF RELIABILITY OF UNREPAIRABLE STRUCTURES

By an *unrepairable structure* we mean a structure whose operation after failure is considered completely impossible or infeasible. However, this does not necessarily mean that a structure of the given type cannot be repaired, as is the case, for example, with electrovacuum devices, meteorological rocket equipment, or ballistic missiles. The very concept of an unrepairable structure (or system) is characterized principally by the specific use of the equipment and not by its form.

Basically, an unrepairable structure should be understood in practice as a structure whose failure during operation leads to irreparable consequences. In this sense, for example, an electronic computer that is being used to control a complicated chemical process, where any interruption in the normal technological process leads to an irreversible disruption of the process, can be thought of as an unrepairable structure. At the same time it is clear that when failure occurs, a computer can be repaired and made fit for further use. However, within the limits of a concrete technological operation, a computer is essentially an unrepairable structure.

In this section we give two determinations for each reliability measure —one probabilistic and the other statistical. We use $\hat{Z}$ to denote the usual unbiased statistical estimator corresponding to the analogous probabilistic measure Z. In general $\hat{Z}$ will be the sample mean of random variables with mathematical expectation Z. It will then follow from the well-known fact from probability theory that as N, the number of trials, increases, this sample mean will converge in probability to Z—that is, for each arbitrarily small ϵ we have

$$\lim_{N \to \infty} \mathcal{P}\{|\hat{Z} - Z| < \epsilon\} = 1.$$

We look at the basic reliability measures of unrepairable structures.

1. Probability of failure-free operation of a structure in the time interval from 0 to t_0.

(a) Probabilistic definition:

$$P(t_0) = P(0; t_0) = \mathcal{P}\{\theta \geq t_0\} = 1 - F(t_0), \quad (1.2.1)$$

where θ is the random duration of failure-free operation of an initially operational structure until failure; $F(t)$ is the distribution function of the random variable θ; and $P(t_0)$ is the probability that the structure operates without failure during the time interval from 0 to t_0—that is, the probability that the random operating-time duration of the structure, starting operation at time $t = 0$, exceeds t_0.

(b) Statistical estimator:

$$\hat{P}(t_0) = \frac{N(t_0)}{N(0)} = 1 - \frac{d(t_0)}{N(0)}, \quad (1.2.2)$$

where $N(t_0)$ is the number of structures still operational at the moment of time t_0—that is, the number of structures that have not failed during the required time interval of length t_0; $N(0)$ is the number of operational structures at the initial moment of time $t = 0$; $d(t_0)$ is the number of structures that have failed up to the moment of time t_0; $\hat{P}(t_0)$ is the ratio of the number of structures operating without failure up to the moment of time t_0 to the number of operational structures at the initial moment of time $t = 0$, or the frequency of the event that the random time interval of failure-free operation of the structure is greater than the required time duration t_0 of failure-free operation (Figure 1.2.1).

2. Probability of failure-free operation of a structure in the time interval from t to $t + t_0$, given failure-free operation up to time t.

(a) Probabilistic definition:

$$P(t; t + t_0) = \mathcal{P}\{0 \geq t + t_0 \mid \theta > t\}$$
$$= \frac{P(0; t + t_0)}{P(0; t)} = \frac{P(t + t_0)}{P(t)}. \quad (1.2.3)$$

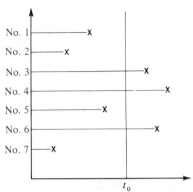

Figure 1.2.1   Time Diagram Illustrating the Definition
of the Statistical Esti-
mators $\hat{P}(t_0)$ and $\hat{Q}(t_0)$.
Realizations Nos. 1–7
comprise $N(0)$ struc-
tures, Nos. 1, 2, 5, 7
comprise $d(t_0)$, Nos. 3,
4, 6 comprise $N(t_0)$.

$P(t;\, t + t_0)$ is the probability that the structure operates
without failure during the required time interval of length t_0
given its operability at time t, or the conditional probability
that the random operating time of the structure up to failure
is greater than the quantity $t + t_0$ given that the structure has
operated without failure during the time interval of length t.

(b)   Statistical estimator:

$$\hat{P}(t;\, t + t_0) = \frac{N(t + t_0)}{N(t)}, \qquad (1.2.4)$$

where $N(t)$ is the number of structures still operational at the
moment of time t.

$\hat{P}(t;\, t + t_0)$ is the ratio of the number of structures operating
without failure up to the moment $t + t_0$ to the number of
structures that are still operational at the moment of time t, or
the frequency of the event that the length of the random time
interval of failure-free operation of the structure is greater
than $t + t_0$, given that it is greater than t (Figure 1.2.2).

3.   Probability of failure of a structure in the time interval from 0 to t_0.

(a)   Probabilistic definition:

$$Q(t_0) = Q(0;\, t_0) = \mathcal{P}\{\theta < t_0\} = F(t_0). \qquad (1.2.5)$$

$Q(t_0)$ is the probability that the structure fails during the required time interval of length t_0 starting operation at the moment of time $t = 0$, or the probability that the random operating time of an initially operational structure up to failure is smaller than the required length of time t_0 of failure-free operation.

It follows from (1.2.1) that

$$Q(t_0) = 1 - P(t_0). \qquad (1.2.6)$$

(b)   Statistical estimator:

$$\hat{Q}(t_0) = \frac{d(t_0)}{N(0)}, \qquad (1.2.7)$$

where $N(0)$ is the number of operational structures at the initial moment of time $t = 0$; $d(t_0)$ is the number of structures that have failed up to the moment of time t_0.

$\hat{Q}(t_0)$ is the ratio of the number of structures that have failed up to the moment of time t_0 to the number of structures

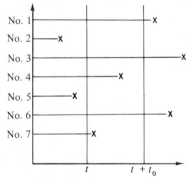

Figure 1.2.2   Time Diagram Illustrating the Definition of the Statistical Estimators $\hat{P}(t,\ t + t_0)$, $\hat{Q}(t,\ t + t_0)$, $\hat{f}(t)$, and $\hat{\lambda}(t)$. Realizations Nos. 1–7 comprise $N(0)$ structures, Nos. 2, 5 comprise $d(t)$, Nos. 2, 4, 5, 7 comprise $d(t + t_0)$, Nos. 1, 3, 4, 6, 7 comprise $N(t)$, Nos. 1, 3, 6 comprise $N(t + t_0)$.

that were operational at the moment of time $t = 0$, or the frequency of the event that the random time interval of failure-free operation of the structure is smaller than the required time interval of length t_0 of failure-free operation (Figure 1.2.1).

It follows from (1.2.2) that

$$\hat{Q}(t_0) = 1 - \hat{P}(t_0). \tag{1.2.8}$$

4.   Probability of failure of a structure in the time interval from t to $t + t_0$, given failure-free operation up to time t.

(a)   Probabilistic definition:

$$Q(t; t + t_0) = 1 - P(t; t + t_0) = 1 - \frac{P(t + t_0)}{P(t)}. \tag{1.2.9}$$

$Q(t; t + t_0)$ is the probability that the structure fails during the required time interval of length t_0 starting from the moment of time t, given its operability at time t, or the conditional probability that the random operating time of the structure up to failure is smaller than the quantity $t + t_0$, given that the structure has operated failure-free during the time interval of length t.

(b)   Statistical estimator:

$$\hat{Q}(t; t + t_0) = \frac{d(t + t_0) - d(t)}{N(t)} = 1 - \frac{N(t + t_0)}{N(t)}, \tag{1.2.10}$$

where $N(t)$ is the number of structures still operational at the moment of time t, and $d(t)$ is the number of structures that have failed up to the moment of time t.

$\hat{Q}(t; t + t_0)$ is the ratio of the number of structures that have failed in the time interval $[t; t + t_0]$ to the number of structures still operational at the moment of time t, or the frequency of the event that the length of the random time interval of failure-free operation of the structure is smaller than $t + t_0$, given that it is greater than t (Figure 1.2.2).

5.   Frequency of failures of a structure initially operational at time 0 at the moment of time t [density of $F(t)$].

(a)   Probabilistic definition:

$$f(t) = \frac{d}{dt} F(t) = \frac{d}{dt} Q(t). \tag{1.2.11}$$

$f(t)$ is the density of the probability that the random time of failure-free operation of the structure is less than t, or the probability density of failure at the moment of time t.

It follows from (1.2.6) that

$$f(t) = -\frac{d}{dt} P(t). \tag{1.2.12}$$

The probability of failure and the probability of failure-free operation of an initially operational structure during a certain time interval are expressed in terms of the failure density as follows:

For the interval from 0 to t_0:

$$Q(t_0) = \int_0^{t_0} f(x) \, dx, \tag{1.2.13}$$

$$P(t_0) = \int_{t_0}^{\infty} f(x) \, dx; \tag{1.2.14}$$

For the interval from t to $t + t_0$:

$$S(t; t + t_0) = \int_t^{t+t_0} f(x) \, dx, \tag{1.2.15}$$

$$R(t; t + t_0) = 1 - \int_t^{t+t_0} f(x) \, dx. \tag{1.2.16}$$

(b) Statistical estimator:

$$\hat{f}(t) = \frac{d(t + \Delta t) - d(t)}{N(0) \, \Delta t}, \tag{1.2.17}$$

where $d(t)$ is the number of failed structures at time t and $N(0)$ is the number of operational structures at the initial time $t = 0$.

$\hat{f}(t)$ is the proportion of failures in a unit of time relative to the number of operational structures at time $t = 0$, or the ratio of the difference between the number of failures up to time $t + \Delta t$ and the number of failures up to time t to the product of the number of operational structures at time $t = 0$ with the length Δt of the time interval. (See Figure 1.2.2, and replace t_0 there by Δt.)

It follows from (1.2.17) that

$$\hat{f}(t) = -\frac{N(t + \Delta t) - N(t)}{N(0) \, \Delta t}. \tag{1.2.18}$$

6. Failure intensity of a structure at time t.
    (a) Probabilistic definition:

$$\lambda(t) = \frac{1}{1 - F(t)} \frac{d}{dt} F(t) = \frac{f(t)}{P(t)}. \qquad (1.2.19)$$

$\lambda(t)$ is the conditional density of the probability of failure of an initially operational structure at time t given that the structure has not failed up to time t.

Using (1.2.12), we can obtain the following expressions for the probability of failure-free operation of the structure

$$P(t_0) = e^{-\int_0^{t_0} \lambda(y)\,dy}, \qquad (1.2.20)$$

and for the conditional probability of failure-free operation from t to $t + t_0$, given operationality at time t_0,

$$P(t; t + t_0) = e^{-\int_t^{t+t_0} \lambda(y)\,dy}. \qquad (1.2.21)$$

(b) Statistical estimator:

$$\hat{\lambda}(t) = \frac{d(t + \Delta t) - d(t)}{N(t)\,\Delta t} = -\frac{N(t + \Delta t) - N(t)}{N(t)\,\Delta t}. \qquad (1.2.22)$$

In order for $\hat{\lambda}(t)$ to provide a good approximation to $\lambda(t)$ in practice, we must have satisfied the conditions

$$N(t) \gg 1 \quad \text{and} \quad \Delta t \text{ is sufficiently small,}$$

where $N(t)$ is the number of structures still operational at time t, and $d(t)$ is the number of structures that have failed up to time t.

$\hat{\lambda}(t)$ is the number of failures per unit time relative to the number of operational structures at time t, or the ratio of the difference between the number of failures by time $t + \Delta t$ and the number of failures up to time t to the product of the number of structures still operational at time t with the length of the time interval Δt (see Figure 1.2.2, and replace t_0 there by Δt).

7. Mean operating time of a structure to failure.
    (a) Probabilistic definition:

$$T_1 = M\{\theta_1\}, \qquad (1.2.23)$$

$$T_1 = \int_0^\infty x f(x)\,dx, \qquad (1.2.24)$$

$$T_1 = \int_0^\infty x\,dQ(x), \qquad (1.2.25)$$

$$T_1 = \int_0^\infty P(x)\,dx. \qquad (1.2.26)$$

T_1 is the mathematical expectation (mean value) of the random operating time of the structure to failure.

(b)   Statistical estimators:

$$\hat{T}_1 = \frac{1}{N(0)} \left(\theta^{(1)} + \theta^{(2)} + \cdots + \theta^{[N(0)]} \right) = \frac{1}{N(0)} \sum_{i=1}^{N(0)} \theta^{(i)},$$

(1.2.27)

or

$$\hat{T}_1 = \theta^{(1)} + \frac{N(0) - 1}{N(0)} [\theta^{(2)} - \theta^{(1)}] + \cdots$$

$$+ \frac{1}{N(0)} \{ \theta^{[N(0)]} - \theta^{[N(0)-1]} \}$$

$$= \sum_{i=1}^{N(0)} \frac{N(0) - i + 1}{N(0)} [\theta^{(i)} - \theta^{(i-1)}],$$

(1.2.28)

where it is assumed that $0 = \theta^{(0)} < \theta^{(1)} < \theta^{(2)} < \cdots < {}^{[N(0)]}$.

$N(0)$ is the number of structures operational at the initial moment of time $t = 0$, and $\theta^{(i)}$ is the ith order statistic formed from the random operating times up to failure of the structures.

$\hat{T}_1$ [of (1.2.27)] is the arithmetic mean of the random operating times of the structures up to failure (Figure 1.2.3).

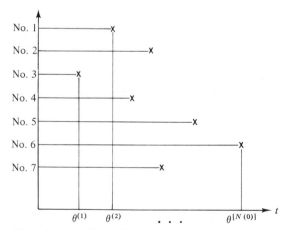

Figure 1.2.3   Time Diagram Illustrating the Definition of the Statistical Estimator $\hat{T}_1$.

Knowledge of any one of the functions $P(t)$, $Q(t)$, $f(t)$, $\lambda(t)$ makes it possible to determine the remaining three (Table 1.1).

1.3 MATHEMATICAL DEFINITIONS OF THE BASIC MEASURES OF RELIABILITY OF REPAIRABLE STRUCTURES

By a repairable structure we mean a structure whose operation after failure can be restored by performance of the needed repair work. It should be kept in mind, however, that the concept of "repairable structure" (or system) is characterized primarily not by the form of the equipment, but by its specific purpose. Basically, a repairable structure should be understood in practice as a structure that can continue to fulfill its function after the elimination of a failure causing the curtailment of its operation. Thus, the restoration* of a structure should not be taken in the narrow sense of the repair of some part or other.

In many cases when one speaks of repair of a structure one has in mind that in case of failure the given structure is completely replaced by an absolutely new one. (In this sense, certain structures that are basically unrepairable can be considered repairable.)

The process of utilization of a structure with repair can be represented as a sequence of intervals of operability alternating with intervals of idleness:

$$\theta_1, \xi_1, \theta_2, \xi_2; \ldots; \theta_n, \xi_n; \ldots.$$

A mathematical model of the process of utilization of such a structure can be taken to be the corresponding random process (Figure 1.3.1).

For structures that will undergo repair, some specific form of random process describing their functioning during the time of utilization is chosen. In general the distributions $F_1(t)$, $F_2(t)$, $F_3(t)$, $\ldots$ of the respective random variables θ_1, θ_2, θ_3, $\ldots$ are taken to be different. The reason is that at the moment of time $t = 0$ the structure is characterized completely

---

* [*Translator's note:* "Renewal" is also used as a synonym for "restoration."]

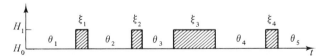

Figure 1.3.1 Random Process Corresponding to a Sequence of Alternating Intervals of Operability and Idleness of a Structure.

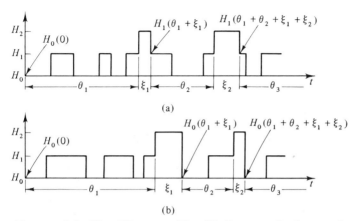

Figure 1.3.2   Time Diagrams of Possible Passages of a System of
Two Redundant Units.

by a certain initial state. For example, if we consider a system consisting of two redundant, independent, repairable units, then its initial
state $H_0(0)$ is generally assumed to be the state where both units are
operational. Likewise, after the very first failure of this system, the
state in which the second duration of failure-free operation θ_2 begins is
$H_1(\theta_1 + \xi_1)$; now only one of the units of the system is operational while
the second is in a state of failure. (We assume that with independent
repair of units the probability that both units will be repaired at the same
moment is practically zero.) A time diagram of passages of such a system from state to state over a period of time is shown in Figure 1.3.2(a).
Of course we can conceive of a maintenance scheme wherein each successive portion of the operation of the system always begins with the
state for which both units of the system are operational [see $H_0(\theta_1 + \xi_1)$
on the time diagram of possible passages in Figure 1.3.2(b)]. Generally
speaking, the initial states may be different before each successive portion
of failure-free operation. However, below we shall deal primarily with
structures whose initial states before the second, third, and further portions of failure-free operation are identical. Consequently all of the random variables θ_2, θ_3, and so on will have identical distributions. This is
the case of greatest practical interest.

Moreover, we shall also assume that all the durations of repair ξ_1, ξ_2, $\ldots$
for the structures under consideration have identical distributions.

We examine the basic reliability measures for repairable structures that
are most frequently used in practice; their formulas are given in this
handbook.

1. Probability of failure-free operation of a structure in the time interval from 0 to t_0.

(a) Probabilistic definition:

$$P(t_0) = P(0; t_0) = \mathcal{P}\{\theta \geq t_0\} = 1 - F(t_0), \quad (1.3.1)$$

where θ is the random operating time of the (initially operational) structure up to the first failure, and $F(t)$ is the distribution function of the random variable θ. $P(t_0)$ is the probability that the structure operates failure-free during the required time interval of length t_0 starting operation at time $t = 0$, or the probability that the random operating time of the structure up to failure is greater than the required time interval of length t_0 of failure-free operation.

(b) Statistical estimator:

$$\hat{P}(t_0) = \frac{N(t_0)}{N(0)} = 1 - \frac{d(t_0)}{N(0)}, \quad (1.3.2)$$

where $N(t_0)$ is the number of structures that have not failed during the required time interval of length t_0; $N(0)$ is the number of structures operational at the initial moment of time $t = 0$; $d(t_0)$ is the number of these structures that have failed prior to time t_0. $\hat{P}(t_0)$ is the ratio of the number of structures operating failure-free up to time t_0 to the number of structures operational at the initial moment of time $t = 0$, or the frequency of the event that the random interval of time of failure-free operation of the structure is greater than the required time t_0 of failure-free operation (Figure 1.2.1).

2. Probability of failure-free operation of a structure during the time interval of length t_0 beginning from the moment of termination of the $(j - 1)$st repair.

(a) Probabilistic definition:

$$P_j(t_0) = \mathcal{P}\{\theta_j \geq t_0\}, \quad (1.3.3)$$
$$P_j(t_0) = 1 - F_j(t_0). \quad (1.3.4)$$

$P_j(t_0)$ is the probability that the structure operates failure-free during the required time interval of length t_0, where the initial moment of this interval coincides with the moment of termination of the $(j - 1)$st repair of the structure (Figure 1.3.3).

Type of favorable event

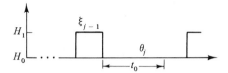

Types of unfavorable events

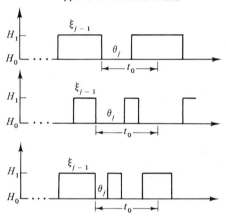

Figure 1.3.3 Time Diagrams Illustrating Probabilistic Definitions of the Quantities $P_j(t_0)$ and $Q_j(t_0)$.

(b) Statistical estimator:

$$\hat{P}_j(t_0) = \frac{N_j(0) - d_j(t_0)}{N_j(0)}, \qquad (1.3.5)$$

$$\hat{P}_j(t_0) = \frac{N_j(t_0)}{N_j(0)}, \qquad (1.3.6)$$

where $N_j(0)$ is the number of (operational) structures that have just undergone their $(j - 1)$st repair within a time t_0 after completion of their $(j - 1)$st repair and $d_j(t_0)$ is the number of structures that have failed within a time t_0 after completion of their $(j - 1)$st repair. $\hat{P}_j(t_0)$ is the ratio of the number of structures whose random time interval of failure-free operation from the moment of termination of the $(j - 1)$st repair to the jth failure is greater than the required value to the over-all quantity of structures (Figure 1.3.4).

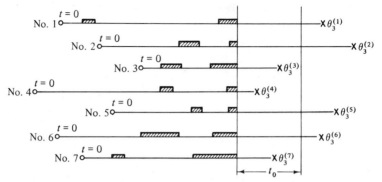

Figure 1.3.4   Time Diagram Illustrating the Definition of the Statistical Estimators $\hat{P}_j(t_0)$ and $\hat{Q}_j(t_0)$ for the case $j = 3$. Realizations Nos. 1–7 comprise $N_3(0)$ structures, Nos. 1, 2, 5, 6 comprise $N_3(t_0)$, Nos. 3, 4, 7 comprise $d_3(t_0)$.

3.   Conditional probability of failure-free operation of a structure in the time interval from t to $t + t_0$, given operationality at time t.

(a)   Probabilistic definition:

$$P(t; t + t_0)$$

$$= \frac{\sum_{j=1}^{\infty} \mathcal{P} \left\{ \sum_{i=1}^{j} (\theta_i + \xi_i) < t < t + t_0 \leq \theta_{j+1} + \sum_{i=1}^{j} (\theta_i + \xi_i) \right\}}{\sum_{j=1}^{\infty} \mathcal{P} \left\{ \sum_{i=1}^{j} (\theta_i + \xi_i) < t \right\}}$$

$$(1.3.7)$$

$P(t; t + t_0)$ is the probability that the structure operates failure-free during the required time interval of length t_0, beginning at time t, given operationality at time t (Figure 1.3.5).

(b)   Statistical estimator:

$$\hat{P}(t; t + t_0) = \frac{N(t) - d(t + t_0 \mid t)}{N(t)}, \qquad (1.3.8)$$

$$\hat{P}(t; t + t_0) = \frac{N(t + t_0 \mid t)}{N(t)}, \qquad (1.3.9)$$

where $d(t + t_0 \mid t)$ is the number of structures operational at the moment t that failed at least once in the interval from t to $t + t_0$; $N(t + t_0 \mid t)$ is the number of structures that are operational at the moment t and never failed in the time interval from t to $t + t_0$; and $N(t)$ is the number of structures operational at time t. $\hat{P}(t, t + t_0)$ is the ratio of the number of structures that are operational at time t and have then operated failure-free up to time $t + t_0$ to the number of structures that are operational at time t (Figure 1.3.6).

4. Probability of failure of a structure in the time interval from 0 to t_0.
    (a) Probabilistic definition:

$$Q(t_0) = Q(0, t_0) = \mathcal{P}\{\theta < t_0\} = F(t_0). \qquad (1.3.10)$$

Type of favorable event

Types of unfavorable events

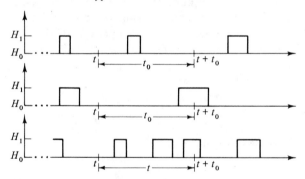

Types of events not considered

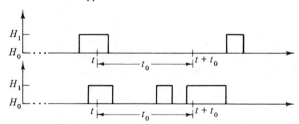

Figure 1.3.5 Time Diagrams Illustrating Probabilistic Definitions of the Quantities $P(t, t + t_0)$ and $Q(t, t + t_0)$.

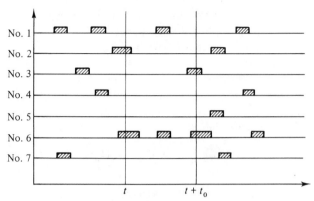

Figure 1.3.6 Time Diagram Illustrating the Definition of the Statistical Estimators $\hat{P}(t, \ t + t_0)$ and $\hat{Q}(t, \ t + t_0)$. Realizations Nos. 1, 3, 4, 5, 7 comprise $N(t)$ structures, Nos. 1, 3 comprise $d(t + t_0 \mid t)$, Nos. 4, 5, 7 comprise $N(t + t_0 \mid t)$, Nos. 2, 6 comprise $r(t)$.

$Q(t_0)$ is the probability that an initially operational structure fails at least once during the required time interval of length t_0, having started to operate at time $t = 0$, or the probability that the random length of time of failure-free operation of the structure is less than the required interval of failure-free operation.

It follows from (1.3.1) that

$$Q(t_0) = 1 - P(t_0). \tag{1.3.11}$$

(b) Statistical estimator:

$$\hat{Q}(t_0) = \frac{d(t_0)}{N_0}, \tag{1.3.12}$$

where $d(t_0)$ is the number of structures that have failed at least once up to time t_0 and N_0 is the number of structures that are operational at the initial moment of time $t = 0$. $\hat{Q}(t_0)$ is the ratio of the number of structures that have failed at least once up to time t_0 to the number of structures operational at time $t = 0$, or the frequency of the event that the random time interval beginning at $t = 0$ of failure-free operation of the structure is smaller than the required interval of length t_0 of failure-free operation (Figure 1.2.1).

It follows from (1.3.2) that

$$\hat{Q}(t_0) = 1 - \hat{P}(t_0). \tag{1.3.13}$$

5. Probability of failure of a structure during the time interval of length t_0 starting from the moment of termination of the $(j - 1)$st repair.

(a) Probabilistic definition:

$$Q_j(t_0) = 1 - P_j(t_0), \tag{1.3.14}$$
$$Q_j(t_0) = \mathcal{P}\{\theta_j < t_0\}, \tag{1.3.15}$$
$$Q_j(t_0) = F_j(t_0). \tag{1.3.16}$$

$Q_j(t_0)$ is the probability that the structure fails in a time interval of length t_0 given that the initial moment of this interval coincides with the moment of the $(j - 1)$st repair of the structure (Figure 1.3.3).

(b) Statistical estimator:

$$\hat{Q}_j(t_0) = \frac{d_j(t_0)}{N_j(0)}, \tag{1.3.17}$$

$$\hat{Q}_j(t_0) = \frac{N_j(0) - N_j(t_0)}{N_j(0)}, \tag{1.3.18}$$

where $N_j(0)$ is the number of structures that have just completed their $(j - 1)$st repair; $N_j(t_0)$ is the number of these structures that have not failed within a time t_0 after completion of their $(j - 1)$st repair; and $d_j(t_0)$ is the number of these structures that have failed within a time t_0 after completion of their $(j - 1)$st repair. $\hat{Q}_j(t_0)$ is the ratio of the number of structures whose random time interval of failure-free operation from the moment of termination of the $(j - 1)$st repair up to the jth failure is less than the required value t_0 to the over-all number of structures considered (Figure 1.3.4).

6. Probability of a structure's being failed at some point in the time interval from t to $t + t_0$.

(a) Probabilistic definition:

$$S(t, t + t_0) = \sum_{j=1}^{\infty} \mathcal{P}\left\{ \sum_{i=1}^{j} (\theta_i + \xi_i) \le t < \theta_{j+1} + \sum_{i=1}^{j} (\theta_i + \xi_j) \right.$$
$$< t + t_0 \right\} + \sum_{j=1}^{\infty} \mathcal{P}\left\{ \theta_j + \sum_{i=1}^{t-1} (\theta_i + \xi_i) \right.$$
$$\left. < t < \sum (\theta_i + \xi_i) \right\}, \tag{1.3.19}$$
$$S(t, t + t_0) = 1 - R(t, t + t_0). \tag{1.3.20}$$

$S(t, t + t_0)$ is the probability that the structure either fails at least once in the required interval of length t_0 after starting at time t, or is already inoperable at time t (Figure 1.3.5).

(b) Statistical estimator:

$$S(t, t + t_0) = \frac{d(t + t_0 \mid t)r(t)}{N_0}, \tag{1.3.21}$$

where N_0 is the total number of available units; $d(t + t_0 \mid t)$ is the number of these structures that were operational at time t and failed at least once in the interval from t to $t + t_0$; and $r(t)$ is the number of units undergoing repair at time t. $\hat{Q}(t, t + t_0)$ is the ratio of the number of structures that are operational at time t and then fail at least once in the required interval of length t_0 to the total number of structures (Figure 1.3.6) plus the proportion of these units undergoing repair at time t.

7. Mean operating time of a structure up to failure.
   (a) Probabilistic definition:

$$T_1 = M\{\theta\}, \tag{1.3.22}$$

$$T_1 = \int_0^\infty xf(x)\, dx = \int_0^\infty x\, dF(x), \tag{1.3.23}$$

$$T_1 = \int_0^\infty x\, dQ(x), \tag{1.3.24}$$

$$T_1 = \int_0^\infty P(x)\, dx. \tag{1.3.25}$$

T_1 is the mathematical expectation (mean value) of the random operating time of the structure up to failure.

   (b) Statistical estimators:

$$
\begin{aligned}
\hat{T}_1 &= \frac{1}{N(0)}(\theta^{(1)} + \theta^{(2)} + \cdots + \theta^{[N(0)]}) \\
&= \frac{1}{N(0)}\sum_{i=1}^{N(0)} \theta^{(i)}
\end{aligned} \tag{1.3.26}
$$

or

$$
\begin{aligned}
\hat{T}_1 &= \theta^{(1)} + \frac{N(0) - 1}{N(0)}[\theta^{(2)} - \theta^{(1)}] + \cdots \\
&\qquad\qquad + \frac{1}{N(0)}\{\theta^{[N(0)]} - \theta^{[N(0)-1]}\} \\
&= \sum_{i=1}^{N(0)} \frac{N(0) - i + 1}{N(0)}[\theta^{(i)} - \theta^{(i-1)}],
\end{aligned} \tag{1.3.27}
$$

where it is assumed that $0 = \theta^{(0)} < \theta^{(1)} < \theta^{(2)} < \cdots < \theta^{[N(0)]}$.

Here $N(0)$ is the over-all number of structures and $\theta^{(i)}$ is the ith order statistic formed from the random operating times up to failure for the structures. $\hat{T}_1$ [of (1.3.26)] is the arithmetic mean of the random intervals of operating times of the structures up to the first failure (Figure 1.2.3).

8.   Mean time of failure-free operation of a structure from the moment of termination of the $(j - 1)$st repair to the jth failure.

(a)   Probabilistic definition:

$$T_j = M\{\theta_j\}, \tag{1.3.28}$$

$$T_j = \int_0^\infty x f_j(x)\, dx, \tag{1.3.29}$$

$$T_j = \int_0^\infty x\, dQ_j(x), \tag{1.3.30}$$

$$T_j = \int_0^\infty P_j(x)\, dx. \tag{1.3.31}$$

T_j is the mathematical expectation (mean value) of the random length of time of failure-free operation of the structure from the moment of termination of the $(j - 1)$st repair to the jth failure.

(b)   Statistical estimators:

$$\hat{T}_j = \frac{1}{N(0)} (\theta_j^{(1)} + \theta_j^{(2)} + \cdots + \theta_j^{[N(0)]}) = \frac{1}{N(0)} \sum_{i=1}^{N(0)} \theta_j^{(i)} \tag{1.3.32}$$

or

$$\hat{T}_j = \theta_j^{(1)} + \frac{N(0) - 1}{N(0)} [\theta_j^{(2)} - \theta_j^{(1)}] + \frac{1}{N(0)} \{\theta_j^{[N(0)]} - \theta_j^{[N(0)-1]}\}$$

$$= \sum_{i=1}^{N(0)} \frac{N(0) - i + 1}{N(0)} [\theta_j^{(i)} - \theta_j^{(i-1)}], \tag{1.3.33}$$

where it is assumed that $0 = \theta_j^{(0)} < \theta_j^{(1)} < \theta_j^{(2)} < \cdots < \theta_j^{[N(0)]}$.

Here $N(0)$ is the over-all number of structures; $\theta_j^{(i)}$ is the ith order statistic formed from the random operating times after the $(j - 1)$st repair up to the jth failure for the structures.

$\hat{T}_j$ [of (1.3.32)] is the arithmetic mean of the random intervals of failure-free operating times of the $N(0)$ structures from the moment of termination of the $(j - 1)$st repair to the jth failure (Figure 1.3.7).

Remark (concerning items 7 and 8).   As noted in Section 1.1, we shall also make use below of the concept of *mean time of failure-free operation*

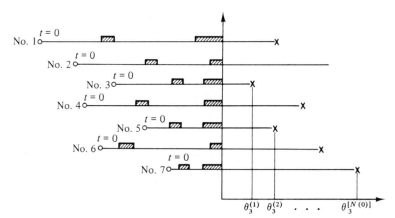

Figure 1.3.7   Time Diagram Illustrating the Definition of the Statistical Estimator $\hat{T}_j$ for the Case $j = 3$.

when mean operating time up to failure and mean operating time between failures are fully equivalent.

9.   Intensity of repair of a structure at time t.

(a)   Probabilistic definition:

$$\mu(t) = \frac{1}{1 - F_r(t)} \frac{d}{dt} F_r(t) = \frac{f_r(t)}{1 - F_r(t)}, \qquad (1.3.34)$$

where f_r is the density of the distribution $F_r(t)$.

$\mu(t)$ is the conditional density of the probability of repair of a structure at time t, given that the structure has not been repaired up to time t.

(b)   Statistical estimator:

$$\hat{\mu}(t) = \frac{d_r(t + \Delta t) - d_r(t)}{N_r(t)\,\Delta t} = -\frac{N_r(t + \Delta t) - N_r(t)}{N_r(t)\,\Delta t}. \qquad (1.3.35)$$

[In order for $\hat{\mu}(t)$ to provide a good approximation to $\mu(t)$ in practice, we must have satisfied the conditions $N_r(t) \gg 1$ and Δt is sufficiently small.] Here $N_r(t)$ is the number of structures failed at time 0 that have not been repaired during the time interval from 0 to t, and $d_r(t)$ is the number of structures for which repair, begun at the moment $t = 0$, terminated by the moment t.

$\hat{\mu}(t)$ is the number of repairs in a unit of time relative to the number of structures not yet repaired by time t, or the ratio of the difference between the number of repairs up to time $t + \Delta t$ and the number of repairs up to time t, to the product of the number of structures not yet repaired by the moment t with the length Δt of the time interval (Figure 1.3.8).

10. Mean repair time of a structure.

(a) Probabilistic definition:

$$\tau = M\{\xi\}, \tag{1.3.36}$$

$$\tau = \int_0^\infty x f_r(x) \, dx = \int_0^\infty x \, dF_r(x), \tag{1.3.37}$$

$$\tau = \int_0^\infty [1 - F_r(x)] \, dx. \tag{1.3.38}$$

τ is the mathematical expectation (mean value) of the random repair time.

(b) Statistical estimators:

$$\hat{\tau} = \frac{1}{N(0)} (\xi^{(1)} + \xi^{(2)} + \cdots + \xi^{[N(0)]}) = \frac{1}{N(0)} \sum_{i=1}^{N(0)} \xi^{(i)}, \tag{1.3.39}$$

or

$$\hat{\tau} = \xi^{(1)} + \frac{N(0) - 1}{N(0)} [\xi^{(2)} - \xi^{(1)}] + \frac{1}{N(0)} \{\xi^{[N(0)]} - \xi^{[N(0)-1]}\}$$

$$= \sum_{i=1}^{N(0)} \frac{N(0) - i + 1}{N(0)} [\xi^{(i)} - \xi^{(i-1)}], \tag{1.3.40}$$

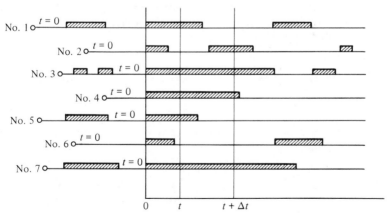

Figure 1.3.8 Time Diagram Illustrating the Definition of the Statistical Estimator $\hat{\mu}(t)$. Realizations Nos. 2, 6 comprise $d_r(t)$ structures, Nos. 1, 3, 4, 5, 7 comprise $N_r(t)$, Nos. 1, 2, 5, 6 comprise $d_r(t + \Delta t)$, Nos. 3, 4, 7 comprise $N_r(t + \Delta t)$.

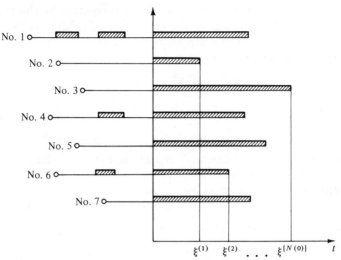

Figure 1.3.9   Time Diagram Illustrating the Definition of the Statistical Estimator $\hat{\tau}$.

where it is assumed that $0 = \xi^{(0)} < \xi^{(1)} < \xi^{(2)} < \cdots < \xi^{[N(0)]}$. Here $N(0)$ is the over-all number of structures, and $\xi^{(i)}$ is the ith order statistic formed from the random repair time for the structures.

$\hat{\tau}$ [of (1.3.39)] is the arithmetic mean of the random repair time intervals of the $N(0)$ structures (Figure 1.3.9).

11.   Nonstationary readiness coefficient of a structure.

(a)   Probabilistic definition:

$$K(t) = \sum_{i=0}^{\infty} \mathcal{P}\left[\sum_{j=0}^{i} (\theta_j + \xi_j) < t < \sum_{j=0}^{i} (\theta_j + \xi_j) + \theta_{i+1} \right].$$

$$(1.3.41)$$

$K(t)$ is the probability that the structure is in an operational state at time t.

(b)   Statistical estimator:

$$\hat{K}(t) = \frac{N_t}{N(0)} = \frac{N(0) - d_t}{N(0)},$$ $$(1.3.42)$$

where $N(0)$ is the over-all number of structures; N_t is the number of structures that are operational at time t, and d_t is the number of structures that are in a state of failure at time t.

$\hat{K}(t)$ is the ratio of the number of structures that are in an operational state at time t to the over-all number of structures (Figure 1.3.10).

12. Readiness coefficient of a structure.

(a) Probabilistic definition:

$$K = \lim_{t \to \infty} K(t). \qquad (1.3.43)$$

K is the probability of finding the structure in an operational state for a stationary random process (that is, for a "sufficiently large" value of the time variable, t), or the asymptotic value of the proportion of time in which the structure is in an operational state for a stationary random process.

For any distributions $F(t)$ and $F_r(t)$, respectively, of the time between failures and of the repair time having finite mean values T and τ, respectively, we can always write

$$K = \frac{T}{T + \tau}. \qquad (1.3.44)$$

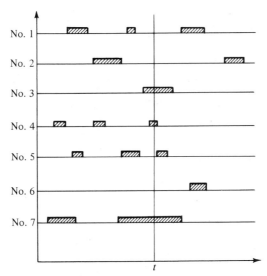

Figure 1.3.10 Time Diagram Illustrating the Definition of the Statistical Estimators $\hat{K}(t)$ and $\hat{k}(t)$. Realizations Nos. 1–7 comprise $N(0)$ structures, Nos. 1, 2, 5, 6 comprise N_t, Nos. 3, 4, 7 comprise d_t.

(b) Statistical estimator:

$$\hat{K} = \frac{N_\infty}{N(0)} = \frac{N(0) - d_\infty}{N(0)}, \qquad (1.3.45)$$

where $N(0)$ is the over-all number of structures; N_∞ is the number of operational structures at an arbitrary "sufficiently distant" moment of time; d_∞ is the number of structures in a state of failure at that same moment of time.

$\hat{K}$ is the ratio of the number of structures in an operational state at a random "sufficiently distant" moment of time to the over-all number of structures (Figure 1.3.11), or the ratio of the time in which the structure is in an operational state to the total time of test, given that the test has started at a random "sufficiently distant" moment of time.

13. Nonstationary idleness coefficient.
(a) Probabilistic definition:

$$k(t) = \sum_{i=1}^{\infty} \mathcal{P} \left[\sum_{j=1}^{i} (\theta_j + \xi_j) + \theta_{i+1} < t < \sum_{j=1}^{j+1} (\theta_j + \xi_j) \right].$$

$$(1.3.46)$$

$k(t)$ is the probability that the structure is in a state of failure (or undergoing repair) at time t.

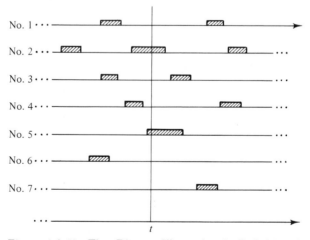

Figure 1.3.11 Time Diagram Illustrating the Definition of the Statistical Estimators $\hat{K}$ and $\hat{k}$. Realizations Nos. 1–7 comprise $N(0)$ structures, Nos. 1, 3, 4, 6, 7 comprise N_∞, Nos. 2, 5 comprise d_∞.

It is obvious that

$$k(t) = 1 - K(t). \qquad (1.3.47)$$

(b) Statistical estimator:

$$\hat{k}(t) = \frac{n_t}{N(0)}, \qquad (1.3.48)$$

where $N(0)$ is the over-all number of structures, and n_t is the number of structures in a state of failure at time t.

$\hat{k}(t)$ is the ratio of the number of structures in a failed state at time t to the over-all number of structures (Figure 1.3.10).

14. Idleness coefficient of a structure.
   (a) Probabilistic definition:

$$k = \lim_{t \to \infty} k(t). \qquad (1.3.49)$$

k is the probability for a stationary random process that the structure is in a state of failure [that is, the probability $k(t)$ evaluated at a "sufficiently distant" moment of time], or the asymptotic value of the proportion of time in which the structure is in a state of failure for a stationary random process.

It is obvious that

$$k = 1 - K. \qquad (1.3.50)$$

   (b) Statistical estimator:

$$\hat{k} = \frac{n_\infty}{N(0)}, \qquad (1.3.51)$$

where $N(0)$ is the over-all number of structures, and n_∞ is the number of structures that are in a failed state at an arbitrary "sufficiently distant" moment of time.

$\hat{k}$ is the ratio of the number of structures that are in a failed state at an arbitrary "sufficiently distant" moment of time to the over-all number of structures (Figure 1.3.11), or the ratio of the time in which the structure is in a state of failure to the total time of test, given that the test has started at a random "sufficiently distant" moment of time.

15. Nonstationary reliability coefficient of a structure in the interval from t to $t + t_0$.

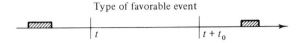

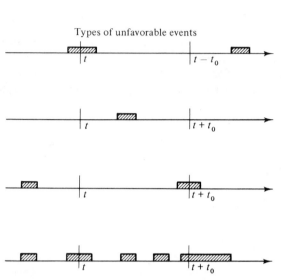

Figure 1.3.12   Time Diagrams Illustrating the Probabilistic Definition of the Quantity $R(t, t_0)$.

(a)   Probabilistic definition:

$$R(t, t_0) = \sum_{i=1}^{\infty} \mathcal{P} \left[\sum_{j=1}^{i} (\theta_j + \xi_j) < t < t + t_0 \right.$$

$$\left. \leq \sum_{j=1}^{i} (\theta_j + \xi_j) + \theta_{i+1} \right]. \quad (1.3.52)$$

$R(t, t_0)$ is the probability that a structure is operational at the moment t and then operates failure-free up to the moment $t + t_0$ (Figure 1.3.12).

(b)   Statistical estimator:

$$\hat{R}(t, t_0) = \frac{N(t + t_0 \mid t)}{N(0)}, \quad (1.3.53)$$

where $N(0)$ is the over-all number of structures; $N(t + t_0 \mid t)$ is the number of structures that were operational at time t and never failed in the time interval from t to $t + t_0$.

$\hat{R}(t, t_0)$ is the ratio of the number of structures that were operational at time t and operated failure-free in the time interval from t to $t + t_0$ to the over-all number of structures (Figure 1.3.13).

16. Reliability coefficient of a structure.

(a) Probabilistic definition:

$$R(t_0) = \lim_{t \to \infty} R(t, t_0). \qquad (1.3.54)$$

$R(t_0)$ is the probability that a structure is operational at an arbitrary "sufficiently distant" moment of time t and will operate failure-free during the succeeding time interval of length t_0.

(b) Statistical estimator:

$$\hat{R}(t_0) = \frac{N(t_\infty + t_0 \mid t_\infty)}{N(0)}, \qquad (1.3.55)$$

where $N(0)$ is the over-all number of structures; $N(T_\infty + t_0 \mid t_\infty)$ is the number of structures that are operational at an arbitrary "sufficiently distant" moment of time t_∞ and operate failure-free during the succeeding time interval of length t_0.

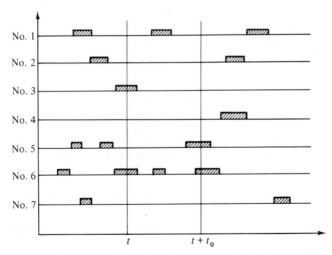

Figure 1.3.13 Time Diagram Illustrating the Definition of the Statistical Estimator $\hat{R}(t, t_0)$. Realizations Nos. 1–7 comprise $N(0)$ structures, Nos. 2, 4, 7 comprise $N(t + t_0 \mid t)$.

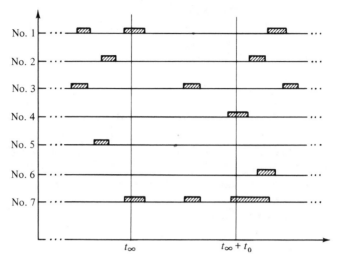

Figure 1.3.14 Time Diagram Illustrating the Definition of the Statistical Estimator $\hat{R}(t_0)$. Realizations Nos. 1-7 comprise $N(0)$ structures, Nos. 2, 5, 6 comprise $N(t_\infty + t_0 \mid t_\infty)$.

$\hat{R}(t_0)$ is the ratio of the number of structures that were operational at an arbitrary "sufficiently distant" moment of time and then operated failure-free during the succeeding time interval of length t_0 to the over-all number of structures (Figure 1.3.14).

BIBLIOGRAPHY

1. GNEDENKO, B. V., BELYAEV, Y. K., AND SOLOV'EV, A. D., *Mathematical Methods in Reliability Theory.* New York: Academic Press, Inc., 1968.
2. MALIKOV, I. M., POLOVKO, A. M., ROMANOV, M. A., AND CHUKREEV, P. A., *Basic Theory and Computation of Reliability*, 2d ed., chap. 2. Moscow: Sudpromgiz, 1960.
3. *Reliability of Technical Systems and Products* (basic concepts; terminology). Moscow: Standartov, 1964.
4. *Reliability Theory in the Domain of Radioelectronics* (terminology), no. 60. Moscow: AN SSR, 1962.
5. SHISHONOK, N. A., REPKIN, V. F., AND BARVINSKII, L. L., *Basic Theory of Reliability and Use of Radioelectronic Equipment*, chap. 3. Moscow: Soviet Radio, 1964.

2

MEASURES OF RELIABILITY
OF A UNIT

PRELIMINARY REMARKS

The concept of a unit is very relative. Often in a reliability analysis it is convenient to isolate some part of the system being considered. Thus in analyzing the reliability of a complicated radioelectronic complex one may take as units an entire radar station, an electronic computer, a power system, and so on; in analyzing the reliability of a radar installation one might consider as units an assembly or chassis of the equipment; in analyzing the reliability of an assembly one can consider an individual module, radio component, tube, transistor, and so forth.

2.1 UNREPAIRABLE UNIT

We assume that we know the distribution law of the operating time of the unit up to failure, $F(t) = \mathcal{P}\{\theta \leq t\}$. The reliability measures of the unit are expressed in terms of the known distribution law or its basic parameters.

In Table 2.1.1 we give the basic reliability measures for an arbitrary distribution law of the time-to-failure. Figures 2.1.1–2.1.4 illustrate the basic formulas in this table.

Table 2.1.2 shows the basic reliability measures of a unit for the exponential distribution law of time-to-failure. [*Translator's note:* In this and future tables in which approximations are given, $\delta_\pm$ will denote the magnitude of the error of the approximation and the subscripted sign will denote the sign of the difference "true value" − "approximate value."]

Table 2.1.1 Unrepairable Unit: Arbitrary Distribution Law of Time Up to Failure, $F(t)$

MEASURE	EXACT VALUE
$P(t_0)$	$1 - F(t_0)$
$Q(t_0)$	$F(t_0)$
$P(t, t + t_0)$	$\dfrac{1 - F(t + t_0)}{1 - F(t)}$
$Q(t, t + t_0)$	$1 - \dfrac{1 - F(t + t_0)}{1 - F(t)} = \dfrac{F(t + t_0) - F(t)}{1 - F(t)}$
T	$\displaystyle\int_0^\infty x f(x)\, dx = \int_0^\infty x\, dF(x) = \int_0^\infty P(x)\, dx$

Table 2.1.2 Unrepairable Unit: Exponential Distribution Law of Time Up to Failure, $F(t) = 1 - e^{-\lambda t}$

MEASURE	EXACT VALUE	APPROXIMATE VALUE	ERROR
$P(t_0)$	$e^{-\lambda t_0}$	$1 - \lambda t_0$	$\delta_+ < \dfrac{(\lambda t_0)^2}{2}$
$Q(t_0)$	$1 - e^{-\lambda t_0}$	λt_0	$\delta_- < \dfrac{(\lambda t_0)^2}{2}$
$P(t, t + t_0)$	$e^{-\lambda t_0}$	$1 - \lambda t_0$	$\delta_+ < \dfrac{(\lambda t_0)^2}{2}$
$Q(t, t + t_0)$	$1 - e^{-\lambda t_0}$	λt_0	$\delta_- < \dfrac{(\lambda t_0)^2}{2}$
T	$\dfrac{1}{\lambda}$	—	—

An illustration of the basic formulas of this table is provided by Figure 2.1.5. Approximate values are given in Table 2.1.2 for the condition

$$\lambda t_0 \ll 1.$$

Example 2.1.1 Suppose the distribution law of the operating time of a unit is given by the following table:

t	0–1	1–2	2–3	3–4	4–5	5–6	6–7	7–8	8–9	9–10	10–
$F(t)$	0	0.03	0.08	0.20	0.45	0.65	0.80	0.90	0.95	0.98	1.00

We want to compute the basic reliability measures of the unit.

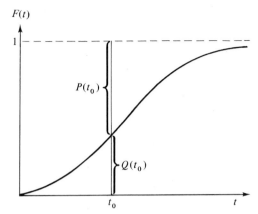

Figure 2.1.1   Illustration of the Formulas for
$P(t_0)$ and $Q(t_0)$.

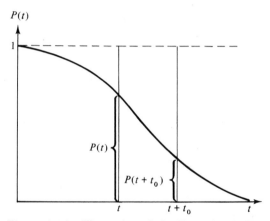

Figure 2.1.2   Illustration of the Formulas for $P(t;$
$t + t_0)$ and $Q(t; t + t_0)$.

Solution.   1.   Probability of failure-free operation for the initial
$t_0 = 4$ hours:

$$P(t_0) = 1 - F(t_0) = 1 - 0.20 = 0.80.$$

2.   Probability of failure prior to $t_0 = 4$ hours:

$$Q(t_0) = F(t_0) = 0.20.$$

3.   Probability of failure-free operation from $t = 2$ hours to $t + t_0 =$
6 hours, given proper operation prior to $t = 2$:

$$P(t, t + t_0) = \frac{1 - F(t + t_0)}{1 - F(t)} = \frac{1 - 0.65}{1 - 0.03} \approx 0.36.$$

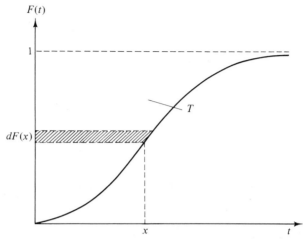

Figure 2.1.3 Illustration of the Principle of Integration
Using the Formula $T = \int_0^\infty x \, dF(x)$.

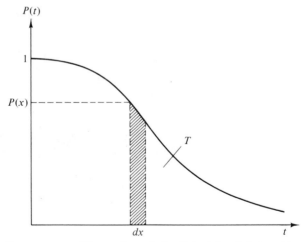

Figure 2.1.4 Illustration of the Principle of Integration
Using the Formula $T = \int_0^\infty P(x) \, dx$.

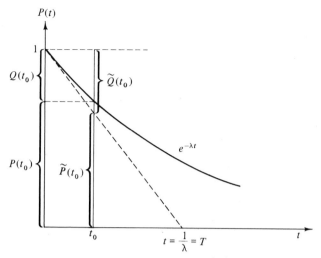

$P(t)$

$Q(t_0)$ $\widetilde{Q}(t_0)$

$e^{-\lambda t}$

$P(t_0)$ $\widetilde{P}(t_0)$

t_0

$t = \dfrac{1}{\lambda} = T$

t

Figure 2.1.5 Illustration of Basic Formulas of Table 2.1.2.

4. Probability of failure in the interval from $t = 2$ hours to $t + t_0 = 6$ hours, given proper operation prior to time $t = 2$:

$$Q(t; t + t_0) = 1 - P(t; t + t_0) = 1 - 0.36 = 0.64.$$

5. Mean operating time up to failure:

$$T = \sum_{i=1}^{10} P(x_i) = \sum_{i=1}^{10} [1 - F(x_i)]$$
$$= 1 + 0.97 + 0.92 + 0.80 + 0.55 + 0.35 + 0.20 + 0.10 + 0.05$$
$$+ 0.02 = 4.96.$$

If $F(t)$ is given in the form of a step function as in this case, the formula for T can be written in the form (Figure 2.1.6)

$$T = \sum_{i=0}^{\infty} [F(t_{i+1}) - F(t_i)]t_i \tag{2.1.1}$$

or in the form (Figure 2.1.7)

$$T = \sum_{i=0}^{\infty} [1 - F(t_i)](t_{i+1} - t_i). \tag{2.1.2}$$

For example, using Formula (2.1.2), we again obtain

$$T = 1 + 0.97 + 0.92 + 0.80 + 0.55 + 0.35 + 0.20 + 0.10 + 0.05$$
$$+ 0.02 = 4.96 \quad \text{hours.}$$

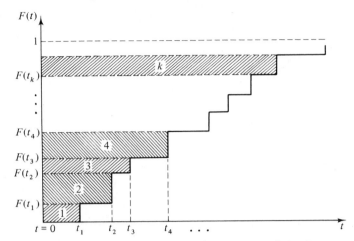

Figure 2.1.6   Illustration of Formula 2.1.1. $1 \div k$ is the area numerically equal to the sum of the first to the kth summands in Formula (2.1.1).

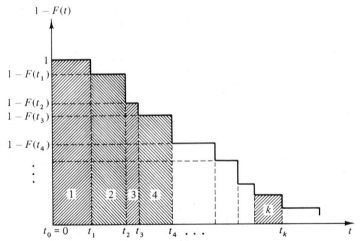

Figure 2.1.7   Illustration of Formula (2.1.2). $1 \div k$ is the area numerically equal to the sum of the first to the kth summands in Formula (2.1.2).

Example 2.1.2   Suppose a unit has an exponential distribution law for the operating time up to failure with distribution parameter

$$\lambda = 2.5 \cdot 10^{-5} \frac{1}{\text{hr}}.$$

We wish to compute the basic reliability measures of the unit.

Solution. 1. Probability of failure-free operation for the initial $t_0 = 2000$ hours:

$$P(t_0) = e^{-\lambda t_0} = e^{-2.5 \cdot 10^{-5} \cdot 2000} = e^{-0.05} = 0.9512.$$

Approximate value:

$$\tilde{P}(t_0) = 1 - \lambda t_0 = 1 - 2.5 \cdot 10^{-5} \cdot 2000 = 0.95.$$

2. Probability of failure before time t_0:

$$Q(t_0) = 1 - e^{-\lambda t_0} = 1 - 0.9512 = 0.0488.$$

Approximate value:

$$\tilde{Q}(t_0) = \lambda t_0 = 2.5 \cdot 10^{-5} \cdot 2000 = 0.05.$$

3. Probability of failure-free operation from $t = 500$ hours to $t + t_0 = 2500$ hours, given the first 500 hours of operation to be failure-free:

$$P(t, t + t_0) = e^{-\lambda t_0} = e^{-2.5 \cdot 10^{-5} \cdot 2000} = e^{-0.05} = 0.9512.$$

Approximate value:

$$\tilde{P}(t, t + t_0) = 1 - \lambda t_0 = 1 - 2.5 \cdot 10^{-5} \cdot 2000 = 0.95.$$

4. Probability of failure in the interval from $t = 500$ hours to $t + t_0 = 2500$ hours, given the first 500 hours to be failure-free:

$$Q(t, t + t_0) = 1 - e^{-\lambda t_0} = 0.9512 = 0.0488.$$

Approximate value:

$$\tilde{Q}(t, t + t_0) = \lambda t_0 = 2.5 \cdot 10^{-5} \cdot 2000 = 0.05.$$

5. Mean operating time to failure:

$$T = \frac{1}{\lambda} = \frac{1}{2.5 \cdot 10^{-5}} = 40,000 \quad \text{hours.}$$

2.2 REPAIRABLE UNIT

We consider repairable units for each of which the intervals of failure-free operation $\theta_1, \theta_2, \ldots, \theta_i, \ldots$ have a common distribution law, $F(t) = \mathcal{P}(\theta \leq t)$, and such that the repair intervals $\xi_1, \xi_2, \ldots, \xi_i, \ldots$ also have a common distribution law $F_r(t) = \mathcal{P}\{\xi \leq t\}$ (Figure 2.2.1). A diagram of the passages of a particular unit from one state to the other is shown in Figure 2.2.2.

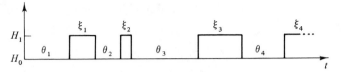

Figure 2.2.1   Random Process Corresponding to a Sequence of Alternating Intervals of Operationality and Idleness of a Unit with Repair.

Figure 2.2.2   Diagram of Transitions of a Unit with Repair from One State to the Other for Repairable Units.

Table 2.2.1 shows the basic reliability measures for arbitrary distribution laws $F(t)$ and $F_r(t)$.

Table 2.2.2 gives the basic reliability measures of a unit for exponential distribution laws of the time of failure-free operation and repair time, $F(t) = 1 - e^{-\lambda t}$, $F_r(t) = 1 - e^{-\mu t}$, respectively.

Approximate values of $P(t_0)$ and $Q(t_0)$ are given when $\lambda t \ll 1$, and $\gamma \ll 1$ for the remaining measures.

The coefficients $K(t)$ and $k(t)$ correspond to the case where the unit is in an operational state at time $t = 0$, and the coefficients $K^0(t)$ and $k^0(t)$ to the case where the unit is in a failed state at time $t = 0$. The formulas for the readiness and idleness coefficients are illustrated in Figures 2.2.3 and 2.2.4.

We do not give a formula for the nonstationary reliability coefficient for the case where the unit is in a failed state at the initial moment of time because, in practice, its use is very restricted.

Remark.   Use of the expression

$$R(t_0) = KP(t_0) \tag{2.2.1}$$

as a precise expression for the reliability coefficient is generally *wrong*. This formula is valid only when the time of failure-free operation of a unit has an exponential distribution—that is, when

$$P(t) = e^{-\lambda t}.$$

Table 2.2.1 Repairable Unit: Arbitrary Distribution Law of Time between Failures, $F(t)$, and Arbitrary Distribution Law of Repair Time, $F_r(t)$

$$\gamma = \frac{\tau}{T}$$

MEASURE	EXACT VALUE	APPROXIMATE VALUE	ERROR
T	$\displaystyle\int_0^\infty x f(x)\, dx = \int_0^\infty x\, dF(x) = \int_0^\infty P(x)\, dx$	—	—
τ	$\displaystyle\int_0^\infty x f_r(x)\, dx = \int_0^\infty x\, dF_r(x) = \int_0^\infty [1 - F_r(x)]\, dx$	—	—
$P(t_0)$	$1 - F(t_0)$	—	—
$Q(t_0)$	$F(t_0)$	—	—
K	$\dfrac{T}{T+\tau} = \dfrac{1}{1+\gamma}$	$1 - \gamma$	$\delta_+ < \gamma^2$
k	$\dfrac{\tau}{T+\tau} = \dfrac{\gamma}{1+\gamma}$	γ	$\delta_- < \gamma^2$
$R(t_0)$	$K\dfrac{1}{T}\displaystyle\int_{t_0}^\infty [1 - F(x)]\, dx$	For units with time-increasing failure intensity $$Ke^{-t_0/T}$$	$\delta_+ < \dfrac{K}{2}\left(\dfrac{t_0}{T}\right)^2$
		For the case $\gamma \ll 1$ and $t_0 \ll T$ $$1 - \dfrac{t_0 + \tau}{T} = 1 - \gamma - \dfrac{t_0}{T}$$	$\delta_+ < 2.5\left[\dfrac{\max(t_0,\tau)}{T}\right]^2$

Table 2.2.2 Repairable Unit: Exponential Distribution Law of Time between Failures, $F(t) = 1 - e^{-\lambda t}$, and Exponential Distribution Law of Repair Time $F_r(t) = 1 - e^{-\mu t}$

$$\gamma = \frac{\lambda}{\mu} = \frac{\tau}{T}$$

MEASURE	EXACT VALUE	APPROXIMATE VALUE	CONDITION ASSUMED SATISFIED	ERROR
$P(t_0)$	$e^{-\lambda t_0}$	$1 - \lambda t_0$	$\lambda t_0 \ll 1$	$\delta_+ < \frac{1}{2}(\lambda t_0)^2$
$Q(t_0)$	$1 - e^{-\lambda t_0}$	λt_0	$\lambda t_0 \ll 1$	$\delta_- < \frac{1}{2}(\lambda t_0)^2$
T	$\dfrac{1}{\lambda}$	—	—	—
τ	$\dfrac{1}{\mu}$	—	—	—
K	$\dfrac{\mu}{\lambda + \mu} = \dfrac{T}{T + \tau} = \dfrac{1}{1 + \gamma}$	$1 - \gamma$	$\gamma \ll 1$	$\delta_+ < \gamma^2$
k	$\dfrac{\lambda}{\lambda + \mu} = \dfrac{\tau}{T + \tau} = \dfrac{\gamma}{1 + \gamma}$	γ	$\gamma \ll 1$	$\delta_- < \gamma^2$
$K(t)$	$K + ke^{-(\lambda+\mu)t}$	$1 - \gamma(1 - e^{-\mu t})$	$\gamma \ll 1$	$\delta_+ \sim \gamma^2(1 - e^{-\mu t})$
$K^0(t)$	$K(1 - e^{-(\lambda+\mu)t})$	$(1 - \gamma)(1 - e^{-\mu t})$	$\gamma \ll 1$	$\delta_- \sim \gamma^2(1 - e^{-\mu t})$

$k(t)$	$k(1 - e^{-(\lambda+\mu)t})$	$\gamma(1 - e^{-\mu t})$	$\gamma \ll 1$	$\delta_- \sim \gamma^2(1 - e^{-\mu t})$
$k^0(t)$	$k + Ke^{-(\lambda+\mu)t}$	$e^{-\mu t} + \gamma(1 - e^{-\mu t})$	$\gamma \ll 1$	$\delta_+ \sim \gamma^2(1 - e^{-\mu t})$
$R(t_0)$	$Ke^{-\lambda t_0}$	$1 - \lambda(\tau + t_0) = 1 - \gamma - \lambda t_0$	$\gamma \ll 1, \lambda t_0 \ll 1$	$\delta_+ < 2.5[\lambda \max(t_0, \tau)]^2$
		$[1 - \gamma(1 - e^{-\mu t})]e^{-\lambda t_0}$	$\gamma \ll 1$	$\delta_+ < \gamma^2(1 - e^{-\mu t})e^{-\lambda t_0}$
		$(K + ke^{-(\lambda+\mu)t})(1 - \lambda t_0)$	$\lambda t_0 \ll 1,$ γ is arbitrary	$\delta_+ < \dfrac{(\lambda t_0)^2}{2}(K + ke^{-(\lambda+\mu)t})$
$R(t, t_0)$	$(K + ke^{-(\lambda+\mu)t})e^{-\lambda t_0}$	$(1 - \lambda t)e^{-\lambda t_0}$	$\gamma \ll 1, \mu t \ll 1$	$\delta_+ < \gamma \lambda t e^{-\lambda t_0}$
		$[1 - \gamma(1 - e^{-\mu t})](1 - \lambda t_0)$	$\gamma \ll 1, \lambda t_0 \ll 1$	$\delta_+ < \gamma^2(1 - e^{-\mu t}) + \dfrac{(\lambda t_0)^2}{2}$
		$1 - \lambda(t_0 + t)$	$\gamma \ll 1, \lambda t_0 \ll 1, \mu t \ll 1$	$\delta_+ < \lambda^2 t_0\left(t + \dfrac{t_0}{2}\right) + \lambda t(\mu t + \gamma)$

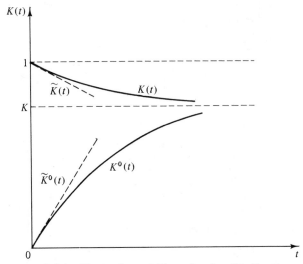

Figure 2.2.3   Illustration of Formulas for the Readiness Coefficient.

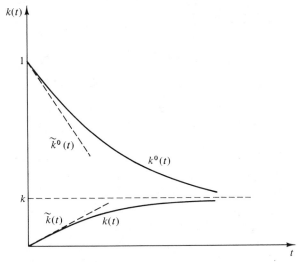

Figure 2.2.4   Illustration of Formulas for the Idleness Coefficient.

Example 2.2.1 Suppose the distribution law of the operating time of a unit up to failure is given by the following table:

t	0–1	1–2	2–3	3–4	4–5	5–6	6–7	7–8	8–9	9–10	10–
$F(t)$	0	0.03	0.08	0.20	0.45	0.65	0.80	0.90	0.95	0.98	1.00

The repair time is a random variable that takes on the value $\xi_1 = 0.2$ hour with probability $p_1 = 0.8$ and the value $\xi_2 = 1.5$ hours with probability $p_2 = 0.2$.

We want to compute the basic reliability measures for failure-free operation of the unit for one hour, and also the readiness and idleness coefficients and the mean values of the time of failure-free operation and the repair time.

Solution. 1. Probability of failure-free operation for the initial $t_0 = 2$ hours:

$$P(t_0) = 1 - F(t_0) = 1 - 0.03 = 0.97.$$

2. Probability of failure before time $t_0 = 2$ hours:

$$Q(t_0) = F(t_0) = 0.03.$$

3. Mean time of failure-free operation (using the result for T in Example 2.1.1):

$$T = 4.96 \quad \text{hours.}$$

4. Mean repair time. In this case the formula for τ can be written as the following sum:

$$\tau = \sum_{i=1}^{n} \xi_i p_i$$

or, under the conditions of the given concrete example,

$$\tau = \xi_1 p_1 + \xi_2 p_2 = 0.2 \cdot 0.8 + 1.5 \cdot 0.2 = 0.46.$$

5. Readiness coefficient:

$$K = \frac{T}{T + \tau} = \frac{4.96}{4.96 + 0.46} = 0.915.$$

Approximate value:

$$\tilde{K} = 1 - \frac{\tau}{T} = 1 - \frac{0.46}{4.96} = 0.908.$$

6. Idleness coefficient:

$$k = \frac{\tau}{T + \tau} = \frac{0.46}{4.96 + 0.46} = 0.085.$$

Approximate value:

$$\tilde{k} = \frac{\tau}{T} = \frac{0.46}{4.96} = 0.092.$$

7. Reliability coefficient for a time interval of length $t_0 = 2$ hours:

$$R(t_0) = Ke^{-t_0/T} = 0.915 \cdot e^{-2/4.96} = 0.915 \cdot 0.670 = 0.613.$$

Approximate value:

$$\tilde{R}(t_0) = 1 - \frac{t_0 + \tau}{T} = 1 - \frac{2 + 0.46}{4.96} = 0.508.$$

Remark. If we had used Formula (2.2.1), we would have obtained the incorrect result

$$R(t_0) = KP(t_0) = 0.915 \cdot 0.97 = 0.885.$$

[Here we have taken the value $P(2) = 1 - F(2) = 0.97$ from the table for the given example.]

Example 2.2.2 Suppose a unit has exponential distribution laws for operating time up to failure and repair time, with parameters $\lambda = 0.04 \frac{1}{\mathrm{hr}}$ and $\mu = 2 \frac{1}{\mathrm{hr}}$, respectively.

We want to compute the basic reliability measures of the unit.

Solution. 1. Probability of failure-free operation for an initial $t_0 = 2$ hours:

$$P(t_0) = e^{-\lambda t_0} = e^{-0.04 \cdot 2} = 0.923.$$

Approximate value:

$$\tilde{P}(t_0) = 1 - \lambda t_0 = 1 - 0.04 \cdot 2 = 0.92.$$

2. Probability of failure before the time $t_0 = 2$ hours:

$$Q(t_0) = 1 - e^{-\lambda t_0} = 1 - 0.923 = 0.077.$$

Approximate value:

$$\tilde{Q}(t_0) = \lambda t_0 = 0.04 \cdot 2 = 0.08.$$

3. Mean time of failure-free operation:

$$T = \frac{1}{\lambda} = \frac{1}{0.04} = 25 \quad \text{hr.}$$

4. Mean repair time:

$$\tau = \frac{1}{\mu} = \frac{1}{2} = 0.5 \quad \text{hr.}$$

5. Readiness coefficient:

$$K = \frac{T}{T + \tau} = \frac{2}{2 + 0.04} = 0.9804.$$

Approximate value:

$$\tilde{K} = 1 - \frac{\tau}{T} = 1 - \frac{0.04}{2} = 0.98.$$

6. Idleness coefficient:

$$k = \frac{\tau}{T + \tau} = \frac{0.04}{2 + 0.04} = 0.0196.$$

Approximate value:

$$\tilde{k} = \frac{\tau}{T} = \frac{0.04}{2} = 0.02.$$

7. Nonstationary readiness coefficient for $t = 2$ hours for the case where the unit is in an operational state at the moment $t = 0$:

$$K(t) = K + ke^{-(\lambda+\mu)t} = 0.9804 + 0.0196 \cdot e^{-(2+0.04)\cdot 2} = 0.9808.$$

Approximate value:

$$\tilde{K}(t) = 1 - \frac{\lambda}{\mu}(1 - e^{-\mu t}) = 1 - \frac{0.04}{2}(1 - e^{-2 \cdot 2}) = 0.9804.$$

8. Nonstationary idleness coefficient for the same case:

$$k(t) = k(1 - e^{-(\lambda+\mu)t}) = 0.0196(1 - e^{-(2+0.04)\cdot 2}) = 0.0192.$$

Approximate value:

$$\tilde{k}(t) = \frac{\lambda}{\mu}(1 - e^{-\mu t}) = \frac{0.04}{2}(1 - e^{-2 \cdot 2}) = 0.0196.$$

9. Reliability coefficient for a time interval of length $t_0 = 2$ hours:

$$R(t_0) = Ke^{-\lambda t_0} = \frac{2}{2 + 0.04} \cdot e^{-0.04 \cdot 2} = 0.9146.$$

Approximate value:

$$\tilde{R}(t_0) = 1 - \frac{\tau + t_0}{T} = 1 - \frac{0.5 + 2}{2.5} = 0.9.$$

10. Nonstationary reliability coefficient for the time interval from $t = 1$ hour to $t + t_0 = 3$ hours:

$$R(t, t + t_0) = (K + ke^{-(\lambda+\mu)t})e^{-\lambda t_0}$$
$$= (0.9804 + 0.0196e^{-(0.04+2)\cdot 1})e^{-0.04\cdot 2} = 0.906.$$

Approximate value:

$$\tilde{R}(t, t + t_0) = 1 - \lambda(t_0 + t) = 1 - 0.04\cdot(2 + 1) = 0.88.$$

BIBLIOGRAPHY

1. GNEDENKO, B. V., BELYAEV, Y. K., AND SOLOV'EV, A. D., *Mathematical Methods in Reliability Theory*, chap. 2. New York: Academic Press, Inc., 1968.
2. MALIKOV, I. M., POLOVKI, A. M., ROMANOV, N. A., AND CHUKREEV, P. A., *Basic Theory and Computation of Reliability*, 2d ed., chap. 3. Moscow: Sudpromgiz, 1960.
3. SHOR, Y. B., *Statistical Methods of Analysis and Control of Quality and Reliability*, chap. 24. Moscow: Soviet Radio, 1962.

3

REDUNDANCY IN NON-REPAIRABLE SYSTEMS

PRELIMINARY REMARKS

In this chapter we present methods for determining the basic reliability measures of redundant systems that either are actually unrepairable, as for example systems of one-time action, or are repairable systems for which, for some reason, it is necessary to use a different reliability measure up to the first failure.

We assume that the time of exchanging a stand-by unit for a failed operating unit is small and can be neglected, and the device accomplishing

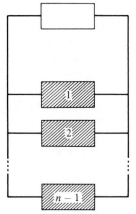

Figure 3.1.1 Block Diagram of the Model Used for the Reliability Computations for a System of One Operating and $n - 1$ Stand-by Units.

the exchange (such as a switch) itself is considered reliable enough so that it has no influence on the over-all system reliability, and therefore we need not take it into account when considering the reliability of the system as a whole. Moreover, we also assume that the failure of individual units in the system is instantaneous.

Below we give the basic reliability characteristics for a model of unrepairable systems with stand-by units based on these assumptions.

3.1 ARBITRARY DISTRIBUTION LAW OF OPERATING TIME OF A UNIT UP TO FAILURE

3.1.1   Stand-by of one unit by $n - 1$ units (Figure 3.1.1) in an operating state*

1.   The probability of failure-free operation of the system in the case of independent operation of all n units is determined from the formula

$$P(t_0) = 1 - \prod_{i=1}^{n} [1 - r_i(t_0)], \qquad (3.1.1)$$

where $r_i(t_0)$ is the probability of failure-free operation of the ith unit.

2.   The mean operating time of the system up to failure is determined from the formula

$$T = \int_0^\infty P(t) \, dt, \qquad (3.1.2)$$

where $P(t)$ is given by (3.1.1).

If the probabilities $r_i(t_0)$ are given graphically or are given numerically at a discrete set of time points, then the mean operating time up to failure can be determined from Formula (3.1.2) by numerical integration.

Example 3.1.1   A system is given that consists of one operating unit and $n - 1 = 2$ stand-by units that are in an operating state. All the units are identical and each has a distribution of operating time up to failure subject to Weibull's law:

$$r(t_0) = e^{-\lambda t_0^\alpha}$$

with parameter $\lambda = 0.01 \, \dfrac{1}{\text{hr}^2}$ and $\alpha = 2$.

---

* [*Editor's note:* In the American literature this would be called parallel operation of the n units.]

We wish to find the probability of failure-free operation of the system for 5 hours, and also to compute the mean operating time of the system up to failure.

Solution.   1.   According to Formula (3.1.1) we have

$$P(t_0) = 1 - \prod_{i=1}^{n} [1 - r_i(t_0)] = 1 - [1 - r(t_0)]^3.$$

First we compute the probability of failure-free operation $r(t_0)$ of one unit for the time $t_0 = 5$ hours:

$$r(t_0) = e^{-\lambda_0 t^\alpha} = e^{-0.01 \cdot 5^2} \approx 0.7788.$$

Then the probability of failure-free operation of the system for an initial period of $t_0 = 5$ hours is

$$P(t_0) = 1 - [1 - r(t_0)]^3 = 1 - (1 - 0.7788)^3 \approx 0.989.$$

2.   To determine the mean operating time of the stand-by system up to failure we first find the general form of the solution. According to Formula (3.1.2) we have

$$T = \int_0^\infty [1 - (1 - e^{-\lambda t^\alpha})^n] \, dt = \int_0^\infty \left[1 - \sum_{i=0}^{n} C_n^i (-1)^i e^{-i\lambda t^\alpha} \right] dt$$

$$= \int_0^\infty \sum_{i=1}^{n} C_n^i (-1)^{i-1} e^{-i\lambda t^\alpha} \, dt = \sum_{i=1}^{n} (-1)^{i-1} C_n^i \int_0^\infty e^{-i\lambda t^\alpha} \, dt$$

$$= \frac{\lambda^{-1/\alpha}}{\alpha} \Gamma \left(\frac{1}{\alpha} \right) \sum_{i=1}^{n} (-1)^{i-1} i^{-1/\alpha} C_n^i,$$

where $\Gamma(x)$ is the gamma function.
Thus we finally obtain

$$T = \lambda^{-1/\alpha} \Gamma \left(1 + \frac{1}{\alpha} \right) \sum_{i=1}^{n} (-1)^{i-1} C_n^i i^{-1/\alpha}.$$

After substituting the numerical values of the given example, we obtain

$$T = \frac{1}{\sqrt{0.01}} \Gamma(1.5) \left[C_3^1 - C_3^2 \frac{1}{\sqrt{2}} + C_3^3 \frac{1}{\sqrt{3}} \right]$$
$$= 10 \cdot 0.886(3 - 2.12 + 0.578) = 12.9 \quad \text{hr.}$$

3.1.2   Stand-by of one unit by $n - 1$ units operating independently that are in a partially operating state

1.   The probability of failure-free operation of the system is defined by the following recursion relation:

$$P_{(n)}(t_0) = 1 + \int_0^{t_0} [1 - r_n^{(0)}(\theta) r_n^{(p)}(\theta, t_0)] \, dP_{(n-1)}(\theta), \qquad (3.1.3)$$

where $r_n^{(0)}(\theta)$ is the probability of failure-free operation of the nth unit in a partial operating state up to time θ; $r_n^{(p)}(\theta, t_0)$ is the probability of failure-free operation of the nth unit in an operating state from time θ to time t_0; and $P_{(i)}(t)$ is the probability of failure-free operation of a system of one operating unit and $(i - 1)$ stand-by units. We take $P_{(1)}(t) = r_1^{(p)}(0, t) = r_1^{(p)}(t)$.

2.   The mean operating time of the system up to failure is given by the formula

$$T = \int_0^\infty P_{(n)}(t) \, dt, \qquad (3.1.4)$$

where $P_{(n)}(t)$ is determined from (3.1.3).

Example 3.1.2 Suppose we have a redundant system in which a single stand-by unit is in a partially operating state. It is assumed that the probability of failure-free operation of each of the two units of the system is subject to Weibull's law,

$$r(t_0) = e^{-\lambda t_0^\alpha},$$

where

$$\lambda = \lambda_p = 0.2 \ \frac{1}{\text{hr}^2}, \qquad \alpha = \alpha_p = 2$$

(for the unit in an operating state),

$$\lambda = \lambda_0 = 0.1 \ \frac{1}{\text{hr}}, \qquad \alpha = \alpha_0 = 1$$

(for the unit in a partially operating state).

We wish to determine an approximate value for the probability of failure-free operation of the redundant system for the time $t_0 = 1$ hour.

Solution. We use an expansion of the formula for the probability of failure-free operation:

$$r(t_0) = e^{-\lambda t_0^\alpha} \approx 1 - \lambda t_0^\alpha, \qquad \text{if } \lambda t_0^\alpha \ll 1.$$

Using Formula (3.1.3), we successively obtain

$$P_{(1)}(t_0) = r_1^{(p)}(t_0) = 1 - \lambda_p t_0^{\alpha_p},$$

$$P_{(2)}(t_0) = 1 + \int_0^{t_0} [1 - r_2^{(0)}(\theta) r_2^{(p)}(\theta, t_0)]\, dP_{(1)}(\theta)$$

$$= 1 - \int_0^{t_0} \{1 - [1 - \lambda_0 \theta^{\alpha_0}][1 - \lambda_p(t_0 - \theta)^{\alpha_p}]\} \lambda_p \alpha_p \theta^{\alpha_p - 1}\, d\theta$$

$$= 1 - \alpha_p \lambda_p \int_0^{t_0} \{1 - (1 - \lambda_0 \theta)[1 - \lambda_p(t_0 - \theta)^2]\} \theta\, d\theta$$

$$= 1 - \frac{\alpha_p \lambda_p}{3} \left(\lambda_0 t_0^3 + \frac{1}{4} \lambda_p t_0^4 - \frac{1}{10} \lambda_0 \lambda_p t_0^5 \right).$$

In this expression we have assumed that $r_2^{(p)}(\theta, t_0) = r_2^{(p)}(t_0 - \theta)$.

Now, substituting the numerical values, we obtain

$$P_{(2)}(t_0 = 1 \text{ hr}) \approx 0.98.$$

3.1.3 Stand-by of one unit by $n - 1$ units in a nonoperating state

1. The probability of failure-free operation of the system is given by the following recursion relation:

$$P_{(n)}(t_0) = 1 + \int_0^{t_0} [1 - r_n(t_0 - \theta)]\, dP_{(n-1)}(\theta), \qquad (3.1.5)$$

where $r_n(t)$ is the probability of failure-free operation of the nth unit; $P_i(t)$ is the probability of failure-free operation of a system of one operating and $i - 1$ stand-by units, with $P_{(1)}(t) = r_1(t)$.

2. The mean operating time up to failure is given by the formula

$$T = \sum_{i=1}^{n} T_i, \qquad (3.1.6)$$

where T_i is the mean operating time up to failure of the ith unit.

Example 3.1.3   We have a system of one operating and two stand-by units in a nonoperating state. The units in an operating state have a normal distribution of operating time up to failure:

$$r_i(t_0) = \frac{1}{\sigma_i \sqrt{2\pi}} \int_{t_0}^{\infty} e^{-(x-T_i)^2/2\sigma_i^2} \, dx$$

with parameters $T_1 = 100$ hours, $\sigma_1 = 10$ hours, $T_2 = 150$ hours, $\sigma_2 = 15$ hours, $T_3 = 100$ hours, $\sigma_3 = 20$ hours.

We want to find the probability of failure-free operation of the stand-by system for $t_0 = 300$ hours and to compute the mean operating time up to failure.

Solution.   In this example we can use the fact that the sum of mutually independent normally distributed random variables also has a normal distribution with mean equal to the sum of the mean values of the summed random variables and variance equal to the sum of the corresponding variances—that is,

$$P(t_0) = 1 - \frac{1}{\sigma_\Sigma \sqrt{2\pi}} \int_{-\infty}^{t_0} e^{-(\alpha-T)^2/2\sigma_\Sigma^2} \, dx,$$

where

$$T = T_1 + T_2 + T_3 = (100 + 150 + 300) = 350 \quad \text{hr},$$
$$\sigma_\Sigma = \sqrt{\sigma_1^2 + \sigma_2^2 + \sigma_3^2} = \sqrt{10^2 + 15^2 + 20^2} \approx 26.93 \quad \text{hr}.$$

Hence we find that for the time $t_0 = 300$ hours,

$$P(t_0) = P(t_0 = 300 \text{ hr}) = 1 - \frac{1}{\sqrt{2\pi}} \int_{-\infty}^{-(350-300)/26.93} e^{-x^2/2} \, dx = \Phi(1.86).$$

The value of $\Phi(u)$ is taken from Table A.3.1 in the Appendix. Then, after finding that $\Phi(u) = \Phi(1.86) = 0.968$, we obtain the required result:

$$P(t_0) = P(t_0 = 300 \text{ hr}) = 0.968.$$

3.1.4   Redundancy of a collection of units with operating stand-by

We consider a system of m operating units in series connection and $(n-1)$ stand-by units (Figure 3.1.2). Failure of the stand-by system occurs at the moment of the nth failure.

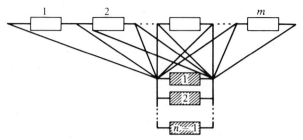

Figure 3.1.2 Block Diagram of the Model Used for the
Reliability Computations for a System of m
Operating and $n - 1$ Stand-by Units.

In order to determine the probability of failure-free operation of this
system when all the units are different, it is convenient to consider a
generating polynomial of the form

$$\pi(x) = (r_1 x + q_1)(r_2 x + q_2) \cdots (r_N x + q_N)$$
$$= B_N x^N + B_{N-1} x^{N-1} + \cdots + B_i x^i + \cdots + B_0 x^0,$$

where

$$r_i = r_i(t);$$
$$q_i = q_i(t) = 1 - r_i(t);$$

$N = m + (n - 1)$ is the total number of units in the system.

It is easy to see that the coefficient $B_i = B_i(t)$ for x^i is the probability
that precisely i of the N units remain unfailed at time t. Then:

1.   The probability of failure-free operation of the system is given by
the formula

$$P(t_0) = \sum_{i=m}^{N} B_i(t_0). \tag{3.1.7}$$

If all units have the same reliability, then

$$P(t_0) = \sum_{i=m}^{N} C_N^i [r(t_0)]^i [1 - r(t_0)]^{N-i}. \tag{3.1.8}$$

2.   The mean operating time up to failure is given by the formula

$$T = \int_0^\infty P(t) \, dt, \tag{3.1.9}$$

where $P(t)$ is given by Formulas (3.1.7) and (3.1.8) for the respective
cases.

Example 3.1.4   Suppose we have a system of two operating and two stand-by units in an operating state. The probability of failure-free operation of the ith unit ($i = 1, 2, 3, 4$) is subject to Weibull's law:

$$r_i(t_0) = e^{-\lambda_i t_0^{\alpha_i}},$$

where

$$\lambda_1 = 0.04 \; \frac{1}{\text{hr}}, \qquad \alpha_1 = 1;$$

$$\lambda_2 = 0.03 \; \frac{1}{\text{hr}^2}, \qquad \alpha_2 = 2;$$

$$\lambda_3 = 0.02 \; \frac{1}{\text{hr}^3}, \qquad \alpha_3 = 3;$$

$$\lambda_4 = 0.01 \; \frac{1}{\text{hr}^4}, \qquad \alpha_4 = 4.$$

We wish to determine the probability of failure-free operation of the system for a period of $t_0 = 1$ hour.

Solution.   By hypothesis $N = 4$, $m = 2$, $n - 1 = 2$.

The generating polynomial has the form

$$
\begin{aligned}
\pi(x) &= (r_1 x + q_1)(r_2 x + q_2)(r_3 x + q_3)(r_4 x + q_4) \\
&= r_1 r_2 r_3 r_4 x^4 + (r_1 r_2 r_3 q_4 + r_1 r_2 r_4 q_3 + r_1 r_3 r_4 q_2 + r_2 r_3 r_4 q_1) \cdot x^3 \\
&\quad + (r_1 r_2 q_3 q_4 + r_1 r_3 q_2 q_4 + r_1 r_4 q_2 q_3 + r_2 r_3 q_1 q_4 + r_2 r_4 q_1 q_3 + r_3 r_4 q_1 q_2) \cdot x^2 \\
&\quad + (r_1 q_2 q_3 q_4 + r_2 q_1 q_3 q_4 + r_3 q_1 q_2 q_4 + r_4 q_1 q_2 q_3) \cdot x + q_1 q_2 q_3 q_4.
\end{aligned}
$$

Using Formula (1.3.7), we obtain

$$
\begin{aligned}
P(t_0) &= \sum_{i=2}^{4} B_i(t_0) = B_2(t_0) + B_3(t_0) + B_4(t_0) \\
&= (r_1 r_2 q_3 q_4 + r_1 r_3 q_2 q_4 + r_1 r_4 q_2 q_3 + r_2 r_3 q_1 q_4 + r_2 r_4 q_1 q_3 + r_3 r_4 q_1 q_2) \\
&\quad + (r_1 r_2 r_3 q_4 + r_1 r_2 r_4 q_3 + r_1 r_3 r_4 q_2 + r_2 r_3 r_4 q_1) + r_1 r_2 r_3 r_4.
\end{aligned}
$$

By hypothesis

$$
\begin{aligned}
r_1(t_0) &= e^{-\lambda_1 t_0^{\alpha_1}} = e^{-0.04} \approx 0.96; & q_1(t_0) &\approx 0.04, \\
r_2(t_0) &= e^{-\lambda_2 t_0^{\alpha_2}} = e^{-0.03} \approx 0.97; & q_2(t_0) &\approx 0.03, \\
r_3(t_0) &= e^{-\lambda_3 t_0^{\alpha_3}} = e^{-0.02} \approx 0.98; & q_3(t_0) &\approx 0.02, \\
r_4(t_0) &= e^{-\lambda_4 t_0^{\alpha_4}} = e^{-0.01} \approx 0.99; & q_4(t_0) &\approx 0.01.
\end{aligned}
$$

Using these values, we obtain as the desired probability

$$P(t_0) \approx 0.997.$$

3.2 GENERAL METHOD FOR DETERMINING THE RELIABILITY OF NONREPAIRABLE REDUNDANT SYSTEMS WITH EXPONENTIALLY DISTRIBUTED OPERATING TIME UP TO FAILURE OF INDIVIDUAL UNITS

We consider a system consisting of m operating and $n - 1$ stand-by units. All $N = m + (n - 1)$ units may fail, where a succession of failures is subject to the following conditions:

1.   If the jth failure has occurred by time t, then, independent of the moment of occurrence of these j failures, the probability that a failure occurs in a small interval of time $(t, t + \Delta t)$ is

$$\Lambda_j \, \Delta t + o(\Delta t),$$

while the probability that no failure occurs in this time interval is

$$1 - \Lambda_j \, \Delta t + o(\Delta t).$$

2.   After the occurrence of the nth failure the nonrepairable system is in a failed state and no further changes occur in the system; therefore

$$\Lambda_n = 0.$$

If the jth failure has occurred at the moment t, then we say that the system has passed into the state H_j at that time. In other words, the system may be in any of a finite number of states corresponding to the number of failed units: $H_0, H_1, \ldots, H_j, \ldots, H_{n-1}, H_n$. By condition 2, the last state, H_n, is the failed or "terminal" state of the system. The state H_j is the state in which the system has j failed units.

A diagram of passages of the given system is shown in Figure 3.2.1, in accordance with which we obtain a system of differential equations of the following form:

$$p_j'(t) = \Lambda_{j-1} p_{j-1}(t) - \Lambda_j p_j(t), \qquad 0 \le j \le n, \Lambda_{-1} = \Lambda_n = 0. \quad (3.2.1)$$

Here $p_j(t)$ is the probability that at the moment of time t the system is in state H_j. We assume that these probabilities satisfy initial conditions of the form

$$p_0(0) = 1, \quad p_j(0) = 0, \quad 1 \le j \le n. \quad (3.2.2)$$

$$H_0 \xrightarrow{\Lambda_0} H_1 \xrightarrow{\Lambda_1} H_2 \xrightarrow{\Lambda_2} \cdots \xrightarrow{\Lambda_{j-1}} H_j \xrightarrow{\Lambda_j} \cdots \xrightarrow{\Lambda_{n-2}} H_{n-1} \xrightarrow{\Lambda_{n-1}} H_n$$

Figure 3.2.1   Diagram of Passages from One State to Another by a System of k Operating and $n - 1$ Stand-by Units.

To find the distribution law $\mathcal{P}\{\theta_n \leq t\} = \mathcal{P}_n(t)$ of the random variable θ_n, the random moment of time when the system enters the failed state H_n [it is easy to see that $\mathcal{P}_n(t) = p_n(t)$], we apply the Laplace transform to the system of differential equations (3.2.1):

$$a_j(s) = \int_0^\infty p_j(t)e^{-st}\, dt. \tag{3.2.3}$$

After integrating by parts, we establish the validity of the following equality:

$$\int_0^\infty p_j'(t)e^{-st}\, dt = -p_j(0) + sa_j(s). \tag{3.2.4}$$

Then the system of differential equations (3.2.1) is transformed into a system of algebraic equations of the form

$$\Lambda_{j-1}a_{j-1}(s) - (\Lambda_j + s)a_j(s) = -p_j(0), \qquad 0 \leq j \leq n,\ \Lambda_{-1} = \Lambda_n = 0. \tag{3.2.5}$$

By Cramer's rule,

$$a_n(s) = \frac{\Delta_n(s)}{\Delta(s)}, \tag{3.2.6}$$

where

$$
\Delta(s) =
\begin{vmatrix}
-(\Lambda_0 + s) & 0 & 0 & \cdots & & & \\
\Lambda_0 & -(\Lambda_1 + s) & 0 & \cdots & & & \\
0 & \Lambda_1 & -(\Lambda_2 + s) & \cdots & & & \\
\vdots & & & & & & \\
0 & 0 & 0 & \cdots & 0 & 0 & 0 \\
0 & 0 & 0 & \cdots & 0 & 0 & 0 \\
0 & 0 & 0 & \cdots & 0 & 0 & 0 \\
& & & & \vdots & & \\
& & & \cdots & -(\Lambda_{n-2} + s) & 0 & 0 \\
& & & \cdots & \Lambda_{n-2} & -(\Lambda_{n-1} + s) & 0 \\
& & & \cdots & 0 & \Lambda_{n-1} & -s
\end{vmatrix}
$$

$$= -s[-(\Lambda_{n-1} + s)][-(\Lambda_{n-2} + s)]\cdots[-(\Lambda_0 + s)]$$

$$= (-1)^n \prod_{j=0}^{n-1} (\Lambda_j + s), \tag{3.2.7}$$

$$\Delta_n(s) = \begin{vmatrix} -(\Lambda_0 + s) & 0 & 0 & \cdots & 0 & 0 & -1 \\ \Lambda_0 & -(\Lambda_1 + s) & 0 & \cdots & 0 & 0 & 0 \\ 0 & \Lambda_1 & -(\Lambda_2 + s) & \cdots & 0 & 0 & 0 \\ \cdot & & & & \cdot & & \\ \cdot & & & & \cdot & & \\ \cdot & & & & \cdot & & \\ 0 & 0 & 0 & \cdots & -(\Lambda_{n-2} + s) & 0 & 0 \\ 0 & 0 & 0 & \cdots & \Lambda_{n-2} & -(\Lambda_{n-1} + s) & 0 \\ 0 & 0 & 0 & \cdots & 0 & \Lambda_{n-1} & 0 \end{vmatrix}$$

$$= (-1)^{n-1} \prod_{j=0}^{n-1} \Lambda_j. \tag{3.2.8}$$

According to (3.2.6) we have

$$a_n(s) = \frac{\prod_{j=0}^{n-1} \Lambda_j}{s \prod_{j=0}^{n-1} (\Lambda_j + s)}. \tag{3.2.9}$$

We find the desired probability $\mathcal{P}_n(t) = p_n(t)$ by inverting the Laplace transform:

$$p_n(t) = \frac{1}{2\pi i} \int_{g-i\infty}^{g+i\infty} \frac{\prod_{j=0}^{n-1} \Lambda_j e^{st}\, ds}{s \prod_{j=0}^{n-1} (\Lambda_j + s)}, \qquad g > 0. \tag{3.2.10}$$

By a theorem of Laurent we find

$$p_n(t) = 1 - \prod_{j=0}^{n-1} \Lambda_j \sum_{i=0}^{n-1} \frac{e^{-\Lambda_i t}}{\Lambda_i \prod_{\substack{l=0 \\ l \neq i}}^{n-1} (\Lambda_l - \Lambda_i)}. \tag{3.2.11}$$

Hence the probability of failure-free operation of the system is

$$P(t_0) = 1 - p_n(t_0) = \prod_{j=0}^{n-1} \Lambda_j \sum_{i=0}^{n-1} \frac{e^{-\Lambda_i t_0}}{\Lambda_i \prod_{\substack{l=0 \\ l \neq i}}^{n-1} (\Lambda_l - \Lambda_i)} . \qquad (3.2.12)$$

Remark. In practice it is not very convenient to compute the probability of failure-free operation by Formula (3.2.12) since, because the summands have different signs, it is necessary to use a large number of significant figures. The following three formulas can be used for approximate computations.

(a)   If $\Lambda_j t_0 \ll 1, 0 \leq j \leq n - 1$, then a lower bound for the probability of failure-free operation is determined by the formula

$$\tilde{P}(t_0) = 1 - \frac{t_0^n}{n!} \prod_{j=0}^{n-1} \Lambda_j. \qquad (3.2.12a)$$

The relative error of this formula does not exceed the quantity

$$\delta_+ < \frac{t_0}{n+1} \sum_{j=0}^{n-1} \Lambda_j.$$

(b)   If $\Lambda_j t_0$ are not small, $0 \leq j \leq n - 1$; and n is large, then the approximate formula for the probability of failure-free operation has the form

$$\tilde{P}(t_0) = 1 - \frac{t_0^n}{n!} \prod_{j=0}^{n-1} \Lambda_j \exp\left(-\frac{t_0}{n} \sum_{i=0}^{n-1} \Lambda_i\right). \qquad (3.2.12b)$$

The main term of the relative error is

$$\delta \sim \frac{t_0^{2n}}{2n^2} \sum_{j=0}^{n-1} \Lambda_j^2.$$

(c)   If

$$\lim_{n \to \infty} \frac{\displaystyle\sum_{j=0}^{n-1} \frac{1}{\Lambda_j^3}}{\left(\displaystyle\sum_{j=0}^{n-1} \frac{1}{\Lambda_j^2}\right)^{3/2}} = 0,$$

then for large n we have the following approximate formula:

$$\tilde{P}(t_0) \approx 1 - \Phi \left(\frac{t_0 - \sum\limits_{j=0}^{n-1} \frac{1}{\Lambda_j}}{\sqrt{\sum\limits_{i=0}^{n-1} \frac{1}{\Lambda_i^2}}} \right), \qquad (3.2.12c)$$

where

$$\Phi(u) = \frac{1}{\sqrt{2\pi}} \int_{-\infty}^{u} e^{-x^2/2} \, dx.$$

The mean operating time of the system up to failure is

$$T = \sum_{j=0}^{n-1} \frac{1}{\Lambda_j}. \qquad (3.2.13)$$

3.3 PARTICULAR CASES OF DETERMINING THE RELIABILITY OF REDUNDANT SYSTEMS WHEN INDIVIDUAL UNITS HAVE AN EXPONENTIAL DISTRIBUTION LAW FOR OPERATING TIME UP TO FAILURE

We investigate the reliability measures of a system of m operating and $n - 1$ stand-by units (Figure 3.1.2), and in particular a system of m operating units and $n - 1 = 1$ stand-by units (Figure 3.3.1) and a system of m operating and $n - 1 = 2$ stand-by units (Figure 3.3.2).

A diagram of passages of the system from state to state is shown in Figure 3.2.1, where H_j is the state of the system in which j ($j = 0, 1, \ldots, n$) of the $N = m + (n - 1)$ units have failed; H_n is the failed state of the system; Λ_j is the intensity of passage (failure of one of its units) of the system from the state where the system had j failed units to the state

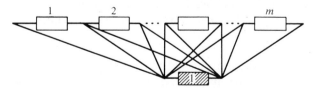

Figure 3.3.1 Block Diagram of the Model Used for Reliability Computations for a System of k Operating Units and One Stand-by Unit.

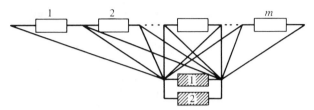

Figure 3.3.2 Block Diagram of the Model Used for Reliability Computations for a System of k Operating and Two Stand-by Units.

where the system has one more failed unit—that is, $j + 1$ ($j = 0, 1, \ldots, n - 1$).

We give the reliability measures of systems with one, two, and $n - 1$ stand-by units for the general case and for three particular cases, namely, the cases of operating, partially operating, and nonoperating stand-by units:

In the operating case:

$$\Lambda_j = m\lambda + [n - (j + 1)]\lambda = (N - j)\lambda, \qquad 0 \leq j \leq n - 1,$$

where λ is the failure intensity of a unit.

For partially operating units:

$$\Lambda_j = m\lambda + [n - (j + 1)]\nu\lambda, \qquad 0 \leq j \leq n - 1,$$

where ν is the load coefficient of the stand-by (operating proportion of a partially operating unit)—that is, the failure intensity of a stand-by unit is $\nu\lambda$, $0 \leq \nu \leq 1$.

In the nonoperating case:

$$\Lambda_j = m\lambda, \qquad 0 \leq j \leq n - 1.$$

In Tables 3.3.1 and 3.3.2 we give exact values for the basic reliability coefficients of a system with $n - 1$ stand-by units.

Next we present examples of the computation of reliability measures.

Example 3.3.1 Consider a system with $m = 4$ operating and $n - 1 = 2$ stand-by units. The intensity of passage of the system from the state with six operational units to a state with at least one failed unit is

$$\Lambda_0 = 6\lambda.$$

If failure of any one unit has occurred, then the remaining operational units begin operating in an overloaded state, where the intensity of passage of the system from the state with one failed unit to the state with two failed units is

$$\Lambda_1 = 10\lambda.$$

Table 3.3.1   System with $n - 1$ Stand-by Units:
General Case

SYSTEM	MEASURE	EXACT VALUE
m operating units and 1 stand-by	T	$\dfrac{\Lambda_0 + \Lambda_1}{\Lambda_0 \Lambda_1}$
	$P(t_0)$	$\dfrac{1}{\Lambda_0 - \Lambda_1}(\Lambda_0 e^{-\Lambda_1 t_0} - \Lambda_1 e^{-\Lambda_0 t_0})$
m operating units and 2 stand-bys	T	$\dfrac{\Lambda_0 \Lambda_1 + \Lambda_0 \Lambda_2 + \Lambda_1 \Lambda_2}{\Lambda_0 \Lambda_1 \Lambda_2}$
	$P(t_0)$	$\dfrac{[\Lambda_1 \Lambda_2(\Lambda_1 - \Lambda_2)e^{-\Lambda_0 t_0} - \Lambda_0 \Lambda_2(\Lambda_0 - \Lambda_2)e^{-\Lambda_1 t_0} + \Lambda_0 \Lambda_1(\Lambda_0 - \Lambda_1)e^{-\Lambda_2 t_0}]}{(\Lambda_0 - \Lambda_1)(\Lambda_1 - \Lambda_2)(\Lambda_0 - \Lambda_2)}$
m operating units and $n - 1$ stand-bys	T	$\displaystyle\sum_{j=0}^{n-1} \frac{1}{\Lambda_j}$
	$P(t_0)$	$\displaystyle\prod_{j=0}^{n-1} \Lambda_j \sum_{i=0}^{n-1} \frac{e^{-\Lambda_i t_0}}{\Lambda_j \displaystyle\prod_{\substack{l=0 \\ l \neq i}}^{n-1} (\Lambda_l - \Lambda_i)}$

The intensity of passage of the system into the failed state—that is, with three failed units—is

$$\Lambda_2 = 12\lambda.$$

Let

$$\lambda = 0.1 \frac{1}{\text{hr}}.$$

We wish to determine the basic reliability measures of such a system.
Solution.   From Table 3.3.1 we find:
1.   The mean operating time of the system up to failure is

$$T = \frac{\Lambda_0 \Lambda_1 + \Lambda_0 \Lambda_2 + \Lambda_1 \Lambda_2}{\Lambda_0 \Lambda_1 \Lambda_2} = \frac{6\lambda \cdot 10\lambda + 6\lambda \cdot 12\lambda + 10\lambda \cdot 12\lambda}{6\lambda \cdot 10\lambda \cdot 12\lambda}$$

$$= \frac{252}{720} \cdot \frac{1}{\lambda} = \frac{252}{720} \cdot \frac{1}{0.1} \quad \text{hr} = 3.5 \quad \text{hr}.$$

Table 3.3.2 System with $n - 1$ Stand-by Units in Various Operating States

SYSTEM	MEASURE	OPERATING STATE	PARTIALLY OPERATING STATE	NONOPERATING STATE
m operating units and 1 stand-by	T	$\dfrac{1}{\lambda}\cdot\dfrac{2m+1}{m(m+1)}$	$\dfrac{1}{\lambda}\cdot\dfrac{2m+v}{m(m+v)}$	$\dfrac{1}{\lambda}\cdot\dfrac{2}{m}$
	$P(t_0)$	$(m+1)e^{-m\lambda t_0} - me^{-(m+1)\lambda t_0}$	$\dfrac{1}{v}(m+v-me^{-v\lambda t_0})e^{-m\lambda t_0}$	$(1+m\lambda t_0)e^{-m\lambda t_0}$
m operating units and 2 stand-bys	T	$\dfrac{1}{\lambda}\cdot\dfrac{[m(m+1)+m(m+2)+(m+1)(m+2)]}{(m+1)(m+2)}$	$\dfrac{1}{\lambda}\cdot\dfrac{[m(m+v)+m(m+2v)+(m+v)(m+2v)]}{m(m+v)(m+2v)}$	$\dfrac{1}{\lambda}\cdot\dfrac{3}{m}$
	$P(t_0)$	$\tfrac{1}{2}e^{-m\lambda t_0}[(m+1)(m+2)-2m(m+2)e^{-\lambda t_0}+m(m+1)e^{-2\lambda t_0}]$	$\dfrac{1}{2v^2}e^{-m\lambda t_0}[(m+v)(m+2v)-2m(m+2v)e^{-v\lambda t_0}+m(m+v)e^{-2v\lambda t_0}]$	$\left[1+m\lambda t_0+\dfrac{(m\lambda t_0)^2}{2}\right]e^{-m\lambda t_0}$
m operating units and $n-1$ stand-bys	T	$\dfrac{1}{\lambda}\sum_{j=0}^{n-1}\dfrac{1}{N-j}$	$\dfrac{1}{\lambda}\sum_{j=0}^{N-m}\dfrac{1}{m+v_i}$	$\dfrac{1}{\lambda}\cdot\dfrac{n}{m}$
	$P(t_0)$	$(N-m+1)C_N^{N-1}\displaystyle\sum_{j=m}^{N}\dfrac{(-1)^{j-m}}{j}\times C_{N-m}^{j-m}e^{-j\lambda t_0}$	$\dfrac{\prod\limits_{j=0}^{N-m}(m+v_j)}{v^{N-m}(N-m)!}\displaystyle\sum_{i=0}^{N-m}\dfrac{(-1)^i C_{N-m}^i}{m+v_i}\times e^{-(m+r_i)\lambda t_0}$	$e^{-m\lambda t_0}\displaystyle\sum_{j=0}^{n-1}\dfrac{(m\lambda t_0)^i}{j!}$

2. The probability of failure-free operation of the system for the initial time of $t_0 = 1$ hour is

$$P(t_0) = \frac{\Lambda_1\Lambda_2(\Lambda_1 - \Lambda_2)e^{-\Lambda_0 t_0} - \Lambda_0\Lambda_2(\Lambda_0 - \Lambda_2)e^{-\Lambda_1 t_0} + \Lambda_0\Lambda_1(\Lambda_0 - \Lambda_1)^{-\Lambda_2 t_0}}{(\Lambda_0 - \Lambda_1)(\Lambda_1 - \Lambda_2)(\Lambda_0 - \Lambda_2)}$$

$$= \frac{10\lambda \cdot 12\lambda(10\lambda - 12\lambda)e^{-6\lambda t_0} - 6\lambda \cdot 12\lambda(6\lambda - 12\lambda)e^{-10\lambda t_0}}{(6\lambda - 10\lambda)(10\lambda - 12\lambda)(6\lambda - 12\lambda)}$$

$$+ \frac{6\lambda \cdot 10\lambda(6\lambda - 10\lambda)e^{-12\lambda t_0}}{(6\lambda - 10\lambda)(10\lambda - 12\lambda)(6\lambda - 12\lambda)}$$

$$= \frac{-240\lambda^3 e^{-6\lambda t_0} + 432\lambda^3 e^{-10\lambda t_0} - 240\lambda^3 e^{-12\lambda t_0}}{-48\lambda^3}$$

$$= \frac{1}{48}(240e^{-6 \cdot 0.1 \cdot 1} - 432e^{-10 \cdot 0.1 \cdot 1} + 240e^{-12 \cdot 0.1 \cdot 1}) \approx 0.91.$$

Example 3.3.2 Suppose we have a system of $m = 6$ operating units and one stand-by unit in an operating state. The failure intensity of each unit is $\lambda = 0.01 \dfrac{1}{\text{hr}}$.

We want to determine the probability of failure-free operation of the system for $t_0 = 10$ hours.

Solution. Using Table 3.3.2, we obtain

$$P(t_0) = (m + 1)e^{-m\lambda t_0} - me^{-(m+1)\lambda t_0} = (6 + 1)e^{-6 \cdot 0.01 \cdot 10}$$
$$- 6 \cdot e^{-(6+1) \cdot 0.01 \cdot 10} \approx 0.862.$$

Example 3.3.3 Suppose a system consists of $m = 12$ operating units and one stand-by in a partially operating state with load coefficient $\nu = 0.1$. The failure intensity of the operating unit is $\lambda = 0.04 \dfrac{1}{\text{hr}}$.

We want to determine the mean operating time up to failure of the system.

Solution. From Table 3.3.2 we find

$$T = \frac{1}{\lambda} \cdot \frac{2m + \nu}{m(m + \nu)} = \frac{1}{0.04} \cdot \frac{2 \cdot 12 + 0.1}{12(12 + 0.1)} \approx 4.15 \quad \text{hr.}$$

Example 3.3.4 Suppose we have a system of $m = 6$ operating units and one stand-by. The failure intensity of an operating unit is $\lambda = 0.001 \dfrac{1}{\text{hr}}$ and the stand-by unit is in a nonoperating state.

We want to compute the probability of failure-free operation for $t_0 = 100$ hours.

Solution.   From Table 3.3.2 we obtain

$$P(t_0) = (1 + m\lambda t_0)e^{-m\lambda t_0} = (1 + 6 \cdot 0.1)e^{-6 \cdot 0.1} \approx 1.6 \cdot 0.5488 \approx 0.878.$$

Example 3.3.5 Suppose we have a system of m units and two stand-bys in an operating state.

We wish to determine the number of units k_0 to be used in the system so that the mean operating time up to failure of the entire system is not less than the mean operating time up to failure of one unit. (Recall that when one unit fails, the remaining units overload.)

Solution.   Using Table 3.3.2, we obtain

$$T = \frac{1}{\lambda} \cdot \frac{m_0(m_0 + 1) + m_0(m_0 + 2) + (m_0 + 1)(m_0 + 2)}{m_0(m_0 + 1)(m_0 + 2)} = \frac{1}{\lambda}.$$

Hence

$$m_0^3 - 4m_0 = 2.$$

Solving this equation graphically, we find

$$m_0 \approx 2.2.$$

Example 3.3.6 Suppose a system consists of $m = 2$ operating units and two stand-by units in a partially operating state with load coefficient $\nu = 0.6$.

The failure intensity of a unit in an operating state is $\lambda = 0.02 \, \frac{1}{\text{hr}}$.

We wish to compute the probability of failure-free operation for $t_0 = 100$ hours.

Solution.   Using Table 3.3.2, we find

$$P(t_0) = \frac{1}{2\nu^2} e^{-m\lambda t_0}[(m + \nu)(m + 2\nu) - 2m(m + 2\nu)e^{-\nu\lambda t_0}$$
$$+ m(m + \nu)e^{-2\nu\lambda t_0}]$$
$$= \frac{1}{2 \cdot (0.6)^2} e^{-2 \cdot 0.02 \cdot 100}[(2 + 0.6)(2 + 2 \cdot 0.6)$$
$$- 4 \cdot (2 + 2 \cdot 0.6)e^{-0.6 \cdot 0.02 \cdot 100} + 2(2 + 0.6)e^{-2 \cdot 0.6 \cdot 0.02 \cdot 100}]$$
$$\approx 0.13.$$

Example 3.3.7 Suppose we have a system of $m = 10$ operating units and two stand-by units in a nonoperating state. The failure intensity of an operating unit is $\lambda = 0.01 \, \frac{1}{\text{hr}}$.

We want to compute the probability of failure-free operation of the system for $t_0 = 100$ hours.

Solution. From Table 3.3.2 we find

$$P(t_0) = \left[1 + m\lambda t_0 + \frac{(m\lambda t_0)^2}{2}\right] e^{-m\lambda t_0}$$

$$= \left[1 + 10 \cdot 0.01 \cdot 100 + \frac{(10 \cdot 0.01 \cdot 100)^2}{2}\right] e^{-10 \cdot 0.01 \cdot 100}$$

$$\approx 0.003.$$

Example 3.3.8 We are given a system of $m = 2$ operating units and $n - 1 = 3$ stand-by units in an operating state. The failure intensity of each unit is $\lambda = 0.0001 \dfrac{1}{\text{hr}}$.

We want to compute the probability of failure-free operation for $t_0 = 100$ hours.

Solution. From Table 3.3.2 we obtain

$$P(t_0) = (N - m + 1)C_N^{m-1} \sum_{j=m}^{N} \frac{(-1)^{j-m}}{j} C_{N-m}^{j-m} e^{-j\lambda t_0}$$

$$= (5 - 2 + 1)C_5^1 \sum_{j=2}^{5} \frac{(-1)^{j-2}}{j} C_{5-2}^{j-2} e^{-j \cdot 0.0001 \cdot 100}$$

$$\approx 20(0.49010 - 0.97045 + 0.72059 - 0.19025) \approx 0.9998.$$

Example 3.3.9 Suppose a system consists of $m = 10$ operating units and $n - 1 = 3$ stand-by units in a partially operating state ($\nu = 0.5$). The failure intensity of an operating unit is $\lambda = 0.01 \dfrac{1}{\text{hr}}$.

We have to determine the approximate probability of failure-free operation for an initial time of $t_0 = 10$ hours.

Solution. Since $\Lambda_j = m\lambda + (n - j - 1)\nu\lambda$ for a partially operating stand-by unit, Formula (3.2.12a) yields

$$\tilde{P}(t_0) = 1 - \frac{t_0^n}{n!} \prod_{j=0}^{n-1} \Lambda_j = 1 - \frac{(\lambda t_0)^n}{n!} \prod_{j=0}^{n-1} [m + (n - 1 - j)\nu]$$

$$= 1 - \frac{(0.1)^4}{24} \prod_{j=0}^{3} [10 + (3 - j) \cdot 0.5] \approx 0.945.$$

Example 3.3.10 Suppose we are given a system of $m = 10$ units and $n - 1 = 3$ stand-by units. The failure intensity of an operating unit is $\lambda = 0.01 \dfrac{1}{\text{hr}}$.

We want to compute the mean operating time up to failure of this system for three different stand-by conditions: operating, partially operating ($\nu = 0.5$), and nonoperating.

Solution. Using the appropriate formulas from Table 3.3.2, we obtain:

for the operating state:

$$T = \frac{1}{\lambda} \sum_{j=0}^{n-1} \frac{1}{N-j} = \frac{1}{\lambda} \sum_{j=0}^{3} \frac{1}{13-j}$$

$$= \frac{1}{0.01} \left(\frac{1}{13} + \frac{1}{12} + \frac{1}{11} + \frac{1}{10} \right) = 35.2 \quad \text{hr};$$

for the partially operating state ($\nu = 0.5$):

$$T = \frac{1}{\lambda} \sum_{j=0}^{N-m} \frac{1}{m+\nu j} = \frac{1}{\lambda} \sum_{j=0}^{3} \frac{1}{10 + 0.5j}$$

$$= \frac{1}{0.01} \left(\frac{1}{10} + \frac{1}{10.5} + \frac{1}{11} + \frac{1}{11.5} \right) = 37.3 \quad \text{hr};$$

for the nonoperating state:

$$T = \frac{1}{\lambda} \cdot \frac{n}{m} = \frac{4}{0.01 \cdot 10} = 40 \quad \text{hr.}$$

BIBLIOGRAPHY

1. BAZOVSKY, I., *Reliability: Theory and Practice*, chaps. 11, 12. Englewood Cliffs, N.J.: Prentice-Hall, Inc., 1961.
2. GNEDENKO, B. V., BELYAEV, Y. K., AND SOLOV'EV, A. D., *Mathematical Methods in Reliability Theory*. New York: Academic Press, Inc., 1968.
3. POLOVKO, A. M., *Foundations of Reliability Theory*, chaps. 5, 6. Moscow: Nauka, 1961.
4. SOLOV'EV, A. D., "On Redundancy without Repair," in the collection *Cybernetics in the Service of Communism*. Moscow: Energiya, 1964.

4

REDUNDANCY WITH REPAIR

PRELIMINARY REMARKS

In this chapter we consider a system consisting of m operating units and $n - 1$ stand-by units (Figure 4.0.1). By unit we mean, as usual, an individual part, node, block, and in some cases an entire system, and so on.

At any moment of time a particular system is considered operational if at least m of all the units $N = m + n - 1$ units are in an operating state. We assume that the moment of failure of any operating unit is discovered immediately, and it is instantaneously replaced by an operational stand-by unit. The stand-by units are also constantly maintained in operational condition. These assumptions correspond in practice to the case where equipment with redundancy has 100 per cent error-free control of the bringing into use of all units by the structures used to switch in stand-by units, and these latter structures are characterized by very short switching time and high reliability.

Repair of failed units (both operating and stand-by) can be done for either unrestricted or restricted service.

Unrestricted repair is a repair scheme in which a failed unit is immediately put in for repair—that is, we assume that the number of repair facilities is sufficient for simultaneous repair of all failed units of the system

Restricted repair is a repair scheme in which not more than one failed unit can undergo repair at any moment of time—that is, we assume there is precisely one repair facility.

The basic reliability measures are found for these two cases. In practice, the number of repair facilities intended for repair of failed units in the system may be more than one but less than the total number of units

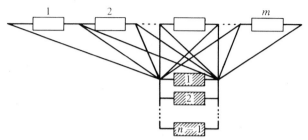

Figure 4.0.1 Block Diagram of the Model Used for the Reliability Computations for a System of m Operating and $n - 1$ Stand-by Units.

in the system. In this case, we have an intermediate situation between the two cases above, and the precise formulas for estimating the various reliability measures are quite cumbersome. Therefore we propose to carry out our computations with the formulas for restricted and unrestricted repair and take the resulting quantities as lower and upper bounds for the true value.

As we did earlier, we assume that a stand-by unit is in a nonoperating state and that its probability of failure in any given time interval is zero. The probability of failure of a stand-by unit in a partially operating state is always less than the probability of failure of an operating unit under otherwise equal conditions.

In our model we assume that m operating units are in series connection —that is, at a moment of failure of all stand-by units and one of the operating units the system is considered to have failed and its further functioning is curtailed until one of the failed units is repaired.

Finally, we assume that both failure-free operation time and repair time have exponential distributions. However, as was noted earlier, if the mean time of failure-free operation of an individual unit is much greater than the corresponding time of idleness (repair), then all the given characteristics still remain in force in the case where only the failure-free operating time is exponentially distributed while the repair time has arbitrary distribution. (Indeed, approximate formulas for many reliability measures are written under this assumption.)

4.1 EXACT VALUES OF RELIABILITY MEASURES OF REDUNDANT SYSTEMS WITH REPAIR

4.1.1 Description of the model

Many actual redundant systems can be described by the following model. We consider a system of $n - 1$ redundant units. Such a system can be found in a finite number of states (corresponding to the number of failed units): H_0, H_1, ..., H_j, ..., H_n, where H_j is the state where there are j failed units in the system.

In accord with the above assumptions the system is characterized by the following conditions: if the system is in state H_j at the moment of time t (j failed units in the system), then in the next time interval Δt it passes to the state H_{j+1} with probability $\Lambda_j \Delta t + o(\Delta t)$ (that is, another one of the good units in the system fails); with probability $M_j \Delta t + o(\Delta t)$ it passes to the state H_{j-1} (that is, one of the previously failed units is restored); it remains in the state H_j with probability $1 - (\Lambda_j + M_j) \Delta t + o(\Delta t)$.

The state H_n, in which there are n failed units in the system (that is, $n - 1$ stand-by units and one of the operating units fail), is obviously the failed state of the system. We shall distinguish two cases: the failed state of the system H_n is *absorbing* or *reflecting*. The first case leads to the problems of finding the probability of failure-free operation of the system, the mean operating time up to failure, the mean operating time between failures, and so on. The second entails the problems of finding nonstationary and stationary readiness and delay coefficients, various stationary characteristics, and so on.

A diagram of the passages of the described system from one state to another is given in Figure 4.1.1, where:

H_j, $0 \leq j \leq n$, is the state of the system where j of the $N = m + (n - 1)$ have failed;

Λ_j, $0 \leq j \leq n - 1$, is the passage intensity of the system [failure of one of its $(N - j)$ remaining good units] from the state with j failed units to the state with one additional failed unit—that is, $j + 1$;

M_j, $1 \leq j \leq n$, is the passage intensity of the system (restoration of one of the j previously failed units) from the state with j failed units to the state with one fewer failed units—that is, $j - 1$.

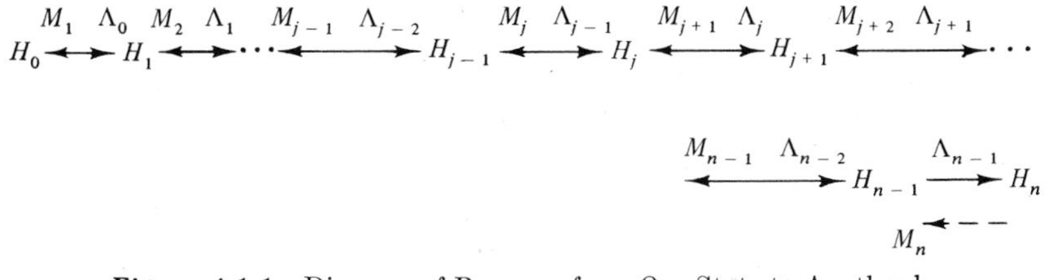

Figure 4.1.1 Diagram of Passages from One State to Another by a System with $n - 1$ Stand-by Units.

4.1.2   State H_n: absorbing

The behavior of a system is described by the following system of differential equations:

$$p_j'(t) = \Lambda_{j-1}p_{j-1}(t) - (\Lambda_j + M_j)p_j(t) + M_{j+1}p_{j+1}(t),$$
$$0 \leq j \leq n, \quad \Lambda_{-1} = \Lambda_n = M_0 = M_n = M_{n+1} = 0, \qquad (4.1.1)$$

where $p_j(t)$ is the probability that the system is in state H_j at time t, and

$$\sum_{j=0}^{n} p_j(t) = 1. \qquad (4.1.2)$$

Suppose the probabilities $p_j(t)$ satisfy initial conditions of the form

$$p_j(0) = p_j^*, \qquad 0 \leq j \leq n, \qquad (4.1.3)$$

where the p_j^*'s are given numbers.

To find the distribution law $\mathcal{P}\{\theta_n \leq t\} = \mathcal{P}_n(t)$ of the random variable θ_n, where θ_n is the random moment of time when the system arrives for the first time at the absorbing failed state H_n [it is easy to see that $\mathcal{P}_n(t) = p_n(t)$], we apply the Laplace transform to the system of differential equations (4.1.1):

$$a_j(s) = \int_0^\infty p_j(t)e^{-st}\, dt. \qquad (4.1.4)$$

Since

$$\int_0^\infty p_j'(t)e^{-st}\, dt = -p_j(0) + sa_j(s), \qquad 0 \leq j \leq n, \qquad (4.1.5)$$

the system of differential equations (4.1.1) is transformed to a system of algebraic equations:

$$\Lambda_{j-1}a_{j-1}(s) - (\Lambda_j + M_j + s)a_j(s) + M_{j+1}a_{j+1}(s) = -p_j(0),$$
$$0 \leq j \leq n, \quad \Lambda_{-1} = \Lambda_n = M_0 = M_n = M_{n+1} = 0.$$

By Cramer's rule,

$$a_n(s) = \frac{D_n(s)}{D(s)}. \qquad (4.1.6)$$

Here

$$
D(s) = \begin{vmatrix}
-(\Lambda_0 + s) & M_1 & 0 & \cdots \\
\Lambda_0 & -(\Lambda_1 + M_1 + s) & M_2 & \cdots \\
0 & \Lambda_1 & -(\Lambda_2 + M_2 + s) & \cdots \\
\cdot & & & \\
\cdot & & & \\
\cdot & & & \\
0 & 0 & 0 & \cdots \\
0 & 0 & 0 & \cdots \\
0 & 0 & 0 & \cdots \\
\end{vmatrix}
$$

$$
\begin{array}{cccc}
\cdots & 0 & 0 & 0 \\
\cdots & 0 & 0 & 0 \\
\cdots & 0 & 0 & 0 \\
& \cdot & & \\
& \cdot & & \\
& \cdot & & \\
\cdots & -(\Lambda_{n-2} + M_{n-2} + s) & M_{n-1} & 0 \\
\cdots & \Lambda_{n-1} & -(\Lambda_{n-1} + M_{n-1} + s) & 0 \\
\cdots & 0 & \Lambda_{n-1} & -s \\
\end{array}
\qquad (4.1.7)
$$

$$
= s\,\Delta_n(s),
$$

where

$$
\Delta_n(s) = \begin{vmatrix}
-(\Lambda_0 + s) & M_1 & 0 & \cdots \\
\Lambda_0 & -(\Lambda_1 + M_1 + s) & M_2 & \cdots \\
0 & \Lambda_1 & -(\Lambda_2 + M_2 + s) & \cdots \\
\cdot & & & \\
\cdot & & & \\
\cdot & & & \\
0 & 0 & 0 & \cdots \\
0 & 0 & 0 & \cdots \\
\end{vmatrix}
$$

$$
\begin{array}{ccc}
\cdots & 0 & 0 \\
\cdots & 0 & 0 \\
\cdots & 0 & 0 \\
& \cdot & \\
& \cdot & \\
& \cdot & \\
\cdots & -(\Lambda_{n-2} + M_{n-2} + s) & M_{n-1} \\
\cdots & \Lambda_{n-2} & -(\Lambda_{n-1} + M_{n-1} + s) \\
\end{array}
\qquad , \quad (4.1.8)
$$

and the determinant $D_n(s)$ is

$$
D_n(s) = \begin{vmatrix}
-(\Lambda_0 + s) & M_1 & 0 & \cdots \\
\Lambda_0 & -(\Lambda_1 + M_1 + s) & M_2 & \cdots \\
0 & \Lambda_1 & -(\Lambda_2 + M_2 + s) & \cdots \\
\vdots & & & \\
\vdots & & & \\
\vdots & & & \\
0 & 0 & 0 & \cdots \\
0 & 0 & 0 & \cdots \\
0 & 0 & 0 & \cdots \\
\end{vmatrix}
$$

$$
\begin{vmatrix}
\cdots & 0 & 0 & -p_0(0) \\
\cdots & 0 & 0 & -p_1(0) \\
\cdots & 0 & 0 & -p_2(0) \\
& \vdots & & \\
& \vdots & & \\
\cdots & -(\Lambda_{n-2} + M_{n-2} + s) & M_{n-1} & -p_{n-2}(0) \\
\cdots & \Lambda_{n-2} & -(\Lambda_{n-1} + M_{n-1} + s) & -p_{n-1}(0) \\
\cdots & 0 & \Lambda_{n-1} & -p_n(0) \\
\end{vmatrix} . \quad (4.1.9)
$$

Expanding the determinant $D_n(s)$ by the elements of the last row, we obtain

$$
D_n(s) = -p_n(0) \, \Delta_n(s) - \Lambda_{n-1} D_{n-1}(s).
$$

Using this relation recursively, it is easy to obtain

$$
D_n(s) = -p_n(0) \cdot \Delta_n(s) + \sum_{i=0}^{n-1} (-1)^{n-i+1} \cdot p_i(0) \cdot \Delta_i(s) \prod_{j=i}^{n-1} \Lambda_j. \quad (4.1.10)
$$

The desired probability $\mathcal{P}_n(t) = p_n(t)$ is obtained by inverse Laplace transformation:

$$
p_n(t) = \frac{1}{2a_i} \int_C a_n(s) e^{st} \, ds = \frac{1}{2\pi i} \int_C \frac{D_n(s) e^{st}}{-s \, \Delta_n(s)} \, ds, \quad (4.1.11)
$$

where the contour C includes inside all the zeros of the denominator of the integrand. Below, in Section 4.2.5, we shall show that $\Delta_n(s)$ is a polynomial of degree n, with all roots $s_1, s_2, \ldots, s_n$ distinct and negative.

Then from (4.1.11), by the theorem with the aid of some computation, we obtain

$$p_n(t) = -\frac{D_n(0)}{\Delta_n(0)} - \sum_{i=1}^{n} \frac{D_n(s_i)e^{s_i t}}{s_i \prod_{\substack{j=1 \\ j \neq i}}^{n} (s_i - s_j)}. \tag{4.1.12}$$

Adding to the first row of the determinant $\Delta_n(0)$, given by expression (4.1.8), all the remaining rows and expanding it by the elements of the first row, and then by the elements of the first column, we easily obtain

$$\Delta_n(0) = (-1)^n \prod_{i=0}^{n-1} \Lambda_i. \tag{4.1.13}$$

But then it follows from (4.1.10) that

$$D_n(0) = -\Delta_n(0) \tag{4.1.14}$$

and (4.1.12) takes the form

$$p_n(t) = 1 - \sum_{i=1}^{n} \frac{D_n(s_i) \cdot e^{s_i t}}{s_i \prod_{\substack{j=1 \\ j \neq i}}^{n} (s_j - s_j)}. \tag{4.1.15}$$

Hence $P(t) = 1 - p_n(t)$, the probability of failure-free operation of the system, is

$$P(t) = \sum_{i=1}^{n} \frac{D_n(s_i)e^{s_i t}}{\prod_{\substack{j=1 \\ j \neq i}}^{n} (s_i - s_j)}. \tag{4.1.16}$$

Exact values of the roots of the polynomial $\Delta_n(s)$ are found easily (in terms of radicals) for $n \leq 3$. For $n > 3$ the roots of the polynomial $\Delta_n(s)$ can be determined approximately with the required accuracy by the usual methods of solution of an nth-degree algebraic equation, where in the computations it is convenient to make use of the fact that for successive values of n the roots of $\Delta_n(s)$ separate (that is, alternate with) those of $\Delta_{n+1}(s)$. (This property will be proved in Section 4.2.5.)

Now we find $M[\theta_n]$, the mathematical expectation of the time the system first falls into the absorbing failed state H_n. Obviously this is the mean time of failure-free operation of a system that began operating under arbitrary initial conditions of the form (4.1.3). We let $M[\theta_n] = T^{(n)}$. By definition

$$T^{(n)} = \int_0^\infty P(t)\, dt = \int_0^\infty t\mathcal{O}'_n(t)\, dt = \int_0^\infty t p'_n(t)\, dt.$$

It follows from (4.1.5) that

$$\int_0^\infty p'_n(t) e^{-st}\, dt = -p_n(0) + s a_n(s).$$

Differentiating this equality in s and then letting $s = 0$, we obtain

$$\int_0^\infty p'_n(t) \cdot (-t)\, dt = -T^{(n)} = \frac{d}{ds}\left[s a_n(s) \right]\Big|_{s=0}.$$

Using (4.1.6), (4.1.7), and (4.1.14) we obtain

$$T^{(n)} = \frac{1}{\Delta_n(0)} \frac{d}{ds}\left[D_n(s) + \Delta_n(s) \right]\Big|_{s=0}. \tag{4.1.17}$$

Adding to the last row of the determinant $D_n(s)$ [Formula (4.1.9)] all the remaining rows and writing in the determinant $\Delta_n(s)$ [Formula (4.1.8)] another (last) row and column, which leave its value unchanged, we obtain

$$D_n(s) + \Delta_n(s) =
\begin{vmatrix}
-(\Lambda_0 + s) & M_1 & 0 & \cdots \\
\Lambda_0 & -(\Lambda_1 + M_1 + s) & M_2 & \cdots \\
0 & \Lambda_1 & -(\Lambda_2 + M_2 + s) & \cdots \\
 & \cdot & & \\
 & \cdot & & \\
 & \cdot & & \\
0 & 0 & 0 & \cdots \\
0 & 0 & 0 & \cdots \\
-s & -s & -s & \cdots \\
\end{vmatrix}$$

$$\begin{array}{cccc}
\cdots & 0 & 0 & -p_0(0) \\
\cdots & 0 & 0 & -p_1(0) \\
\cdots & 0 & 0 & -p_2(0) \\
 & & \cdot & \\
 & & \cdot & \\
 & & \cdot & \\
\cdots & -(\Lambda_{n-2} + M_{n-2} + s) & M_{n-1} & -p_{n-2}(0) \\
\cdots & \Lambda_{n-2} & -(\Lambda_{n-1} + M_{n-1} + s) & -p_{n-1}(0) \\
\cdots & -s & -s & -1
\end{array}$$

$$-\begin{vmatrix}
-(\Lambda_0+s) & M_1 & 0 & \cdots & 0 & 0 & 0 \\
\Lambda_0 & -(\Lambda_1+M_1+s) & M_2 & \cdots & 0 & 0 & 0 \\
0 & \Lambda_1 & -(\Lambda_2+M_2+s) & \cdots & 0 & 0 & 0 \\
\vdots & & & & & & \\
0 & 0 & 0 & \cdots & -(\Lambda_{n-2}+M_{n-2}+s) & M_{n-1} & 0 \\
0 & 0 & 0 & \cdots & \Lambda_{n-2} & -(\Lambda_{n-1}+M_{n-1}+s) & 0 \\
-s & -s & -s & \cdots & -s & -s & -1
\end{vmatrix}$$

$$= -sD^{(n)}(s),$$

where

$$D^{(n)}(s) = \begin{vmatrix}
-(\Lambda_0+s) & M_1 & 0 & \cdots & 0 & 0 & -p_0(0) \\
\Lambda_0 & -(\Lambda_1+M_1+s) & M_2 & \cdots & 0 & 0 & -p_1(0) \\
0 & \Lambda_1 & -(\Lambda_2+M_2+s) & \cdots & 0 & 0 & -p_2(0) \\
\vdots & & & & & & \\
0 & 0 & 0 & \cdots & -(\Lambda_{n-2}+M_{n-2}+s) & M_{n-1} & -p_{n-2}(0) \\
0 & 0 & 0 & \cdots & \Lambda_{n-2} & -(\Lambda_{n-1}+M_{n-1}+s) & -p_{n-1}(0) \\
1 & 1 & 1 & \cdots & 1 & 1 & 0
\end{vmatrix}. \tag{4.1.18}$$

Then from (4.1.17) it follows that

$$T^{(n)} = -\frac{D^{(n)}(0)}{\Delta_n(0)}, \qquad (4.1.19)$$

where, according to (4.1.18), the determinant $D^{(n)}(0)$ is

$$
D^{(n)}(0) = \begin{vmatrix}
-\Lambda_0 & M_1 & 0 & \cdots \\
\Lambda_0 & -(\Lambda_1 + M_1) & M_2 & \cdots \\
0 & \Lambda_1 & -(\Lambda_2 + M_2) & \cdots \\
& \cdot & & \\
& \cdot & & \\
& \cdot & & \\
0 & 0 & 0 & \cdots \\
0 & 0 & 0 & \cdots \\
1 & 1 & 1 & \cdots \\
\end{vmatrix}
$$

$$
\begin{matrix}
\cdots & 0 & 0 & -p_0(0) \\
\cdots & 0 & 0 & -p_1(0) \\
\cdots & 0 & 0 & -p_2(0) \\
& \cdot & & \\
& \cdot & & \\
\cdots & -(\Lambda_{n-2} + M_{n-2}) & M_{n-1} & -p_{n-2}(0) \\
\cdots & \Lambda_{n-2} & -(\Lambda_{n-1} + M_{n-1}) & -p_{n-1}(0) \\
\cdots & 1 & 1 & 0 \\
\end{matrix} \qquad (4.1.20)
$$

and the determinant $\Delta_n(0)$ is given by Formula (4.1.13).

Now we find a convoluted expression for the determinant $D^{(n)}(0)$. We multiply the first row of the determinant $D^{(n)}(0)$ by $1/\Lambda_0$ and add it to the last row. Now we add the first row to the second. Then, expanding the obtained determinant by the elements of the first column (actually by the single element "$-\Lambda_0$") and multiplying the last row of the new determinant by Λ_0, we obtain

$$
D^{(n)}(0) = (-1)^1 \cdot \begin{vmatrix}
-\Lambda_1 & M_2 & \cdots \\
\Lambda_1 & -(\Lambda_2 + M_2) & \cdots \\
\cdot & & \\
\cdot & & \\
\cdot & & \\
0 & 0 & \cdots \\
0 & 0 & \cdots \\
\Lambda_0 + M_1 & \Lambda_0 & \cdots \\
\end{vmatrix}
$$

$$
\begin{vmatrix}
\cdots & 0 & 0 & -[p_0(0) + p_1(0)] \\
\cdots & 0 & 0 & -p_2(0) \\
 & \cdot & & \\
 & \cdot & & \\
 & \cdot & & \\
\cdots & -(\Lambda_{n-2} + M_{n-2}) & M_{n-1} & -p_{n-2}(0) \\
\cdots & \Lambda_{n-2} & -(\Lambda_{n-1} + M_{n-1}) & -p_{n-1}(0) \\
\cdots & \Lambda_0 & \Lambda_0 & -p_0(0)
\end{vmatrix}
$$

Again we multiply the first row of the determinant by $(\Lambda_0 + M_1)/\Lambda_1$ and add it to the last row, then add the first row to the second. Now, expanding the obtained determinant by the elements of the first column (actually by the single element "$-\Lambda_1$") and multiplying the last row of the new determinant by Λ_1, we obtain

$$
D^{(n)}(0) = (-1)^2 \cdot
\begin{vmatrix}
-\Lambda_2 & M_3 & \cdots & 0 \\
\cdot & & & \\
\cdot & & & \\
\cdot & & & \\
0 & 0 & \cdots & -(\Lambda_{n-2} + M_{n-2}) \\
0 & 0 & \cdots & \Lambda_{n-2} \\
[\Lambda_0\Lambda_1 + M_2(\Lambda_0 + M_1)] & \Lambda_0\Lambda_1 & \cdots & \Lambda_0\Lambda_1
\end{vmatrix}
$$

$$
\begin{vmatrix}
\cdots & 0 & -[p_0(0) + p_1(0) + p_2(0)] \\
 & \cdot & \\
 & \cdot & \\
 & \cdot & \\
\cdots & M_{n-1} & -p_{n-2}(0) \\
\cdots & -(\Lambda_{n-1} + M_{n-1}) & p_{n-1}(0) \\
\cdots & \Lambda_0\Lambda_1 & -\{p_0(0)\Lambda_1 + [p_0(0) + p_1(0)](\Lambda_0 + M_1)\}
\end{vmatrix}
$$

Again we multiply the first row of the determinant by

$$
\frac{[\Lambda_0\Lambda_1 + M_2(\Lambda_0 + M_1)]}{\Lambda_2}
$$

and add it to the last row. We also add the first row to the second and expand the obtained determinant by the elements of the first column (actually by the single element "$-\Lambda_2$") and multiply the last row of the new determinant by Λ_2. Carrying out this procedure k times, we obtain

$$D^{(n)}(0) = (-1)^k \cdot
\begin{vmatrix}
-\Lambda_k & M_{k+1} & 0 & \cdots \\
\Lambda_k & -(\Lambda_{k+1} + M_{k+1}) & M_{k+2} & \cdots \\
0 & \Lambda_{k+1} & -(\Lambda_{k+2} + M_{k+2}) & \cdots \\
\cdot & & & \\
\cdot & & & \\
\cdot & & & \\
0 & 0 & 0 & \cdots \\
0 & 0 & 0 & \cdots \\
r_k & \prod_{i=0}^{k-1} \Lambda_i & \prod_{i=0}^{k-1} \Lambda_i & \cdots \\
\end{vmatrix}$$

$$
\begin{vmatrix}
\cdots & 0 & 0 & -\pi_k(0) \\
\cdots & 0 & 0 & -p_{k+1}(0) \\
\cdots & 0 & 0 & -p_{k+2}(0) \\
 & & & \cdot \\
 & & & \cdot \\
 & & & \cdot \\
\cdots & -(\Lambda_{n-2} + M_{n-2}) & M_{n-1} & -p_{n-2}(0) \\
\cdots & \Lambda_{n-2} & -(\Lambda_{n-1} + M_{n-1}) & -p_{n-1}(0) \\
\cdots & \prod_{i=0}^{k-1} \Lambda_i & \prod_{i=0}^{k-1} \Lambda_i & S_k \\
\end{vmatrix}. \quad (4.1.21)
$$

Here

$$\pi_k(0) = \sum_{i=0}^{k} p_i(0), \tag{4.1.22}$$

$$r_k = \prod_{i=0}^{k-1} \Lambda_i + M_k r_{k-1}, \qquad r_0 = 1, \tag{4.1.23}$$

$$S_k = \Lambda_{k-1} S_{k-1} - \pi_{k-1}(0) r_{k-1}, \qquad s_0 = 0,\ s_1 = -p_0(0). \tag{4.1.24}$$

Using the recursion relation (4.1.23), we obtain

$$r_k = \prod_{i=1}^{k} M_i \cdot \sum_{j=0}^{k} \Theta_j, \tag{4.1.25}$$

where

$$\Theta_j = \frac{\Lambda_0 \Lambda_1 \cdots \Lambda_{j-1}}{M_1 M_2 \cdots M_j}, \qquad \Theta_0 = 1. \tag{4.1.26}$$

Using (4.1.25), from the second recursion relation (4.1.24) we obtain

$$S_k = -\prod_{i=0}^{k-1} \Lambda_i \cdot \sum_{j=0}^{k-1} \frac{\pi_j(0) \cdot \sum_{l=0}^{j} \Theta_l}{\Lambda_j \Theta_j}. \tag{4.1.27}$$

But it follows from (4.1.21) that

$$D^{(n)}(0) = (-1)^n S_n \tag{4.1.28}$$

or

$$D^{(n)}(0) = (-1)^{n+1} \cdot \prod_{i=0}^{n-1} \Lambda_i \sum_{j=0}^{n-1} \frac{\pi_j(0) \sum_{l=0}^{j} \Theta_l}{\Lambda_j \Theta_j}. \tag{4.1.29}$$

Then from (4.1.18), (4.1.20) we finally obtain

$$T^{(n)} = \sum_{j=0}^{n-1} \frac{\pi_j(0) \sum_{l=0}^{j} \Theta_l}{\Lambda_j \Theta_j}. \tag{4.1.30}$$

From the general formula (4.1.30) we obtain as special cases the corresponding formulas for the mean operating time of the system up to failure, $T_1^{(n)}$, and for the mean operating time of the system between failures, $T_2^{(n)}$. Indeed in the first case

$$\pi_j(0) = \sum_{i=0}^{j} p_i(0) = 1, \qquad 0 \le j \le n - 1, \tag{4.1.31}$$

since $p_0(0) = 1$. Then from (4.1.30) we obtain

$$T_1^{(n)} = \sum_{j=0}^{n-1} \frac{\sum_{l=0}^{j} \Theta_l}{\Lambda_j \Theta_j}. \tag{4.1.32}$$

In the second case

$$\pi_j(0) = \sum_{i=0}^{j} p_i(0) = 0, \qquad 0 \le j \le n - 2,$$

$$\pi_{n-1}(0) = \sum_{i=0}^{n-1} p_i(0) = 1, \tag{4.1.33}$$

since $p_{n-1}(0) = 1$. Then from (4.1.30) we have

$$T_2^{(n)} = \frac{\sum_{l=0}^{n-1} \Theta_l}{\Lambda_{n-1}\Theta_{n-1}} . \qquad (4.1.34)$$

Yet another useful relationship follows from (4.1.32), (4.1.34):

$$T_2^{(n)} = T_1^{(n)} - T_1^{(n-1)}. \qquad (4.1.35)$$

It is easy to see that for the given model of a system, the mean operating time between the ith and $(i + 1)$st failures (falling into an absorbing failed state) does not depend on i for $i \geq 1$.

4.1.3 State H_n: reflecting

Now we consider a system whose failed state H_n isreflecting. The behavior of such a system is described by a system of differential equations of the same form as (4.1.1) except for the last two equations, which have the form

$$\begin{aligned} p'_{n-1}(t) &= \Lambda_{n-2}p_{n-2}(t) - (\Lambda_{n-1} + M_{n-1})p_{n-1}(t) + M_n p_n(t), \\ p'_n(t) &= \Lambda_{n-1}p_{n-1}(t) - M_n p_n(t). \end{aligned} \qquad (4.1.36)$$

We find the quantities $p_n(t)$ and $\lim_{t \to \infty} p_n(t)$ determining the length of time during which the system is in the failed state H_n; that is, $p_n(t)$ and $\lim_{t \to \infty} p_n(t)$ obviously characterize the time dependence and the stationary value of the idleness coefficient of the given system. The nonstationary and stationary readiness coefficients are

$$K(t) = 1 - p_n(t), \qquad (4.1.37)$$
$$K = 1 - \lim_{t \to \infty} p_n(t). \qquad (4.1.38)$$

Applying the Laplace transform (4.1.4) to the altered system of differential equations [(4.1.1), (4.1.36)] under arbitrary initial conditions (4.1.3), by Cramer's rule we obtain the following expression for $a_n(s)$:

$$a_n(s) = \frac{D_n(s)}{A_n(s)} . \qquad (4.1.39)$$

The determinant $D_n(s)$ is given, as before, by Formula (4.1.9), and the determinant $A_n(s)$ is equal to

$$
A_n(s) = \begin{vmatrix}
-(\Lambda_0 + s) & M_1 & 0 & \cdots \\
\Lambda_0 & -(\Lambda_1 + M_1 + s) & M_2 & \cdots \\
0 & \Lambda_1 & -(\Lambda_2 + M_2 + s) & \cdots \\
 & & \cdot & \\
 & & \cdot & \\
 & & \cdot & \\
0 & 0 & 0 & \cdots \\
0 & 0 & 0 & \cdots \\
0 & 0 & 0 & \cdots \\
\cdots & 0 & 0 & 0 \\
\cdots & 0 & 0 & 0 \\
\cdots & 0 & 0 & 0 \\
 & \cdot & & \\
 & \cdot & & \\
 & \cdot & & \\
-(\Lambda_{n-2} + M_{n-2} + s) & M_{n-1} & 0 \\
\cdots \quad \Lambda_{n-2} & -(\Lambda_{n-1} + M_{n-1} + s) & M_n \\
\cdots \quad 0 & \Lambda_{n-1} & -(M_n + s)
\end{vmatrix} . \quad (4.1.40)
$$

Expanding the determinant $A_n(s)$ by the elements of the last column and taking (4.1.8) into account, we obtain

$$
A_n(s) = -(M_n + s)\Delta_n(s) - M_n\Lambda_{n-1}\Delta_{n-1}(s). \qquad (4.1.41)
$$

It is easy to show that

$$
\Delta_n(s) + \Lambda_{n-1}\Delta_{n-1}(s) = A_{n-1}(s),
$$

whence

$$
A_n(s) = -s\Delta_n(s) - M_n A_{n-1}(s). \qquad (4.1.42)
$$

Making successive use of this recursion relation, we obtain

$$
A_n(s) = -s\left[\Delta_n(s) + \sum_{i=0}^{n-1} (-1)^{n-i}\Delta_i(s) \prod_{j=i+1}^{n} M_j\right]. \qquad (4.1.43)
$$

Then, in view of (4.1.10), we obtain

$$
a_n(s) = \frac{D_n(s)}{A_n(s)} = \frac{p_n(0)\Delta_n(s) + \displaystyle\sum_{i=0}^{n-1} (-1)^{n-i}p_i(0)\Delta_i(s) \prod_{j=i}^{n-1} \Lambda_j}{s\left[\Delta_n(s) + \displaystyle\sum_{i=0}^{n-1} (-1)^{n-i}\Delta_i(s) \cdot \prod_{j=i+1}^{n} M_j\right]}. \qquad (4.1.44)
$$

The desired probability $p_n(t)$ is found by inverse Laplace transformation:

$$p_n(t) = \frac{1}{2\pi i} \int_C \frac{D_n(s)e^{st}\,ds}{A_n(s)} \tag{4.1.45}$$

and can be computed in the same way as was (4.1.12), using the residue theorem, if we know the roots of the polynomial $A_n(s)$.

From the form of the integrand in (4.1.45) it follows that $\lim\limits_{t\to\infty} p_n(t)$ can be found simply as the residue at the point $s = 0$:

$$p_n = \lim_{t\to\infty} p_n(t) = \frac{p_n(0)\Delta_n(0) + \sum\limits_{i=0}^{n-1}(-1)^{n-i}p_i(0)\Delta_i(0)\prod\limits_{j=i}^{n-1}\Lambda_j}{\Delta_n(0) + \sum\limits_{i=0}^{n-1}(-1)^{n-i}\Delta_i(0)\prod\limits_{j=i+1}^{n}M_j}.$$

Using (4.1.14), (4.1.13), and (4.1.26), we easily obtain

$$p_n = \frac{1}{1 + \dfrac{1}{\Theta_n}\sum\limits_{i=0}^{n-1}\Theta_i}. \tag{4.1.46}$$

Hence

$$K = 1 - p_n = \frac{1}{1 + \dfrac{\Theta_n}{\sum\limits_{i=0}^{n-1}\Theta_i}}. \tag{4.1.47}$$

It follows from Formula (4.1.46) that the quantity $p_n = \lim\limits_{t\to\infty} p_n(t)$ does not depend on the quantities $\{p_i(0), 0 \le i \le n\}$; therefore the fact that we have used initial conditions of the form (4.1.3) in deriving Formula (4.1.46) is not essential.

Since the mean repair time τ of the system in our case is

$$\tau = \frac{1}{M_n}, \tag{4.1.48}$$

the expression for the readiness coefficient K can also be obtained directly [as $\lim\limits_{t\to\infty} p_n(t)$] by the following formula if we know the values of the mean repair time τ and $T_2^{(n)}$:

$$K = \frac{T_2^{(n)}}{\tau + T_2^{(n)}} = \frac{1}{1 + \dfrac{\tau}{T_2^{(n)}}} = \frac{1}{1 + \dfrac{1}{M_n T_2^{(n)}}} = \frac{1}{1 + \dfrac{\Theta_n}{\sum\limits_{i=0}^{n-1} \Theta_i}},$$

which coincides with (4.1.47).

Finally, the stationary reliability characteristics can be computed by the usual methods applicable in queuing theory. Indeed, as $t \to \infty$, instead of the systems of differential equations (4.1.1) and (4.1.36), using a technique that may be found in [1], we may write

$$(\Lambda_j + M_j)p_j = \Lambda_{j-1}p_{j-1} + M_{j+1}p_{j+1},$$
$$0 \le j \le n, \quad \Lambda_{n-1} = M_0 = \Lambda_n = M_{n+1} = 0.$$

Hence we easily obtain

$$p_j = \Theta_j p_0,$$

where Θ_j is determined from Formula (4.1.26). Since

$$\sum_{j=0}^{n} p_j = 1 = p_0 \sum_{j=0}^{n} \Theta_j,$$

then

$$p_0 = \frac{1}{\sum\limits_{j=0}^{n} \Theta_j}$$

and finally

$$p_j = \frac{\Theta_j}{\sum\limits_{i=0}^{n} \Theta_i}, \qquad 0 \le j \le n. \tag{4.1.49}$$

Then

$$K = \sum_{j=0}^{n-1} p_j = \frac{\sum\limits_{j=0}^{n-1} \Theta_j}{\sum\limits_{i=0}^{n} \Theta_i} = \frac{1}{1 + \dfrac{\Theta_n}{\sum\limits_{j=0}^{n-1} \Theta_j}},$$

which coincides with (4.1.47).

Finally we determine the mean time of failure-free stationary operation, $T_R^{(n)}$; that is, the mean operating time up to failure of the system, which began operating under initial conditions

$$p_j(0) = \lim_{t \to \infty} p_j(t) = p_j, \qquad 0 \le j \le n, \qquad (4.1.50)$$

where p_j is determined by Formula (4.1.49).

Using Formulas (4.1.49), (4.1.30), and (4.1.22), we obtain

$$T_R^{(n)} = \frac{1}{\sum\limits_{s=0}^{n} \Theta_s} \cdot \sum_{j=0}^{n-1} \frac{\left(\sum\limits_{l=0}^{j} \Theta_l \right)^2}{\Lambda_j \Theta_j}$$

where the quantity Θ_i is determined from Formula (4.1.26).

4.2 APPROXIMATE ESTIMATES OF RELIABILITY MEASURES AND THEIR ERROR

In deriving approximate estimates we shall make use of the following basic assumption:

$$\min M_\alpha \gg N \cdot \max \Lambda_\beta, \qquad 0 \le (\alpha, \beta) \le n, \qquad (4.2.1)$$

where N is the total number of units in the system (operating and stand-by).

4.2.1 Mean time of failure-free operation of a system

It follows from (4.1.30) that when condition (4.2.1) is satisfied, an approximate estimate $\tilde{T}^{(n)}$ for the mean time of failure-free operation of a system $T^{(n)}$ is the quantity

$$\tilde{T}^{(n)} = \frac{\pi_{n-1}(0)}{\Lambda_{n-1} \Theta_{n-1}}, \qquad (4.2.2)$$

where π_{n-1} and Θ_{n-1} are determined by Formulas (4.1.22) and (4.1.26), respectively, and

$$\delta_+ = T^{(n)} - \tilde{T}^{(n)} \approx \frac{1}{\Lambda_{n-2} \Theta_{n-2}} \left[\pi_{n-2}(0) + \frac{\Lambda_0 M_{n-1}}{\Lambda_{n-1} M_1} \pi_{n-1}(0) \right]. \qquad (4.2.3)$$

Since all the quantities Λ_s, Θ_s, $\pi_s(0)$ are nonnegative, Formula (4.2.2) gives a lower bound for the mean time of failure-free operation of the system.

4.2.2 Mean operating time of a system up to failure

It follows from (4.1.32) [or from (4.2.2) and (4.2.3)] that when condition (4.2.1) is satisfied, an approximate estimate $\tilde{T}_1^{(n)}$ for the mean operating time $T_1^{(n)}$ of a system up to failure is the quantity

$$\tilde{T}_1^{(n)} = \frac{1}{\Lambda_{n-1}\Theta_{n-1}} = \frac{M_1 M_2 \cdots M_{n-1}}{\Lambda_0 \Lambda_1 \cdots \Lambda_{n-1}}, \tag{4.2.4}$$

and the error satisfies

$$\delta_+ = T_1^{(n)} - \tilde{T}_1^{(n)} \sim \frac{1}{\Lambda_{n-2}\Theta_{n-2}} \left(1 + \frac{\Lambda_0 M_{n-1}}{\Lambda_{n-1} M_1} \right). \tag{4.2.5}$$

4.2.3 Mean operating time of a system between failures

From (4.1.34) [or from (4.2.2) and (4.2.3)] it follows that when condition (4.2.1) is satisfied, an approximate estimate $\tilde{T}_2^{(n)}$ for the mean operating time of the system between failures, $T_2^{(n)}$, is the quantity

$$\tilde{T}_2^{(n)} = \frac{1}{\Lambda_{n-1}\Theta_{n-1}} = \frac{M_1 M_2 \cdots M_{n-1}}{\Lambda_0 \Lambda_1 \cdots \Lambda_{n-1}} = \tilde{T}_1^{(n)}, \tag{4.2.6}$$

where

$$\delta_+ = T_2^{(n)} - \tilde{T}_2^{(n)} \sim \frac{\Lambda_0}{\Lambda_{n-1} M_1 \Theta_{n-1}}. \tag{4.2.7}$$

4.2.4 Mean time of failure-free stationary operation of a system

It follows from (4.1.51) [or from (4.2.2) and (4.2.3)] that when condition (4.2.1) is satisfied, an approximate estimate $\tilde{T}_R^{(n)}$ for the mean time of failure-free stationary operation of the system $T_R^{(n)}$ is the quantity

$$\tilde{T}_R^{(n)} = \frac{1}{\Lambda_{n-1}\Theta_{n-1}} = \frac{M_1 M_2 \cdots M_{n-1}}{\Lambda_0 \Lambda_1 \cdots \Lambda_{n-1}} = \tilde{T}_1^{(n)} = \tilde{T}_2^{(n)}, \tag{4.2.8}$$

and the error satisfies

$$\delta_+ = T_R^{(n)} - \tilde{T}_R^{(n)} \sim \frac{1}{\Lambda_{n-2}\Theta_{n-2}} \left(1 + \frac{\Lambda_0 M_{n-1}}{\Lambda_{n-1} M_1} \right). \tag{4.2.9}$$

4.2.5 Probability of failure-free operation of a system

In deriving an approximate estimate of the probability of failure-free operation of a system we need to use certain properties of the polynomials $\Delta_n(s)$, which we now proceed to establish.

Expanding the determinant $\Delta_n(s)$ [Formula (4.1.8)] by the elements of the last column, we obtain a recursion relation of the form

$$\Delta_n(s) = -(\Lambda_{n-1} + M_{n-1} + s)\Delta_{n-1}(s) - M_{n-1}\Lambda_{n-2}\Delta_{n-2}(s), \quad (4.2.10)$$

where $\Delta_0(s) = 1$ and $\Delta_1(s) = -(\Lambda_0 + s)$.

Using (4.2.10), we establish the following two properties of the polynomials $\Delta_j(s)$:

1.   All roots of the polynomial $\Delta_j(s)$ are negative and distinct;
2.   for each j, the roots of $\Delta_j(s)$ separate or alternate with those of $\Delta_{j+1}(s)$.

The method of proof for these assertions is due to A. D. Solov'ev [2].
In accord with (4.2.10) we have

$$\Delta_0(s) = 1,$$
$$\Delta_1(s) = -(\Lambda_0 + s),$$
$$\Delta_2(s) = s^2 + s(\Lambda_0 + \Lambda_1 + M_1) + \Lambda_0\Lambda_1,$$
$$\Delta_j(s) = -(\Lambda_{j-1} + M_{j-1} + s)\Delta_{j-1}(s) - \Lambda_{j-2}M_{j-1}\Delta_{j-2}(s).$$

It follows from (4.1.13) that at the point $s = 0$ all even polynomials are positive, all odd ones are negative. From (4.1.8) we find that as $s \to \pm\infty$ the polynomial $\Delta_j(s)$ behaves like the quantity

$$\Delta_j(s) \sim (-1)^j s^j;$$

that is, as $s \to -\infty$ all the polynomials are positive.
The polynomial $\Delta_1(s)$ always has one root

$$s_1^{(1)} = -\Lambda_0;$$

that is (see Figure 4.2.1), the polynomial $\Delta_1(s)$ behaves as follows:
At the point $s = 0$

$$\text{the polynomial } \Delta_1(0) < 0;$$

At the point $s = s_1^{(1)} = -\Lambda_0$

$$\text{the polynomial } \Delta_1(s_1^{(1)}) = 0;$$

At the points s lying immediately to the left of the point $s_1^{(1)}$

$$\text{the polynomial } \Delta_1(s) > 0.$$

The roots of the polynomial $\Delta_2(s)$ are

$$s_{1,2}^{(2)} = \tfrac{1}{2}\Big[-(\Lambda_0 + \Lambda_1 + M_1) \pm \sqrt{(\Lambda_0 + \Lambda_1 + M_1)^2 - 4\Lambda_0\Lambda_1}\,\Big].$$

The fact that $s_2^{(2)} < 0$ is obvious. In order to show that $s_1^{(2)} < 0$, it suffices to show that

$$\Lambda_0 + \Lambda_1 + M_1 > \sqrt{(\Lambda_0 + \Lambda_1 + M_1)^2 - 4\Lambda_0\Lambda_1}.$$

Squaring both sides of this inequality, we immediately obtain the needed result. Thus both roots $s_{1,2}^{(2)}$ of the polynomial $\Delta_2(s)$ are negative. Now we prove that the root $s_1^{(2)}$ lies to the right of $s_1^{(1)}$, and $s_2^{(2)}$ to the left, i.e., we prove that

$$s_1^{(2)} > s_1^{(1)}, \qquad s_2^{(2)} < s_1^{(1)}.$$

Writing these inequalities in explicit form, we obtain:

$$\tfrac{1}{2}\Big[-(\Lambda_0 + \Lambda_1 + M_1) + \sqrt{(\Lambda_0 + \Lambda_1 + M_1)^2 - 4\Lambda_0\Lambda_1}\,\Big] > -\Lambda_0,$$

$$\tfrac{1}{2}\Big[-(\Lambda_0 + \Lambda_1 + M_1) - \sqrt{(\Lambda_0 + \Lambda_1 + M_1)^2 - 4\Lambda_0\Lambda_1}\,\Big] < -\Lambda_0,$$

and the last inequality is conveniently rewritten in the form

$$\tfrac{1}{2}\Big[(\Lambda_0 + \Lambda_1 + M_1) + \sqrt{(\Lambda_0 + \Lambda_1 + M_1)^2 - 4\Lambda_0\Lambda_0}\,\Big] > \Lambda_0.$$

Thus we need to show that

$$\sqrt{(\Lambda_0 + \Lambda_1 + M_1)^2 - 4\Lambda_0\Lambda_1} > \Lambda_1 + M_1 - \Lambda_0,$$

$$\sqrt{(\Lambda_0 + \Lambda_1 + M_1)^2 - 4\Lambda_0\Lambda_1} > \Lambda_0 - \Lambda_1 - M_1.$$

Squaring these inequalities and making simple transformations, we obtain the equivalent conditions

$$4\Lambda_0 M_1 > 0, \qquad 4\Lambda_0 M_1 > 0,$$

which are obviously satisfied. Thus we have shown that indeed the root $s_1^{(2)}$ lies to the right of $s_1^{(1)}$, and $s_2^{(2)}$ to the left of it. Then the polynomial $\Delta_2(s)$ behaves as follows (see Figure 4.2.1):

At the point $s = 0$

the polynomial $\Delta_2(0) > 0$;

At the point $s_2^{(1)}$, lying to the right of the point $s_1^{(1)}$,

the polynomial $\Delta_2(s_2^{(1)}) = 0$;

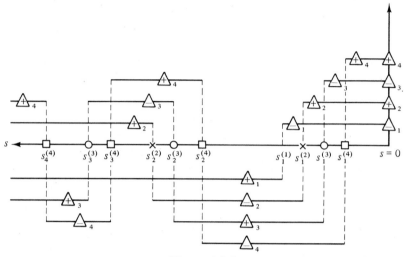

Figure 4.2.1.

At points s lying between the points $s_2^{(1)}$, $s_2^{(2)}$

the polynomial $\Delta_2(s) < 0$;

At the point $s_2^{(2)}$, lying to the left of the point $s_1^{(1)}$,

the polynomial $\Delta_2(s_2^{(2)}) = 0$;

Finally, at points s lying to the left of the point $s_2^{(2)}$

the polynomial $\Delta_2(s) > 0$.

The polynomial $\Delta_3(s)$ is written, by virtue of (4.2.10), as

$$\Delta_3(s) = -(\Lambda_2 + M_2 + s)\Delta_2(s) - \Lambda_1 M_2 \Delta_1(s).$$

The behavior of this polynomial is described as follows:
At the point $s = 0$

the polynomial $\Delta_3(0) < 0$;

At the point $s_1^{(2)}$

the polynomial $\Delta_3(s_1^{(2)}) = \Lambda_1 M_2 \Delta_1(s_1^{(2)})$;

but $\Delta_1(s_1^{(2)}) < 0$; therefore $\Delta_3(s_1^{(2)}) > 0$; that is, from the point $s = 0$ to the point $s = s_1^{(2)}$ the polynomial $\Delta_3(s)$ changes sign;

At the point $s_2^{(2)}$

$$\text{the polynomial } \Delta_3(s_2^{(2)}) = -\Lambda_1 M_2 \Delta_1(s_2^{(2)});$$

but $\Delta_1(s_2^{(2)}) > 0$, therefore $\Delta_3(s_2^{(2)}) < 0$; that is, between the points $s = s_1^{(2)}$ and $s = s_2^{(2)}$ the polynomial $\Delta_3(s)$ again changes sign;
At the points $s \to -\infty$

$$\text{the polynomial } \Delta_3(s) > 0;$$

that is, between the points $s = s_2^{(2)}$ and $s \to -\infty$ the polynomial $\Delta_3(s)$ again (for the third time) passes through zero.

On the basis of the reasoning above, we have established the mode of behavior of the polynomial $\Delta_3(s)$ and the arrangement of its roots as plotted in Figure 4.2.1.

Similarly we can also describe the behavior of the polynomials $\Delta_4(s)$, $\Delta_5(s), \ldots, \Delta_j(s), \ldots$, and it would follow that the two properties of the polynomials $\Delta_j(s)$ formulated above are indeed satisfied for each j.

Since the method of proof is now obvious, we shall not give it for the general case.

Now with the use of Formula (4.1.8) we can represent the determinant (or the polynomials) $\Delta_n(s)$ in the form

$$\Delta_n(s) = \Delta_n(0) + s^0 \left. \frac{d\Delta_n(s)}{ds} \right|_{s=0} + \frac{1}{2!} s^2 \left. \frac{d^2\Delta_n(s)}{ds^2} \right|_{s=0} + \cdots$$
$$+ \frac{1}{n!} s^n \left. \frac{d^n\Delta_n(s)}{ds^n} \right|_{s=0}.$$

Letting

$$\bar{\Delta}_n(s) = \frac{\Delta_n(s)}{\Delta_n(0)}, \tag{4.2.11}$$

we obtain

$$\bar{\Delta}_n(s) = 1 + s \left. \frac{d\bar{\Delta}_n(s)}{ds} \right|_{s=0} + \frac{1}{2!} s^2 \left. \frac{d^2\bar{\Delta}_n(s)}{ds^2} \right|_{s=0} + \cdots + \frac{1}{m!} s^m \left. \frac{d^m\bar{\Delta}_n(s)}{ds^m} \right|_{s=0}$$
$$+ \cdots + \frac{1}{n!} s^n \left. \frac{d^n\bar{\Delta}_n(s)}{ds^n} \right|_{s=0}. \tag{4.2.12}$$

It follows from (4.1.8) and (4.1.9) that for the initial conditions

$$p_0(0) = 1, \quad p_j(0) = 0, \quad 1 \le j \le n, \tag{4.2.13}$$

the following equality is satisfied:

$$\frac{d}{ds} \left[D_n^{(1)}(s) + \Delta_n(s) \right] \bigg|_{s=0} = \frac{d}{ds} \Delta_n(s) \bigg|_{s=0}. \tag{4.2.14}$$

Indeed,

$$D_n^{(1)}(s) + \Delta_n(s) = \begin{vmatrix} -(\Lambda_0 + s) & M_1 & 0 & \cdots \\ \Lambda_0 & -(\Lambda_1 + M_1 + s) & M_2 & \cdots \\ 0 & \Lambda_1 & -(\Lambda_2 + M_2 + s) \\ \cdot \\ \cdot \\ \cdot \\ 0 & 0 & 0 & \cdots \\ 0 & 0 & 0 & \cdots \\ -s & -s & -s & \cdots \end{vmatrix}$$

$$\begin{matrix} \cdots & 0 & 0 & -1 \\ \cdots & 0 & 0 & 0 \\ \cdots & 0 & 0 & 0 \end{matrix}$$

$$\begin{matrix} \cdots & -(\Lambda_{n-2} + M_{n-2} + s) & M_{n-1} & 0 \\ \cdots & \Lambda_{n-2} & -(\Lambda_{n-1} + M_{n-1} + s) & 0 \\ \cdots & -s & -s & 0 \end{matrix}$$

. (4.2.15)

Subtracting from the last row of the determinant (4.2.15) all the preceding rows and expanding the resulting determinant by the elements of the last column, we obtain

$$D_n^{(1)}(s) + \Delta_n(s) = \Delta_n(s) + (-1)^{n+1} \prod_{i=0}^{n-1} \Lambda_i, \qquad (4.2.16)$$

from which the validity of Equation (4.2.14) also follows. But then from (4.1.17) and (4.2.11) we obtain

$$T_1^{(n)} = \frac{1}{\Delta_n(0)} \cdot \frac{d}{ds} \Delta_n(s) \Big|_{s=0} = \bar{\Delta}_n(0). \qquad (4.2.17)$$

Introducing the new variable

$$y = T_1^{(n)} s, \qquad (4.2.18)$$

instead of (4.2.12), we can write

$$\bar{\Delta}_n(y) = 1 + y + \frac{1}{2!(T_1^{(n)})^2} \cdot \frac{d^2 \bar{\Delta}_n(s)}{ds} \Big|_{s=0} \cdot y^2 + \cdots$$
$$+ \frac{1}{m!(T_1^{(n)})^m} \cdot \frac{d^m \bar{\Delta}_n(s)}{ds^m} \Big|_{s=0} \cdot y^m + \cdots + \frac{1}{n!(T_1^{(n)})^n} \cdot \frac{d^n \bar{\Delta}_n(s)}{ds^n} \Big|_{s=0} \cdot y^n$$
$$= 1 + y + \beta_n y^2 + \beta_{n3} y^3 + \cdots + \beta_{nl} y^l + \cdots + \beta_{nn} y^n. \qquad (4.2.19)$$

Above we showed that the roots of the polynomial $\Delta_n(s)$ [and consequently also $\bar{\Delta}_n(s)$] are distinct and negative. Going on, instead of (4.2.19) we can write

$$\bar{\Delta}_n(y) = (1 + x_0 y)(1 + x_1 y)(1 + x_2 y) \cdots (1 + x_{n-1} y), \quad (4.2.20)$$

where $x_0 > x_1 > x_2 > \cdots > x_{n-1} > 0$.

From (4.2.19) and (4.2.20) we find:

$$x_0 + x_1 + x_2 + \cdots + x_{n-1} = 1, \quad (4.2.21)$$

$$(x_0 x_1 + x_0 x_2 + \cdots + x_0 x_{n-1}) + (x_1 x_2 + x_1 x_3 + \cdots + x_1 x_{n-1}) + \cdots$$
$$+ (x_{n-3} x_{n-2} + x_{n-3} x_{n-1}) + x_{n-2} x_{n-1} = \beta_n. \quad (4.2.22)$$

Squaring both sides of Equation (4.2.21) and taking (4.2.22) into account, we obtain

$$x_0^2 + x_1^2 + x_2^2 + \cdots + x_{n-1}^2 + 2\beta_n = 1.$$

Thus

$$x_0 + x_1 + x_2 + \cdots + x_{n-1} = 1,$$
$$x_0^2 + x_1^2 + x_2^2 + \cdots + x_{n-1}^2 = 1 - 2\beta_n. \quad (4.2.23)$$

Hence, first

$$1 - 2\beta_n < x_0(x_0 + x_1 + x_2 + \cdots + x_{n-1}) = x_0 \quad (4.2.24)$$

and second

$$1 - 2\beta_n = x_0^2 + (x_1^2 + x_2^2 + \cdots + x_{n-1}^2) < x_0^2 + (x_1 + x_2 + \cdots + x_{n-1})^2$$
$$= x_0^2 + (1 - x_0)^2;$$

that is,

$$x_0^2 - x_0 + \beta_n > 0. \quad (4.2.25)$$

Somewhat later we shall show that if condition (4.2.1) is satisfied, then $\beta_n \ll 1$. Then it follows from (4.2.24) that the quantity x_0 is close to unity, and from (4.2.25) we obtain:

$$x_0 > \frac{1 + \sqrt{1 - 4\beta_n}}{2}. \quad (4.2.26)$$

Thus

$$1 - x_0 = x_1 + x_2 + \cdots + x_{n-1} < \frac{1 - \sqrt{1 - 4\beta_n}}{2}. \quad (4.2.27)$$

Since it follows from (4.2.23) that

$$x_0 < \sqrt{1 - 2\beta_n} < 1 - \beta_n, \quad (4.2.28)$$

then obviously both estimates [Formulas (4.2.26) and (4.2.27)] are asymptotically exact, since from (4.2.26) and (4.2.27) we obtain

$$x_0 = 1 - \beta_n + O(\beta_n^2)$$

which coincides with (4.2.28).

From (4.2.16) and (4.1.13) with initial conditions (4.2.13) it follows that

$$\mathfrak{D}_n^{(1)}(s) = (-1)^{n+1} \prod_{i=0}^{n-1} \Lambda_i = -\Delta_n(0).$$

Then in view of (4.2.11) we can write in place of (4.1.11):

$$p_n^{(1)}(t) = \frac{1}{2\pi i} \int_C \frac{e^{st}\, ds}{s\bar{\Delta}_n(s)}. \tag{4.2.29}$$

Considering (4.2.18) and (4.2.20), we obtain

$$p_n^{(1)}(t) = \frac{1}{2\pi i} \int_C \frac{e^{xy}\, dy}{y(1 + x_0 y)(1 + x_1 y)\cdots(1 + x_{n-1}y)}, \tag{4.2.30}$$

where we define

$$x = \frac{t}{T_1^{(n)}}.$$

Obviously the function $p_n^{(1)}(t)$ defined by Formula (4.2.29) is the distribution function of the normalized quantity

$$\frac{\theta_n}{T_1^{(n)}} = \bar{\theta}_n,$$

where the quantity θ_n, as before, is the random moment of time when the system first goes into state H_n.

Thus

$$p_n^{(1)}(t) = p_n^{(1)}(T_1^{(n)} x) = \mathcal{P}\{\bar{\theta}_n < x\}.$$

It follows from (4.2.30) (see, for example, [3]) that

$$\bar{\theta}_n = \bar{\theta}_{n0} + \bar{\theta}_{n1} + \bar{\theta}_{n2} + \cdots + \bar{\theta}_{n,n-1},$$

where the $\bar{\theta}_{ni}$ are independent and $\mathcal{P}\{\bar{\theta}_{ni} > x\} = e^{-x/x_i}$.

Letting

$$\mathcal{P}\{\bar{\theta}_{n1} + \bar{\theta}_{n2} + \cdots + \bar{\theta}_{n,n-1} > x\} = \varphi_n(x),$$

we can write

$$p_n^{(1)}(T_1^{(n)}x) = \mathcal{P}\{[\bar{\theta}_{n0} + (\bar{\theta}_{n1} + \bar{\theta}_{n2} + \cdots + \bar{\theta}_{n,n-1})] < x\}$$

$$= \int_0^x \frac{1}{x_0} e^{-(x-u)/x_0} \cdot [1 - \varphi_n(u)]\, du$$

$$= 1 - e^{-x/x_0} - \frac{1}{x_0} \int_0^x e^{-(x-u)/x_0}\varphi_n(u)\, du. \qquad (4.2.31)$$

From (4.2.31) we form a difference of the following form:

$$p_n^{(1)}(T_1^{(n)}x) - 1 + e^{-x} = e^{-x} - e^{-x/x_0} - \frac{1}{x_0} \int_0^x e^{-(x-u)/x_0}\varphi_n(u)\, du. \qquad (4.2.32)$$

We bound this difference from above as follows:

$$p_n^{(1)}(T_1^{(n)}x) - 1 + e^{-x} < e^{-x} - e^{-x/x_0}.$$

It is easy to show that

$$e^{-x} - e^{-x/x_0} < (1 - x_0)e^{-x_{\max}},$$

where

$$x_{\max} = - \frac{x_0}{1 - x_0} \ln x_0 > x_0,$$

since $x_0 < 1$, even though close to unity. Then

$$e^{-x_{\max}} < e^{-x_0} < \frac{1}{e x_0}$$

and

$$e^{-x} - e^{-x/x_0} < \frac{1 - x_0}{e x_0}.$$

Thus an upper bound for the difference (4.2.32) is the quantity

$$p_n^{(1)}(T_1^{(n)}x) - 1 + e^{-x} < \frac{1 - x_0}{e x_0}. \qquad (4.2.33)$$

We obtain a lower bound for the difference defined by Formula (4.2.32) as follows:

$$p_n^{(1)}(T_1^{(n)}x) - 1 + e^{-x} > - \frac{1}{x_0} \int_0^x e^{-(x-u)/x_0}\varphi_n(u)\, du$$

$$> - \frac{1}{x_0} \int_0^x \varphi_n(u)\, du > - \frac{1}{x_0} \int_0^\infty \varphi_n(u)\, du$$

$$= - \frac{\mathrm{M}[\bar{\theta}_{n1} + \bar{\theta}_{n2} + \cdots + \bar{\theta}_{n,n-1}]}{x_0} = - \frac{1 - x_0}{x_0}. \qquad (4.2.34)$$

The last equality follows from (4.2.23).

From (4.2.33) and (4.2.34) we have

$$\max_{0\leq x<\infty} |p_n^{(1)}(T_1^{(n)}x) - 1 + e^{-x}| < \frac{1 - x_0}{x_0}$$

or, using (4.2.26) and (4.2.27) for $\beta_n < \frac{1}{4}$, we obtain

$$\max_{0\leq t<\infty} |p_n^{(1)}(t) - 1 + e^{-t/T_1^{(n)}}| < \frac{1 - x_0}{x_0} < \frac{1 - \sqrt{1 - 4\beta_n}}{1 + \sqrt{1 - 4\beta_n}}$$

$$< \frac{1 - (1 - 4\beta_n)}{1 + (1 - 4\beta_n)}$$

$$= \frac{2\beta_n}{1 - 2\beta_n} < 2\beta_n(1 + 4\beta_n). \quad (4.2.35)$$

Thus we have shown that an approximate estimate $\tilde{P}^{(1)}(t)$ for the probability of failure-free operation of the system $P^{(1)}(t)$ [where $P^{(1)}(t)$ is the exact value of the probability of failure-free operation of the system for initial conditions (4.2.13)] is the function

$$\tilde{P}^{(1)}(t) = e^{-t/T_1^{(n)}}, \quad (4.2.36)$$

where the following inequality is satisfied for the error δ of the estimate (4.2.36) when $\beta_n < \frac{1}{4}$ *for all moments of time t:*

$$\max_{0\leq t<\infty} \delta = \max_{0\leq t<\infty} |P^{(1)}(t) - \tilde{P}^{(1)}(t)|$$

$$= \max_{0\leq t<\infty} |P^{(1)}(t) - e^{-t/T_1^{(n)}}|$$

$$< 2\beta_n(1 + 4\beta_n). \quad (4.2.37)$$

The satisfaction of condition (4.2.1) is not needed, generally speaking, for validity of the estimate (4.2.36) and the bound (4.2.37). It suffices that $\beta_n < \frac{1}{4}$, since here the quantities Λ_α and M_β are related, and n is irrelevant. Since

$$\max_{0\leq t<\infty} |P_n^{(1)}(t) - 1 + e^{-t/T_1^{(n)}}| \sim \beta_n$$

as $\beta_n \to 0$, then the estimate (4.2.36) is asymptotically exact.

Now we find an explicit expression for the quantity β_n. It follows from (4.2.19) that

$$\beta_n = \frac{1}{2!(T_1^{(n)})^2} \cdot \frac{d^2\bar{\Delta}_n(s)}{ds^2}\bigg|_{s=0}.$$

Using (4.2.10), (4.2.11), we obtain

$$\frac{d^2\bar{\Delta}_n(s)}{ds^2}\bigg|_{s=0} = \frac{2}{\Lambda_{n-1}} \cdot \frac{d\bar{\Delta}_{n-1}(s)}{ds}\bigg|_{s=0} + \left(1 + \frac{M_{n-1}}{\Lambda_{n-1}}\right)\frac{d^2\bar{\Delta}_{n-1}(s)}{ds^2}\bigg|_{s=0}$$
$$- \frac{M_{n-1}}{\Lambda_{n-1}} \cdot \frac{d^2\bar{\Delta}_{n-2}(s)}{ds^2}\bigg|_{s=0}.$$

Letting

$$\frac{d^2\bar{\Delta}_n(s)}{ds^2}\bigg|_{s=0} = d_n, \qquad \frac{d\bar{\Delta}_n(s)}{ds}\bigg|_{s=0} = \omega_n, \qquad (4.2.38)$$

we obtain

$$\alpha_n = \frac{2}{\Lambda_{n-1}}\omega_{n-1} + \left(1 + \frac{M_{n-1}}{\Lambda_{n-1}}\right)\alpha_{n-1} - \frac{M_{n-1}}{\Lambda_{n-1}}\alpha_{n-1}. \qquad (4.2.39)$$

Using (4.2.39) and introducing the notation

$$\alpha_n - \alpha_{n-1} = b_n, \qquad 2\omega_{n-1} = \delta_{n-1}, \qquad (4.2.40)$$

we obtain

$$b_n = \frac{1}{\Lambda_{n-1}}\delta_{n-1} + \frac{M_{n-1}}{\Lambda_{n-1}}b_{n-1}. \qquad (4.2.41)$$

Making successive use of this recursion relation, we show easily that

$$b_n = \frac{1}{\Lambda_{n-1}\Theta_{n-1}} \cdot \sum_{j=0}^{n-1} \delta_j\Theta_j, \qquad (4.2.42)$$

where Θ_j, as before, is defined by Formula (4.1.26). Since, according to Formulas (4.2.40), (4.2.38), (4.2.17), and (4.1.32),

$$\delta_{n-1} = 2\omega_{n-1} = 2\frac{d\bar{\Delta}_{n-1}(s)}{ds}\bigg|_{s=0}$$

$$= 2T_1^{(n-1)} = 2\sum_{i=0}^{n-2}\frac{\sum_{j=0}^{i}\Theta_j}{\Lambda_i\Theta_i},$$

then

$$\delta_j = 2\sum_{k=0}^{j-1}\frac{\sum_{i=0}^{k}\Theta_i}{\Lambda_k\Theta_k}. \qquad (4.2.43)$$

It follows that

$$b_n = \frac{2}{\Lambda_{n-1}\Theta_{n-1}} \sum_{j=0}^{n-1} \Theta_j \sum_{k=0}^{j-1} \frac{\sum_{i=0}^{k} \Theta_i}{\Lambda_k \Theta_k} . \tag{4.2.44}$$

Since, by (4.2.40),

$$b_n = \alpha_n - \alpha_{n-1},$$

using (4.2.44), we obtain

$$\alpha_n = 2 \sum_{l=0}^{n-1} \frac{1}{\Lambda_l \Theta_l} \sum_{j=0}^{l} \Theta_j \sum_{k=0}^{j-1} \frac{\sum_{i=0}^{k} \Theta_i}{\Lambda_k \Theta_k} . \tag{4.2.45}$$

It follows from (4.2.19) and (4.2.38) that

$$\beta_n = \frac{1}{2!(T_1^{(n)})^2} \cdot \frac{d^2 \bar{\Delta}_n(s)}{ds^2}\bigg|_{s=0} = \frac{\alpha_n}{2(T_1^{(n)})^2} .$$

Thus, the exact value of the coefficient β_n is

$$\beta_n = \frac{\displaystyle\sum_{l=0}^{n-1} \frac{1}{\Lambda_l \Theta_l} \sum_{j=0}^{l} \Theta_j \sum_{k=0}^{j-1} \frac{\sum_{i=0}^{k} \Theta_i}{\Lambda_k \Theta_k}}{\left(\displaystyle\sum_{l=0}^{n-1} \frac{\sum_{j=0}^{l} \Theta_j}{\Lambda_l \Theta_l}\right)^2} . \tag{4.2.46}$$

Using (4.1.32), we write

$$\beta_n = \frac{1}{(T_1^{(n)})^2} \cdot \sum_{l=0}^{n-1} \frac{1}{\Lambda_l \Theta_l} \sum_{j=0}^{l} T_1^{(j)} \Theta_j.$$

We denote the approximate value of the coefficient β_n by $\tilde{\beta}_n$. From condition (4.2.1) it follows that

$$\sum_{k=0}^{j-1} \frac{\sum_{i=0}^{k} \Theta_i}{\Lambda_k \Theta_k} \sim \frac{1}{\Lambda_{j-1}\Theta_{j-1}}, \qquad \sum_{l=0}^{n-1} \frac{\sum_{j=0}^{l} \Theta_j}{\Lambda_l \Theta_l} \sim \frac{1}{\Lambda_{n-1}\Theta_{n-1}} .$$

Since

$$\sum_{j=0}^{l} \Theta_j \frac{1}{\Lambda_{j-1}\Theta_{j-1}} = \sum_{j=0}^{l} \frac{\Lambda_0\Lambda_1\cdots\Lambda_{j-1}\cdot M_1 M_2\cdots M_{j-1}}{M_1 M_2\cdots M_j\cdot\Lambda_0\Lambda_1\cdots\Lambda_{j-2}\Lambda_{j-1}} = \sum_{j=1}^{l} \frac{1}{M_j},$$

then

$$\sum_{l=0}^{n-1} \frac{1}{\Lambda_l\Theta_l}\cdot\sum_{j=1}^{l}\frac{1}{M_j} \sim \frac{1}{\Lambda_{n-1}\Theta_{n-1}}\sum_{j=1}^{n-1}\frac{1}{M_j}$$

and

$$\tilde{\beta}_n = \frac{\dfrac{1}{\Lambda_{n-1}\Theta_{n-1}}\displaystyle\sum_{j=1}^{n-1}\frac{1}{M_j}}{\left(\dfrac{1}{\Lambda_{n-1}\Theta_{n-1}}\right)^2} = \Lambda_{n-1}\Theta_{n-1}\sum_{j=1}^{n-1}\frac{1}{M_j}.$$

Thus

$$\tilde{\beta}_n = \Lambda_{n-1}\Theta_{n-1}\sum_{j=1}^{n-1}\frac{1}{M_j} = \Lambda_0\prod_{i=1}^{n-1}\frac{\Lambda_i}{M_i}\sum_{j=1}^{n-1}\frac{1}{M_j}. \qquad (4.2.47)$$

Since it follows from condition (4.2.1) that $\Lambda_{max} \ll M_{min}/N$, using (4.2.47) and the given conditions $n - 1 \geq 1$, $k \geq 1$, we obtain

$$\tilde{\beta}_n < (n-1)\left(\frac{\Lambda_{max}}{M_{min}}\right)^n \ll \frac{n-1}{N^n} = \frac{n-1}{(n-1+k)^n} \leq \tfrac{1}{4}.$$

Thus when condition (4.2.1) is satisfied, we always have

$$\tilde{\beta}_n \ll \tfrac{1}{4}.$$

As has already been noted, the estimate (4.2.36) and its error (4.2.37) hold for all moments of time t. We can also obtain another estimate of the probability of failure-free operation of the system, which will hold for all times $t \geq t_\varphi$, where t_φ is a completely determined quantity. In Solov'ev's work [2] it is shown that for the coefficients β_{nl} (Formula (4.2.19)) the following bound holds:

$$\beta_{nl} < \frac{\beta_n^{l-1}}{(l-1)!}, \qquad 3 \leq l \leq n, \qquad (4.2.48)$$

and, in particular,

$$\beta_{n3} < \frac{\beta_n^2}{2!}.$$

Then, to within terms of order $O(\beta_n^2)$, we easily obtain from (4.2.19) that

$$\bar{\Delta}_n(y) \approx \beta_n(y - y_1)(y - y_2),$$

where

$$y_1 = -(1 + \beta_n), \qquad y_2 = -\frac{1}{\beta_n}(1 - \beta_n - \beta_n^2). \qquad (4.2.49)$$

Since, in accordance with (4.2.29),

$$p_n^{(1)}(t) = \frac{1}{2\pi i} \int_C \frac{e^{st}\, ds}{s\bar{\Delta}_n(s)}, \qquad (4.2.50)$$

making the substitution (4.2.18) in (4.2.50) and substituting (4.2.49), we obtain

$$p_n^{(1)}(t) \approx \frac{1}{2\pi i} \int_C \frac{e^{t/T_1^{(n)}}\, dy}{\beta_n y(y - y_1)(y - y_2)}. \qquad (4.2.51)$$

Hence by the residue theorem we obtain

$$p_n^{(1)}(t) \approx \frac{1}{\beta_n}\left[\frac{1}{y_1 y_2} + \frac{e^{(t/T_1^{(n)})y_1}}{y_1(y_1 - y_2)} + \frac{e^{(t/T_1^{(n)})y_2}}{y_2(y_2 - y_1)}\right]. \qquad (4.2.52)$$

Taking account of (4.2.49), to within terms of order $O(\beta_n^2)$ we obtain

$$p_n^{(1)}(t) \approx 1 - [(1 + \beta_n)e^{-(1+\beta_n)(t/T_1^{(n)})} - \beta_n e^{-((1-\beta_n-\beta_n^2)/\beta_n)\cdot(t/T_1^{(n)})}]. \qquad (4.2.53)$$

Note that both of the summands in the square brackets in (4.2.53) contain terms of order $O(\beta_n^2)$ and smaller. Since it follows from (4.2.19) that the expansion (4.2.49) holds only with an accuracy up to terms of order $O(\beta_n^2)$, then in (4.2.53) we must retain only terms of order β_n. Taking account of (4.2.49), we obtain:

$$p_n^{(1)}(t) \approx 1 - \left[1 + \beta_n\left(1 - \frac{t}{T_1^{(n)}}\right)\right]e^{-t/T_1^{(n)}}. \qquad (4.2.54)$$

Then the probability of failure-free operation of the system can be represented in the form

$$P^{(1)}(t) - 1 - p_n^{(1)}(t) \approx \left[1 + \beta_n\left(1 - \frac{t}{T_1^{(n)}}\right)\right]e^{-t/T_1^{(n)}}, \qquad (4.2.55)$$

where $P^{(1)}(t)$ is the probability of failure-free operation of the system that began working in a completely operational state. [Formula (4.2.50) holds for initial conditions (4.2.13).]

In both estimates, (4.2.36) and (4.2.55), the quantity $T_1^{(n)}$ in the exponent [Formula (4.1.32)], although not complicated, is cumbersome to compute in the majority of cases. An approximate value for this quantity (the mean value of the time up to failure) is easily determined from Formula (4.2.4). Therefore for crude estimates of the probability of failure-free operation we can use a less precise, though essentially simpler, estimate depending on the approximate value of the mean operating time up to failure $\tilde{T}_1^{(n)}$.

It follows from (4.2.4) and (4.2.5) that

$$\frac{T_1^{(n)} - \tilde{T}_1^{(n)}}{\tilde{T}_1^{(n)}} - \frac{\Delta \tilde{T}_1^{(n)}}{\tilde{T}_1^{(n)}} \sim \frac{\Lambda_0}{M_1} + \frac{\Lambda_{n-1}}{M_{n-1}}. \qquad (4.2.56)$$

Then from (4.2.55) we can obtain

$$|P^{(1)}(t) - e^{-t/\tilde{T}_1^{(n)}}| \sim \frac{1}{e}\left(\frac{\Lambda_0}{M_1} + \frac{\Lambda_{n-1}}{M_{n-1}}\right). \qquad (4.2.57)$$

Thus for the probability of failure-free operation we can propose the following estimates, arranged in decreasing order of precision:

NO.	ESTIMATE FOR THE PROBABILITY OF FAILURE-FREE OPERATION	ERROR
1	$\left[1 + \beta_n\left(1 - \dfrac{t}{T_1^{(n)}}\right)\right]e^{-t/T_1^{(n)}}$	$\delta \sim \beta_n^2$
2	$e^{-t/T_1^{(n)}}$	$\delta < 2\beta_n(1 + 4\beta_n)$
3	$e^{-t/\tilde{T}_1^{(n)}}$	$\delta \sim \dfrac{1}{e}\left(\dfrac{\Lambda_0}{M_1} + \dfrac{\Lambda_{n-1}}{M_{n-1}}\right)$

According to the considerations above the preferable estimate is the "second," Formula (4.2.36), the error of which for all moments of time t does not exceed $2\beta_n(1 + 4\beta_n)$, where β_n is given by Formula (4.2.46).

4 2.6 Readiness coefficient

It follows from (4.1.47) that when condition (4.2.1) is satisfied, an approximate estimate $\tilde{K}$ of the readiness coefficient of the system K is the quantity

$$\tilde{K} = 1 - \Theta_n \qquad (4.2.58)$$

and

$$\delta_+ = K - \tilde{K} \sim \Theta_1\Theta_n, \qquad (4.2.59)$$

where Θ_j, as before, is defined by Formula (4.1.26).

4.2.7 Reliability coefficient

Now we find an approximate expression for the reliability coefficient of a system. According to a known formula from renewal theory [3] the precise value of the reliability coefficient is

$$R(t) = \frac{K}{T_2^{(n)}} \int_t^\infty P^{(2)}(\theta)\, d\theta \qquad (4.2.60)$$

where in our case K and $T_2^{(n)}$ are defined by Formulas (4.1.47) and (4.1.34), and the function $P^{(2)}(\theta)$ by Formula (4.1.11), which under the initial conditions (4.1.33) takes the form

$$P^{(2)}(t) = 1 - p_n^{(2)}(t) = 1 - \frac{1}{2\pi i} \int_C \frac{\Lambda_{n-1}\Delta_{n-1}(s)e^{st}\, ds}{-s\Delta_n(s)}. \qquad (4.2.61)$$

Since, according to (4.1.13),

$$\Delta_n(0) = -\Lambda_{n-1}\Delta_{n-1}(0),$$

dividing the numerator and denominator of the integrand in expression (4.2.61) by $\Delta_n(0)$ and taking (4.2.11) into account, we obtain

$$p_n^{(2)}(t) = \frac{1}{2\pi i} \int_C \frac{\bar{\Delta}_{n-1}(s)e^{st}\, ds}{s\bar{\Delta}_n(s)}, \qquad (4.2.62)$$

where the $\bar{\Delta}_j(s)$ are given by the recursion relation (4.2.10).

According to (4.2.19) and (4.2.18), to within terms of order $O(\beta_n^2)$ we have

$$\bar{\Delta}_n(s) = \bar{\Delta}_n(y) \approx 1 + y + \beta_n y^2,$$

$$\bar{\Delta}_{n-1}(s) = \bar{\Delta}_{n-1}(y) \approx 1 + \frac{T_1^{(n-1)}}{T_1^{(n)}}\, y + \beta_{n-1}\left(\frac{T_1^{(n-1)}}{T_1^{(n)}}\right)^2 y^2. \qquad (4.2.63)$$

From (4.1.11), finding the inverse Laplace transform for $R(t)$ directly in the form (4.2.60), we obtain

$$R(t) = \frac{K}{T_2^{(n)}} \cdot \frac{1}{2\pi i} \int_C \frac{T_2^{(n)}s\bar{\Delta}_n(s) + \bar{\Delta}_{n-1}(s) - \bar{\Delta}_n(s)}{s^2\bar{\Delta}_n(s)}\, e^{st}\, ds. \qquad (4.2.64)$$

Now, using (4.2.63) and (4.2.49), we obtain

$$R(t) \approx K[(1 - \beta_n\epsilon_n)e^{-(1+\beta_n)t/T_1^{(n)}} + \beta_n\epsilon_n e^{-(1-\beta_n-\beta_n^2)/\beta_n T_1^{(n)}}], \qquad (4.2.65)$$

where

$$\epsilon_n = \frac{T_1^{(n-1)}}{T_2^{(n)}}\left(1 - \frac{\beta_{n-1}}{\beta_n} \cdot \frac{T_1^{(n-1)}}{T_1^{(n)}}\right). \qquad (4.2.66)$$

Since the expansion in (4.2.49) took account only of terms with accuracy up to order $O(\beta_n^2)$, then in (4.2.65) we must keep terms whose orders are not higher than β_n. Then instead of (4.2.65) we can write

$$R(t) \approx K\left[1 - \beta_n\left(\epsilon_n + \frac{t}{T_1^{(n)}}\right)\right]e^{-t/T_1^{(n)}}. \tag{4.2.67}$$

A cruder estimate than (4.2.67) is one of the form

$$\tilde{R}(t) = Ke^{-t/T_1^{(n)}}.$$

The error of this estimate can be found approximately as the maximum of the function

$$A(t) = K\left|\left[1 - \beta_n\left(\epsilon_n + \frac{t}{T_1^{(n)}}\right)\right]e^{-t/T_1^{(n)}} - e^{-t/T_1^{(n)}}\right|.$$

This maximum is achieved for

$$\frac{t_m}{T_1^{(n)}} = 1 - \epsilon_n$$

and is

$$A_m = A(t_m) = \frac{K\beta_n}{e^{1-\epsilon_n}} \sim \frac{\beta_n}{e}, \tag{4.2.68}$$

since it follows from (4.2.66) that

$$\epsilon_n \sim \frac{\Lambda_{n-1}}{M_{n-1}} \cdot \frac{\dfrac{1}{M_{n-1}}}{\displaystyle\sum_{j=1}^{n-1}\dfrac{1}{M_j}} \ll 1.$$

Thus, we have obtained two estimates for the reliability coefficient of a system: the first,

$$R(t) \approx K\left[1 - \beta_n\left(\epsilon_n + \frac{t}{T_1^{(n)}}\right)\right]e^{-t/T_1^{(n)}} \tag{4.2.69}$$

has error of order $O(\beta_n^2)$, and the second,

$$\tilde{R}(t) = Ke^{-t/T_1^{(n)}}, \tag{4.2.70}$$

has error of order $O(\beta_n)$. Below we shall use only the estimate (4.2.70).

We note that we can also find an estimate and its error analogously for the function $p_n^{(2)}(t)$ [Formula (4.2.62)], the distribution function of the operating time of the system between failures. Indeed, substituting

Table 4.2.1 Basic Reliability Measures for Systems
Obtained in Sections 4.1 and 4.2

MEASURE	EXACT VALUE	APPROXIMATE VALUE	ERROR
T	$\displaystyle\sum_{j=0}^{n-1} \frac{\pi_j(0) \sum_{l=0}^{j} \Theta_l}{\Lambda_j \Theta_j}$	$\dfrac{\pi_{n-1}(0)}{\Lambda_{n-1}\Theta_{n-1}}$	$\delta_+ \sim \dfrac{\pi_{n-2}(0) + \dfrac{\Lambda_0 M_{n-1}}{\Lambda_{n-1}M_1}\pi_{n-1}(0)}{\Lambda_{n-2}\Theta_{n-2}}$
T_1	$\displaystyle\sum_{j=0}^{n-1} \frac{\sum_{l=0}^{j} \Theta_l}{\Lambda_j \Theta_j}$	$\dfrac{1}{\Lambda_{n-1}\Theta_{n-1}}$	$\delta_+ \sim \dfrac{1 + \dfrac{\Lambda_0 M_{n-1}}{\Lambda_{n-1}M_1}}{\Lambda_{n-2}\Theta_{n-2}}$
T_2	$\dfrac{\sum_{l=0}^{n-1} \Theta_l}{\Lambda_{n-1}\Theta_{n-1}}$	$\dfrac{1}{\Lambda_{n-1}\Theta_{n-1}}$	$\delta_+ \sim \dfrac{\Lambda_0}{\Lambda_{n-1}M_1\Theta_{n-1}}$
T_R	$\dfrac{1}{\sum_{s=0}^{n}\Theta_s}\displaystyle\sum_{j=0}^{n-1}\frac{\left(\sum_{l=0}^{j}\Theta_l\right)^2}{\Lambda_j\Theta_j}$	$\dfrac{1}{\Lambda_{n-1}\Theta_{n-1}}$	$\delta_+ \sim \dfrac{1 + \dfrac{\Lambda_0 M_{n-1}}{\Lambda_{n-1}M_1}}{\Lambda_{n-2}\Theta_{n-2}}$
P_j	$\dfrac{\Theta_j}{\sum_{i=0}^{n}\Theta_i}$	—	—
τ	$\dfrac{1}{M_n}$	—	—
K	$\left[1 + \dfrac{\Theta_n}{\sum_{i=0}^{n-1}\Theta_i}\right]^{-1}$	$1 - \Theta_n$	$\delta_+ \sim \Theta_1\Theta_n$
$P^{(1)}(t_0)$	—	e^{-t_0/T_1}	$\delta \sim \beta_n$
$P^{(2)}(t_0)$	—	$\dfrac{T_2}{T_1}e^{-t_0/T_1}$	$\delta \sim \beta_n$
$R(t_0)$	—	Ke^{-t_0/T_1}	$\delta \sim \beta_n$

(4.2.18) into (4.2.62) and taking account of (4.2.49), with accuracy up to terms of order $O(\beta_n^2)$ we may write

$$p_n^{(2)}(t) \approx \frac{1}{2\pi i} \int_C \frac{\bar{\Delta}_{n-1}(y)e^{yt/T_1^{(n)}}\,dy}{\beta_n y(y-y_1)(y-y_2)},$$

whence, by the residue theorem, using (4.2.61), we obtain

$$
\begin{aligned}
P^{(2)}(t) &= \frac{T_2^{(n)}}{T_1^{(n)}}\left\{1 + \beta_n\left[1 - \frac{T_1^{(n-1)}}{T_2^{(n)}}\left(1 - \frac{\beta_{n-1}}{\beta_n}\frac{T_1^{(n-1)}}{T_1^{(n)}}\right)\right]\right\}e^{-(1+\beta_n)t/T_1^{(n)}} \\
&+ \left\{\frac{T_1^{(n-1)}}{T_1^{(n)}}\left(1 - \frac{\beta_{n-1}}{\beta_n}\frac{T_1^{(n-1)}}{T_1^{(n)}}\right) - \beta_n\left[1 + \frac{T_1^{(n-1)}}{T_1^{(n)}} - 2\frac{\beta_{n-1}}{\beta_n}\left(\frac{T_1^{(n-1)}}{T_1^{(n)}}\right)^2\right]\right\} \\
&\qquad\qquad\qquad\qquad \times e^{-(1-\beta_n-\beta_n^2)t/\beta_n T_1^{(n)}}, \qquad (4.2.71)
\end{aligned}
$$

where, according to (4.1.35),

$$T_2^{(n)} = T_1^{(n)} - T_1^{(n-1)}.$$

In Formula (4.2.71) we should keep only terms of order β_n.

Thus, using (4.2.66) instead of (4.2.71), with accuracy up to terms of order $O(\beta_n^2)$ we obtain

$$P^{(2)}(t) \approx \frac{T_2^{(n)}}{T_1^{(n)}}\left[1 + \beta_n\left(1 - \epsilon_n - \frac{t}{T_1^{(n)}}\right)\right]e^{-t/T_1^{(n)}}. \qquad (4.2.72)$$

With accuracy up to terms of order $O(\beta_n)$

$$\tilde{P}^{(2)}(t) = \frac{T_2^{(n)}}{T_1^{(n)}}e^{-t/T_1^{(n)}}. \qquad (4.2.73)$$

In Table 4.2.1 we present all the basic reliability measures of a system obtained in Sections 4.1 and 4.2.

4.3 ANOTHER CRITERION FOR FAILURE OF A SYSTEM

A system with a somewhat more general criterion for failure than the one considered in Sections 4.1 and 4.2 is a system of k operating and $n-1$ stand-by units that has not one, but several, failed states: H_n, H_{n+1}, $\ldots$, $H_{N=k+(n-1)}$. This model describes the operation of a system in which, after failure of all stand-by and one of the k operating units, the remaining operational units continue to be in a "turned-on" state, awaiting the end of repair of the system. This situation is especially typical of complex systems, such as various types of computing complexes, which as a rule are not "shut down" during the repair time of the failed units.

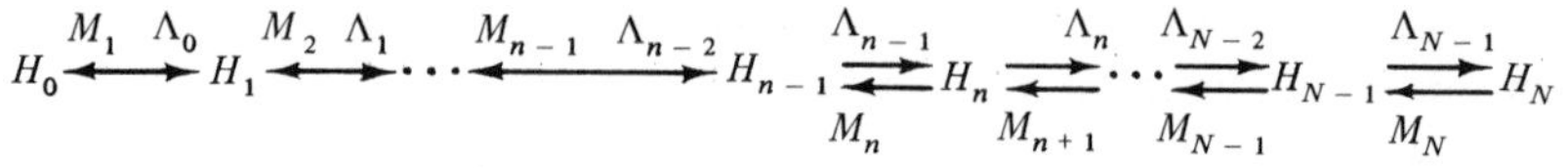

$$H_0 \xrightleftharpoons[\ ]{M_1\ \ \Lambda_0} H_1 \xleftarrow{M_2\ \ \Lambda_1} \cdots \xleftarrow{M_{n-1}\ \ \Lambda_{n-2}} H_{n-1} \underset{M_n}{\overset{\Lambda_{n-1}}{\rightleftarrows}} H_n \underset{M_{n+1}}{\overset{\Lambda_n\ \ \Lambda_{N-2}}{\rightleftarrows}} \cdots \underset{M_{N-1}}{\overset{}{\rightleftarrows}} H_{N-1} \underset{M_N}{\overset{\Lambda_{N-1}}{\rightleftarrows}} H_N$$

Figure 4.3.1.

A diagram of passages of such a system from one state to another is shown in Figure 4.3.1 where, starting with the state H_n (n units have failed), all successive states are failed states.

It is easy to see that in this case such reliability measures for a system as mean operating time T_1 of the system up to failure, mean operating time T_2 of the system between failures, and the probability $P(t)$ of failure-free operation of the system are the same as for the systems considered in Sections 4.1 and 4.2.

It is no less obvious [see (4.1.49)] that in this case

$$p_l = \lim_{t \to \infty} p_l(t) = \frac{\Theta_l}{\displaystyle\sum_{j=0}^{N} \Theta_j}, \qquad 0 \le l \le N, \tag{4.3.1}$$

where Θ_j is defined by Formula (4.1.26). Then the readiness coefficient of the system is

$$K = \sum_{l=0}^{n-1} p_l = \frac{\displaystyle\sum_{l=0}^{n-1} \Theta_l}{\displaystyle\sum_{j=0}^{N} \Theta_j}. \tag{4.3.2}$$

Knowing T_2 and K, we can easily find

$$\tau = \frac{\displaystyle\sum_{j=n}^{N} \Theta_j}{\Lambda_{n-1}\Theta_{n-1}}. \tag{4.3.3}$$

The reliability coefficient of such a system is defined by the general formula (4.2.70):

$$\tilde{R}(t) = K e^{-t/T_1},$$

where K is defined by (4.3.2) and T_1 by (4.1.32). The error δ of this formula, as before, is of order

$$\delta \sim \beta_n,$$

where β_n is defined by Formula (4.2.46).

4.4 SYSTEMS WITH ONE STAND-BY UNIT

We examine the reliability measures of a system with m operating units and one stand-by unit (Figure 4.4.1). The case of a redundant system with one operating and one stand-by unit is considered separately (Figure 4.4.2).

A diagram of passages from one state to another for a system with one stand-by unit is given in Figure 4.4.3, where H_j is the state of the system for which j of its units have failed ($N = m + 1$) and H_2 is the failed state of the system ($j = 0, 1, 2$); Λ_j is the intensity of passage (failure of one of its operational units) of the system from the state with j failed units to the state with one additional failed unit—that is, $j + 1$ ($j = 0, 1$); M_j is the intensity of passage of the system (repair of one of its failed units) from the state with j failed units to the state with one less failed unit—that is, $j - 1$ ($j = 1, 2$).

The reliability measures of a system with one stand-by unit are given for the general case and three particular cases, namely, operating, partially operating, and nonoperating stand-by unit.

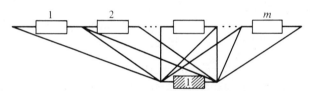

Figure 4.4.1   Block Diagram of the Model Used for Reliability Computations for a System of m Operating Units and One Stand-by Unit.

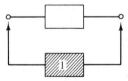

Figure 4.4.2   Block Diagram of the Model Used for Reliability Computations for a Redundant System.

$$H_0 \xleftrightarrow[M_1]{\Lambda_0} H_1 \underset{M_2}{\overset{\Lambda_1}{\rightleftarrows}} H_2$$

Figure 4.4.3   Diagram of Passages from One State to Another by a System with One Stand-by Unit.

For an operating stand-by:

$$\Lambda_0 = m\lambda + \lambda = (m + 1)\lambda.$$

For a partially operating stand-by:

$$\Lambda_0 = m\lambda + \nu\lambda = (m + \nu)\lambda, \qquad \Lambda_1 = m\lambda,$$

where ν is the load coefficient of the stand-by—that is, the failure intensity of the stand-by unit is $\nu\lambda$, $0 \le \nu \le 1$.

For a nonoperating stand-by:

$$\Lambda_0 = m\lambda; \qquad \Lambda_1 = m\lambda.$$

For each of these three cases we examine two forms of repair (service) of failed units:

(1)   unrestricted repair, where two failed units can be repaired simultaneously and independently;

(2)   restricted repair, where not more than one failed unit can undergo repair at any moment of time.

Formally,

$$\mathrm{M}_2 = \begin{cases} 2\mu & \text{if repair is unrestricted,} \\ \mu & \text{if repair is restricted.} \end{cases}$$

In Tables 4.4.1–4.4.13 we show both exact and approximate values for the basic reliability measures of a system with one stand-by unit, with error estimates. The approximate values of the basic reliability measures were obtained for $\gamma = \lambda/\mu \ll 1/(m + 1)$.

After the tables we give examples of the computation of these reliability measures.

Example 4.4.1   Suppose we have a redundant system where the intensity of passage from the state with both units operational to the state of failure of at least one is

$$\Lambda_0 = \lambda + \lambda = 2\lambda.$$

If failure of (any) one of the two units has already occurred, then the remaining operational unit begins operating in an overloaded condition, and its failure intensity becomes

$$\Lambda_1 = 2\lambda.$$

The failed unit is repaired with intensity

$$\mathrm{M}_1 = \mu.$$

Since use of the unit remaining operational may lead to more severe failure of this unit, we assume that the repair intensity of the system (that is, of at least one of the two failed units) is

$$M_2 = \mu + 0.5\mu = 1.5\mu.$$

For this system we set the quantities λ and μ equal to

$$\lambda = 0.05 \, \frac{1}{\text{hr}}, \qquad \mu = 2 \, \frac{1}{\text{hr}}.$$

We wish to determine the reliability measures for such a redundant system of units.

Solution. We make use of Table 4.4.1.

1. The mean operating time up to failure is

$$T_1 = \frac{\Lambda_0 + \Lambda_1 + M_1}{\Lambda_0 \Lambda_1} = \frac{2\lambda + 2\lambda + \mu}{2\lambda \cdot 2\lambda} = \frac{4\lambda + \mu}{4\lambda^2}$$

$$= \frac{4 \cdot 0.05 + 2}{4 \cdot (0.05)^2} = 220 \quad \text{hr.}$$

The approximate value is

$$\tilde{T}_1 = \frac{M_1}{\Lambda_0 \Lambda_1} = \frac{\mu}{2\lambda \cdot 2\lambda} = \frac{2}{4 \cdot (0.05)^2} = 200 \quad \text{hr.}$$

2. The mean operating time between failures is

$$T_2 = \frac{\Lambda_0 + M_1}{\Lambda_0 \Lambda_1} = \frac{2\lambda + \mu}{2\lambda \cdot 2\lambda} = \frac{2 \cdot 0.05 + 2}{4 \cdot (0.05)^2} = 210 \quad \text{hr.}$$

The approximate value is

$$\tilde{T}_2 = \tilde{T}_1 = 200 \quad \text{hr.}$$

3. The mean repair time is

$$\tau = \frac{1}{M_2} = \frac{1}{1.5\mu} = \frac{1}{1.5 \cdot 2} = 0.33 \quad \text{hr.}$$

4. The probability of failure-free operation of the system for an initial time interval of length t_0 is

$$P(t_0) = \frac{1}{s_1 - s_2} (s_1 e^{-s_2 t} - s_2 e^{-s_1 t}),$$

where

$$s_{1,2} = \tfrac{1}{2}\Big[(\Lambda_0 + \Lambda_1 + M_1) \pm \sqrt{(\Lambda_0 + \Lambda_1 + M_1)^2 - 4\Lambda_0 \Lambda_1} \Big].$$

Table 4.4.1   System with One Stand-by Unit: General Case

$$s_{1,2} = \tfrac{1}{2}[(\Lambda_0 + \Lambda_1 + M_1) \pm \sqrt{(\Lambda_0 + \Lambda_1 + M_1)^2 - 4\Lambda_0\Lambda_1}],$$

$$\epsilon_{1,2} = \tfrac{1}{2}[(\Lambda_0 + \Lambda_1 + M_1 + M_2) \pm \sqrt{(\Lambda_0 + \Lambda_1 + M_1 + M_2)^2 - 4(\Lambda_0\Lambda_1 + \Lambda_0 M_2 + M_1 M_2)}],$$

$$\gamma_c = \frac{\Lambda_0\Lambda_1}{M_2(\Lambda_0 + M_1)}.$$

MEASURE	EXACT VALUE	APPROXIMATE VALUE	ERROR
T_1	$\dfrac{\Lambda_0 + \Lambda_1 + M_1}{\Lambda_0\Lambda_1}$	$\dfrac{M_1}{\Lambda_0\Lambda_1}$	$\delta_+ = \dfrac{\Lambda_0 + \Lambda_1}{\Lambda_0\Lambda_1}$
T_2	$\dfrac{\Lambda_0 + M_1}{\Lambda_0\Lambda_1}$	$\dfrac{M_1}{\Lambda_0\Lambda_1}$	$\delta_+ = \dfrac{1}{\Lambda_1}$
τ	$\dfrac{1}{M_2}$	—	—
$P(t_0)$	$\dfrac{1}{s_1 - s_2}\cdot(s_1 e^{-s_2 t_0} - s_2 e^{-s_1 t_0})$	$\exp\left(-\dfrac{\Lambda_0\Lambda_1}{\Lambda_0 + \Lambda_1 + M_1}\, t_0\right)$	$\delta \sim \dfrac{\Lambda_0\Lambda_1}{M_1^2}$
$K(t)$	$1 - \dfrac{\Lambda_0\Lambda_1}{\epsilon_1\epsilon_2}\left[1 - \dfrac{1}{\epsilon_1 - \epsilon_2}(\epsilon_1 e^{-\epsilon_2 t} - \epsilon_2 e^{-\epsilon_1 t})\right]$	$1 - \dfrac{\Lambda_0\Lambda_1}{M_1 M_2}\left[1 - \dfrac{1}{M_2 - M_1}(M_2 e^{-M_1 t} - M_1 e^{-M_2 t})\right]$; if $M_1 = M_2 = M$, then $1 - \dfrac{\Lambda_0\Lambda_1}{M^2}[1 - (1 + Mt)e^{-Mt}]$	$\delta \sim \left(\dfrac{\Lambda_0}{M_1}\right)^3$ $\delta \sim \left(\dfrac{\Lambda_0}{M}\right)^3$
K	$\dfrac{1}{1 + \gamma_c}$	$1 - \dfrac{\Lambda_0\Lambda_1}{M_1 M_2}$	$\delta_+ \sim \dfrac{\Lambda_0^2\Lambda_1}{M_1^2 M_2}$
$R(t_0)$	$\dfrac{M_2[s_2(\Lambda_1 - s_2)e^{-s_1 t_0} - s_1(\Lambda_1 - s_1)e^{-s_2 t_0} + M_1 M_2]}{(s_1 - s_2)(\Lambda_0\Lambda_1 + \Lambda_0 M_2 + M_1 M_2)}$	$\dfrac{1}{1 + \gamma_c}\exp\left(-\dfrac{\Lambda_0\Lambda_1}{\Lambda_0 + \Lambda_1 + M_1}\, t_0\right)$	$\delta \sim \dfrac{\Lambda_0\Lambda_1}{M_1^2}$

First we compute the quantity $s_{1,2}$:

$$s_{1,2} = \tfrac{1}{2}\Big[(2\lambda + 2\lambda + \mu) \pm \sqrt{(2\lambda + 2\lambda + \mu)^2 - 4 \cdot 2\lambda \cdot 2\lambda} \Big]$$
$$= \tfrac{1}{2}\Big[(4 \cdot 0.05 + 2) \pm \sqrt{(4 \cdot 0.05 + 2)^2 - 16 \cdot (0.05)^2} \Big]$$
$$= \tfrac{1}{2}[2.20 \pm 2.19] \; \frac{1}{\text{hr}} \cdot$$

Hence,

$$s_1 = 2.195 \, \frac{1}{\text{hr}}; \qquad s_2 = 0.005 \, \frac{1}{\text{hr}} \cdot$$

Thus

$$P(t_0) = \frac{1}{2.19} \, (2.195 \cdot e^{-0.005 \cdot t_0} - 0.005 \cdot e^{-2.195 \cdot t_0}).$$

5.   The nonstationary readiness coefficient is

$$K(t) = 1 - \frac{\Lambda_0 \Lambda_1}{\epsilon_1 \epsilon_2} \left[1 - \frac{1}{\epsilon_1 - \epsilon_2} \, (\epsilon_1 e^{-\epsilon_2 t} - \epsilon_2 e^{-\epsilon_1 t}) \right],$$

where

$$\epsilon_{1,2} = \tfrac{1}{2}\Big[(\Lambda_0 + \Lambda_1 + M_1 + M_2)$$
$$\pm \sqrt{(\Lambda_0 + \Lambda_1 + M_1 + M_2)^2 - 4(\Lambda_0 \Lambda_1 + \Lambda_0 M_2 + M_1 M_2)} \,\Big].$$

First we compute the quantity $\epsilon_{1,2}$:

$$\epsilon_{1,2} = \tfrac{1}{2}\Big[(2\lambda + 2\lambda + \mu + 1.5\mu)$$
$$\pm \sqrt{(2\lambda + 2\lambda + \mu + 1.5\mu)^2 - 4(2\lambda \cdot 2\lambda + 2\lambda \cdot 1.5\mu + \mu \cdot 1.5\mu)} \,\Big]$$
$$= \tfrac{1}{2}\Big[(4\lambda + 2.5\mu) \pm \sqrt{(4\lambda + 2.5\mu)^2 - 4(4\lambda^2 + 3\lambda\mu + 1.5\mu^2)} \,\Big]$$
$$= \tfrac{1}{2}\Big[(4 \cdot 0.05 + 2.5 \cdot 2)$$
$$\pm \sqrt{(4 \cdot 0.05 + 2.5 \cdot 2)^2 - 4[4 \cdot (0.05)^2 + 3 \cdot 0.05 \cdot 2 + 1.5 \cdot 2^2]} \,\Big]$$
$$= \tfrac{1}{2}(5.2 \pm 1.3) \; \frac{1}{\text{hr}},$$

hence

$$\epsilon_1 = 3.25 \, \frac{1}{\text{hr}}, \qquad \epsilon_2 = 1.95 \, \frac{1}{\text{hr}} \cdot$$

Then

$$K(t) = 1 - \frac{\Lambda_0 \Lambda_1}{\epsilon_1 \epsilon_2} \left[1 - \frac{1}{\epsilon_1 - \epsilon_2} \, (\epsilon_1 e^{-\epsilon_2 t} - \epsilon_2 e^{-\epsilon_1 t}) \right]$$
$$= 1 - \frac{2\lambda \cdot 2\lambda}{\epsilon_1 \epsilon_2} \left[1 - \frac{1}{\epsilon_1 - \epsilon_2} \, (\epsilon_1 e^{-\epsilon_2 t} - \epsilon_2 e^{-\epsilon_1 t}) \right]$$
$$= 1 - \frac{2 \cdot 0.05 \cdot 2 \cdot 0.05}{3.25 \cdot 1.95} \left[1 - \frac{1}{3.25 - 1.95} \, (3.25 e^{-1.95 \cdot t} - 1.95 e^{-3.25 \cdot t}) \right]$$
$$= 0.99842 + 0.00121(3.25 e^{-1.95 \cdot t} - 1.95 e^{-3.25 \cdot t}).$$

6. The readiness coefficient is

$$K = \frac{1}{1 + \gamma_c},$$

where

$$\gamma_c = \frac{\Lambda_0 \Lambda_1}{M_2(\Lambda_0 + M_1)}.$$

In our case,

$$\gamma_c = \frac{2\lambda \cdot 2\lambda}{1.5\mu(2\lambda + \mu)} = \frac{2 \cdot 0.05 \cdot 2 \cdot 0.05}{1.5 \cdot 2 \cdot (2 \cdot 0.05 + 2)} \approx 0.00158.$$

Then

$$K = \frac{1}{1 + \gamma_c} = \frac{1}{1 + 0.00158} \approx 0.99842.$$

7. The approximate value of the reliability coefficient is

$$\tilde{R}(t_0) = \frac{1}{1 + \gamma_c} \exp\left(-\frac{\Lambda_0 \Lambda_1}{\Lambda_0 + \Lambda_1 + M_1} t_0\right)$$

$$= \frac{1}{1 + \dfrac{\Lambda_0 \Lambda_1}{M_2(\Lambda_0 + M_1)}} \exp\left(-\frac{\Lambda_0 \Lambda_1}{\Lambda_0 + \Lambda_1 + M_1} t_0\right)$$

$$= \frac{1}{1 + \dfrac{2\lambda \cdot 2\lambda}{1.5\mu(2\lambda + \mu)}} \exp\left(-\frac{2\lambda \cdot 2\lambda}{2\lambda + 2\lambda + \mu} t_0\right)$$

$$= \frac{1}{1 + \dfrac{2 \cdot 0.05 \cdot 2 \cdot 0.05}{1.5 \cdot 2 \cdot (2 \cdot 0.05 + 2)}} \exp\left(-\frac{2 \cdot 0.05 \cdot 2 \cdot 0.05}{2 \cdot 0.05 + 2 \cdot 0.05 + 2} t_0\right)$$

$$= 0.99842 e^{-0.0046 t_0}.$$

Example 4.4.2 We have a system of $m = 10$ units and one stand-by unit in an operating state. The failure intensity of the operating units is $\lambda = 0.01 \dfrac{1}{\text{hr}}$, the repair intensity of a unit is

$$\mu = 1 \frac{1}{\text{hr}}, \quad \text{and} \quad \gamma = \frac{\lambda}{\mu} = 0.01.$$

We want to determine the basic reliability measures of this system.

Solution.   From Tables 4.4.2 and 4.4.3 we find:

1.   The exact and approximate values of the mean operating time up to failure of the system are

$$T_1 = \frac{1}{m\lambda} \cdot \frac{1 + (2m + 1)\gamma}{(m + 1)\gamma} = \frac{1}{10 \cdot 0.01} \cdot \frac{1 + (2 \cdot 10 + 1) \cdot 0.01}{(10 + 1) \cdot 0.01} = 110 \quad \text{hr},$$

$$\tilde{T}_1 = \frac{1}{m\lambda} \cdot \frac{1}{(m + 1)\gamma} = \frac{1}{10 \cdot 0.01} \cdot \frac{1}{(10 + 1) \cdot 0.01} \approx 99 \quad \text{hr}.$$

2.   The exact value of the mean operating time between failures is

$$T_2 = \frac{1}{m\lambda} \cdot \frac{1 + (m + 1)\gamma}{(m + 1)\gamma} = \frac{1}{10 \cdot 0.01} \cdot \frac{1 + (10 + 1) \cdot 0.01}{(10 + 1) \cdot 0.01} \approx 101 \quad \text{hr}.$$

3.   The mean repair times for unrestricted and restricted repair, respectively, are

$$\tau = \frac{1}{2\mu} = \frac{1}{2 \cdot 1} = 0.5 \quad \text{hr},$$

$$\tau = \frac{1}{\mu} = \frac{1}{1} = 1 \quad \text{hr}.$$

4.   An approximate value of the probability of failure-free operation for an initial time of $t_0 = 1$ hour is

$$\tilde{P}(t_0) = \exp\left[-\frac{m(m + 1)\gamma\lambda t_0}{1 + (2m + 1)\gamma} \right] = \exp\left[-\frac{10(10 + 1) \cdot 0.01 \cdot 0.01 \cdot 1}{1 + (2 \cdot 10 + 1) \cdot 0.01} \right]$$
$$= 0.991.$$

5.   An approximate value for the nonstationary readiness coefficient with unrestricted repair at the moment of time $t = 1$ hour is

$$\tilde{K}(t) = 1 - \frac{m(m + 1)}{2} \gamma^2 [1 - (2 - e^{-(\lambda/\gamma)t})e^{-(\lambda/\gamma)t}]$$

$$= 1 - \frac{10 \cdot (10 + 1)}{2} \cdot (0.01)^2 \cdot [1 - (2 - e^{-(0.01/0.01) \cdot 1})e^{-(0.01/0.01) \cdot 1}]$$

$$\approx 0.9978.$$

6.   The exact value of the readiness coefficient with restricted repair is

$$K = \frac{1}{1 + \gamma_c} = \frac{1}{1 + \dfrac{m(m + 1)\gamma^2}{1 + (m + 1)\gamma}} = \frac{1}{1 + \dfrac{10(10 + 1) \cdot (0.01)^2}{1 + (10 + 1) \cdot 0.01}} \approx 0.9902.$$

Table 4.4.2   System of m Operating Units and One Stand-by Unit: Operating Stand-by, Unrestricted Repair

$$s_{1,2} = \frac{\lambda}{2\gamma}\{[1+(2m+1)\gamma] \pm \sqrt{1+2(2m+1)\gamma+\gamma^2}\},$$

$$\epsilon_{1,2} = \frac{\lambda}{2\gamma}\{[3+(2m+1)\gamma] \pm \sqrt{1+2(2m-1)\gamma+\gamma^2}\},$$

$$\gamma_c = \frac{m(m+1)\gamma^2}{2[1+(m+1)\gamma]}, \qquad \gamma = \frac{\lambda}{\mu}.$$

MEASURE	EXACT VALUE	APPROXIMATE VALUE	ERROR
T_1	$\dfrac{1}{m\lambda}\dfrac{1+(2m+1)\gamma}{(m+1)\gamma}$	$\dfrac{1}{m\lambda}\dfrac{1}{(m+1)\gamma}$	$\delta_+ = \dfrac{1}{m\lambda}\dfrac{1+2m}{m+1}$
T_2	$\dfrac{1}{m\lambda}\dfrac{1+(m+1)\gamma}{(m+1)\gamma}$	$\dfrac{1}{m\lambda}\dfrac{1}{(m+1)\gamma}$	$\delta_+ = \dfrac{1}{m\lambda}$
τ	$\dfrac{1}{2\mu}$	—	—
$P(t_0)$	$\dfrac{1}{s_1-s_2}(s_1 e^{-s_2 t_0} - s_2 e^{-s_1 t_0})$	$\exp\left[-\dfrac{(m+1)\gamma}{1+(2m+1)\gamma}m\lambda t_0\right]$	$\delta \sim m(m+1)\gamma^2$
$K(t)$	$1 - \dfrac{m(m+1)\lambda^2}{\epsilon_1\epsilon_2}\left[1 - \dfrac{1}{\epsilon_1-\epsilon_2}(\epsilon_1 e^{-\epsilon_2 t} - \epsilon_2 e^{-\epsilon_1 t})\right]$	$1 - \dfrac{m(m+1)}{2}\gamma^2\left[1 - (2 - e^{-(\lambda/\gamma)t})e^{-(\lambda/\gamma)t}\right]$	$\delta \sim (m+1)^3\gamma^3$
K	$\dfrac{1}{1+\gamma_c}$	$1 - \dfrac{m(m+1)}{2}\gamma^2$	$\delta_+ \sim \dfrac{m(m+1)^2}{2}\gamma^3$
$R(t_0)$	$\dfrac{2[s_2(m\lambda - s_2)e^{-s_1 t_0} - s_1(m\lambda - s_1)e^{-s_2 t_0}]}{\mu(s_1 - s_2)[2 + 2(m+1)\gamma + m(m+1)\gamma^2]}$	$\dfrac{1}{1+\gamma_c}\exp\left[-\dfrac{(m+1)\gamma}{1+(2m+1)\gamma}m\lambda t_0\right]$	$\delta \sim m(m+1)\gamma^2$

Table 4.4.3 System of m Operating Units and One Stand-by Unit: Operating Stand-by, Restricted Repair

$$s_{1,2} = \frac{\lambda}{2\gamma}\{[1 + (2m+1)\gamma] \pm \sqrt{1 + 2(2m+1)\gamma + \gamma^2}\},$$

$$\epsilon_{1,2} = \frac{\lambda}{2\gamma}\{[2 + (2m+1)\gamma] \pm \sqrt{4m\gamma + \gamma^2}\},$$

$$\gamma_c = \frac{m(m+1)\gamma^2}{1 + (m+1)\gamma}, \qquad \gamma = \frac{\lambda}{\mu}.$$

MEASURE	EXACT VALUE	APPROXIMATE VALUE	ERROR
T_1	$\dfrac{1}{m\lambda}\dfrac{1 + (2m+1)\gamma}{(m+1)\gamma}$	$\dfrac{1}{m\lambda}\dfrac{1}{(m+1)\gamma}$	$\delta_+ = \dfrac{1}{m\lambda}\dfrac{2m+1}{m+1}$
T_2	$\dfrac{1}{m\lambda}\dfrac{1 + (m+1)\gamma}{(m+1)\gamma}$	$\dfrac{1}{m\lambda}\dfrac{1}{(m+1)\gamma}$	$\delta_+ = \dfrac{1}{m\lambda}$
τ	$\dfrac{1}{\mu}$		—
$P(t_0)$	$\dfrac{1}{s_1 - s_2}(s_1 e^{-s_2 t_0} - s_2 e^{-s_1 t_0})$	$\exp\left[-\dfrac{(m+1)\gamma}{1 + (2m+1)\gamma}\,k\lambda t_0\right]$	$\delta \sim m(m+1)\gamma^2$
$K(t)$	$1 - \dfrac{m(m+1)\lambda^2}{\epsilon_1\epsilon_2}\left[1 - \dfrac{1}{\epsilon_1 - \epsilon_2}(\epsilon_1 e^{-\epsilon_2 t} - \epsilon_2 e^{-\epsilon_1 t})\right]$	$1 - m(m+1)\gamma^2\left[1 - \left(1 + \dfrac{1}{\gamma}\,t\right)e^{-(\lambda/\gamma)t}\right]$	$\delta \sim (m+1)^3\gamma^3$
K	$\dfrac{1}{1 + \gamma_c}$	$1 - m(m+1)\gamma^2$	$\delta_+ \sim m(m+1)^2\gamma^3$
$R(t_\theta)$	$\dfrac{s_2(m\lambda - s_2)e^{-s_1 t_0} - s_1(m\lambda - s_1)e^{s_2 t_0}}{\mu(s_1 - s_2)[1 + (m+1)\gamma + m(m+1)\gamma^2]}$	$\dfrac{1}{1 + \gamma_c}\exp\left[-\dfrac{(m+1)\gamma}{1 + (2m+1)\gamma}\,m\lambda t_0\right]$	$\delta \sim m(m+1)\gamma^2$

7.   An approximate value of the reliability coefficient for a time interval of length $t_0 = 1$ hour with restricted repair is given by

$$\tilde{R}(t_0) = \frac{1}{1 + \gamma_c} \exp\left[-\frac{m(m + 1)\gamma\lambda t_0}{1 + (2m + 1)\gamma} \right] = K \exp\left[-\frac{m(m + 1)\gamma\lambda t_0}{1 + (2m + 1)\gamma} \right].$$

Making use of the results obtained in parts 4 and 6 of this example, we obtain

$$\tilde{R}(t_0) = 0.9902 \cdot 0.991 \approx 0.981.$$

Example 4.4.3   Suppose we have a system of $m = 5$ operating units and one stand-by unit. The failure intensity of each operating structure is $\lambda = 0.01 \dfrac{1}{\text{hr}}$, and the stand-by structure is operating with load coefficient $\nu = 0.4$. The repair intensity of any failed structure is $\mu = 1 \dfrac{1}{\text{hr}}$, and $\gamma = \lambda/\mu = 0.01/1 = 0.01$.

We want to determine the basic reliability measures of the system for unrestricted and restricted repair.

Solution.   From Tables 4.4.4 and 4.4.5 we find:

1.   The exact value of the mean operating time of the system up to failure is

$$T_1 = \frac{1}{m\lambda} \cdot \frac{1 + (2m + \nu)\gamma}{(m + \nu)\gamma} = \frac{1}{5 \cdot 0.01} \cdot \frac{1 + (2 \cdot 5 + 0.4) \cdot 0.01}{(5 + 0.4) \cdot 0.01} \approx 409 \quad \text{hr.}$$

2.   The exact value of the mean operating time between failures is

$$T_2 = \frac{1}{m\lambda} \cdot \frac{1 + (m + \nu)\gamma}{(m + \nu)\gamma} = \frac{1}{5 \cdot 0.01} \cdot \frac{1 + (5 + 0.4) \cdot 0.01}{(5 + 0.4) \cdot 0.01} \approx 390 \quad \text{hr.}$$

3.   The mean repair times for unrestricted and restricted repair, respectively, are

$$\tau = \frac{1}{2\mu} = 0.5 \quad \text{hr,}$$

$$\tau = \frac{1}{\mu} = 1 \quad \text{hr.}$$

4.   An approximate value of the probability of failure-free operation of the system for an initial time of $t_0 = 10$ hours is

$$\tilde{P}(t_0) = \exp\left[-\frac{m(m + \nu)\gamma\lambda t_0}{1 + (2m + \nu)\gamma} \right] = \exp\left[-\frac{5 \cdot (5 + 0.4) \cdot 0.01 \cdot 0.01 \cdot 10}{1 + (2 \cdot 5 + 0.4) \cdot 0.01} \right]$$
$$\approx 0.976.$$

Table 4.4.4 System of m Operating Units and One Stand-by Unit: Partially Operating Stand-by, Unrestricted Repair

$$s_{1,2} = \frac{\lambda}{2\gamma}\{[1 + (2m + \nu)\gamma] \pm \sqrt{1 + 2(2m + \nu)\gamma + \nu^2\gamma^2}\},$$

$$\epsilon_{1,2} = \frac{\lambda}{2\gamma}\{[3 + (2m + \nu)\gamma] \pm \sqrt{1 + 2(2m - \nu)\gamma + \nu^2\gamma^2}\},$$

$$\gamma_c = \frac{m(m + \nu)\gamma^2}{2[1 + (m + \nu)\gamma]}, \qquad \gamma = \frac{\lambda}{\mu}.$$

MEASURE	EXACT VALUE	APPROXIMATE VALUE	ERROR
T_1	$\dfrac{1}{m\lambda}\dfrac{1 + (2m + \nu)\gamma}{(m + \nu)\gamma}$	$\dfrac{1}{m\lambda}\dfrac{1}{(m + \nu)\gamma}$	$\delta_+ = \dfrac{2m + \nu}{m\lambda(m + \nu)}$
T_2	$\dfrac{1}{m\lambda}\dfrac{1 + (m + \nu)\gamma}{(m + \nu)\gamma}$	$\dfrac{1}{m\lambda}\dfrac{1}{(m + \nu)\gamma}$	$\delta_+ = \dfrac{1}{m\lambda}$
τ	$\dfrac{1}{2\mu}$	—	—
$P(t_0)$	$\dfrac{1}{s_1 - s_2}(s_1 e^{-s_2 t_0} - s_2 e^{-s_1 t_0})$	$\exp\left[-\dfrac{(m + \nu)\gamma}{1 + (2m + \nu)\gamma}\, m\lambda t_0\right]$	$\delta \sim m(m + \nu)\gamma^2$
$K(t)$	$1 - \dfrac{m(m + \nu)\lambda^2}{\epsilon_1\epsilon_2}\left[1 - \dfrac{1}{\epsilon_1 - \epsilon_2}(\epsilon_1 e^{-\epsilon_2 t} - \epsilon_2 e^{-\epsilon_1 t})\right]$	$1 - \dfrac{m(m + \nu)}{2}\gamma^2[1 - (2 - e^{-(\lambda/\nu)t})e^{-(\lambda/\nu)\mu t}]$	$\delta \sim (m + \nu)^3\gamma^3$
K	$\dfrac{1}{1 + \gamma_c}$	$1 - \dfrac{m(m + \nu)}{2}\gamma^2$	$\delta_+ \sim \dfrac{m(m + \nu)^2}{2}\gamma^3$
$R(t_0)$	$\dfrac{2[s_2(m\lambda - s_2)e^{-s_1 t_0} - s_1(m\lambda - s_1)e^{-s_2 t_0}]}{\mu(s_1 - s_2)[2 + 2(m + \nu)\gamma + m(m + \nu)\gamma^2]}$	$\dfrac{1}{1 + \gamma_c}\exp\left[-\dfrac{(m + \nu)\gamma}{1 + (2m + \nu)\gamma}\, m\lambda t_0\right]$	$\delta \sim m(m + \nu)\gamma^2$

Table 4.4.5 System of m Operating Units and One Stand-by Unit: Partially Operating Stand-by, Restricted Repair

$$s_{1,2} = \frac{\lambda}{2\gamma}\left\{[1 + (2m+\nu)\gamma] \pm \sqrt{1 + 2(2m+\nu)\gamma + \nu^2\gamma^2}\right\},$$

$$\epsilon_{1,2} = \frac{\lambda}{2\gamma}\left\{[2 + (2m+\nu)\gamma] \pm \sqrt{4m\gamma + \nu^2\gamma^2}\right\},$$

$$\gamma_c = \frac{m(m+\nu)\gamma^2}{1+(m+\nu)\gamma}, \qquad \gamma = \frac{\lambda}{\mu}.$$

MEASURE	EXACT VALUE	APPROXIMATE VALUE	ERROR
T_1	$\dfrac{1}{m\lambda}\dfrac{1+(2m+\nu)\gamma}{(m+\nu)\gamma}$	$\dfrac{1}{m\lambda}\dfrac{1}{(m+\nu)\gamma}$	$\delta_+ = \dfrac{2m+\nu}{m\lambda(m+\nu)}$
T_2	$\dfrac{1}{m\lambda}\dfrac{1+(m+\nu)\gamma}{(m+\nu)\gamma}$	$\dfrac{1}{m\lambda}\dfrac{1}{(m+\nu)\gamma}$	$\delta_+ = \dfrac{1}{m\lambda}$
τ	$\dfrac{1}{\mu}$	—	—
$P(t_0)$	$\dfrac{1}{s_1-s_2}(s_1 e^{-s_2 t_0} - s_2 e^{-s_1 t_0})$	$\exp\left[-\dfrac{(m+\nu)\gamma}{1+(2m+\nu)\gamma}m\lambda t_0\right]$	$\delta \sim m(m+\nu)\gamma^2$
$K(t)$	$1 - \dfrac{m(m+\nu)\lambda^2}{\epsilon_1\epsilon_2}\left[1 - \dfrac{1}{\epsilon_1-\epsilon_2}(\epsilon_1 e^{-\epsilon_2 t} - \epsilon_2 e^{-\epsilon_1 t})\right]$	$1 - m(m+\nu)\gamma^2\left[1 - \left(1+\left(\dfrac{\lambda}{\nu}-t\right)\right)e^{-(\lambda/\nu)\gamma t}\right]$	$\delta \sim (m+\nu)^3\gamma^3$
K	$\dfrac{1}{1+\gamma_c}$	$1 - m(m+\nu)\gamma^2$	$\delta_+ \sim m(m+\nu)^2\gamma^3$
$R(t_0)$	$\dfrac{s_2(m\lambda - s_2)e^{-s_1 t_0} - s_1(m\lambda - s_1)e^{-s_2 t_0}}{\mu(s_1 - s_2)[1 + (m+\nu)\gamma + m(m+\nu)\gamma^2]}$	$\dfrac{1}{1+\gamma_c}\exp\left[-\dfrac{(m+\nu)\gamma}{1+(2m+\nu)\gamma}m\lambda t_0\right]$	$\delta \sim m(m+\nu)\gamma^2$

5. An approximate value for the nonstationary readiness coefficient at $t = 10$ hours with unrestricted repair is

$$\tilde{K}(t) = 1 - \frac{m(m + \nu)}{2} \gamma^2[1 - (2 - e^{-(\lambda/\gamma)t})e^{-(\lambda/\gamma)t}]$$

$$= 1 - \frac{5(5 + 0.4)}{2} \cdot (0.01)^2[1 - (2 - e^{-(0.01/0.01) \cdot 10})e^{-(0.01/0.01) \cdot 10}]$$

$$\approx 0.99865.$$

6. The exact value of the readiness coefficient with restricted repair is

$$K = \frac{1}{1 + \gamma_c} = \frac{1}{1 + \dfrac{m(m + \nu)\gamma^2}{1 + (m + \nu)\gamma}} = \frac{1}{1 + \dfrac{5 \cdot (5 + 0.4) \cdot (0.01)^2}{1 + (5 + 0.4) \cdot 0.01}} \approx 0.9974.$$

7. An approximate value of the reliability coefficient of the system for a time interval of length $t_0 = 10$ hours for restricted repair is

$$\tilde{R}(t_0) = K \cdot \tilde{P}(t_0).$$

Using the results obtained in parts 4 and 6 of this example, we obtain:

$$\tilde{R}(t_0) = 0.9974 \cdot 0.976 \approx 0.973.$$

Example 4.4.4   We are given a system of $m = 10$ operating units and one stand-by unit. The failure intensity of an operating unit is $\lambda = 0.01\,\dfrac{1}{hr}$, the repair intensity of a failed unit is $\mu = 1\,\dfrac{1}{hr}$, and $\gamma = \lambda/\mu = 0.01$.

We wish to determine the basic reliability measures of the system for unrestricted and restricted repair.

Solution.   From Tables 4.4.6 and 4.4.7 we find:

1. The exact value of the mean operating time of the system up to failure is

$$T_1 = \frac{1}{m\lambda}\left(2 + \frac{1}{m\gamma}\right) = \frac{1}{10 \cdot 0.01}\left(2 + \frac{1}{10 \cdot 0.01}\right) = 120 \quad hr.$$

2. The exact value of the mean operating time between failures is

$$T_2 = \frac{1}{m\lambda}\left(1 + \frac{1}{m\gamma}\right) = \frac{1}{10 \cdot 0.01}\left(1 + \frac{1}{10 \cdot 0.01}\right) = 110 \quad hr.$$

3. The mean repair time for unrestricted repair is

$$\tau = \frac{1}{2\mu} = \frac{1}{2 \cdot 1} = 0.5 \quad hr.$$

Table 4.4.6 System of m Operating Units and One Stand-by Unit: Nonoperating Stand-by, Unrestricted Repair

$$s_{1,2} = \frac{\lambda}{2\gamma}[(1+2m\gamma) \pm \sqrt{1+4m\gamma}],$$

$$\epsilon_{1,2} = \frac{\lambda}{2\gamma}[(3+2m\gamma) \pm \sqrt{1+4m\gamma}],$$

$$\gamma_c = \frac{m^2\gamma^2}{2(1+m\gamma)}, \qquad \gamma = \frac{\lambda}{\mu}.$$

MEASURE	EXACT VALUE	APPROXIMATE VALUE	ERROR
T_1	$\dfrac{1}{m\lambda}\left(2+\dfrac{1}{m\gamma}\right)$	$\dfrac{1}{m\lambda}\dfrac{1}{m\gamma}$	$\delta_+ = \dfrac{2}{m\lambda}$
T_2	$\dfrac{1}{m\lambda}\left(1+\dfrac{1}{m\gamma}\right)$	$\dfrac{1}{m\lambda}\dfrac{1}{m\gamma}$	$\delta_+ = \dfrac{1}{m\lambda}$
τ	$\dfrac{1}{2\mu}$	—	—
$P(t_0)$	$\dfrac{1}{s_1 - s_2}(s_1 e^{-s_2 t_0} - s_2 e^{-s_1 t_0})$	$\exp\left(-\dfrac{1}{2+\dfrac{1}{m\gamma}}\,m\lambda t_0\right)$	$\delta \sim m^2\gamma^2$
$K(t)$	$1 - \dfrac{m^2\lambda^2}{\epsilon_1 \epsilon_2}\left[1 - \dfrac{1}{\epsilon_1 - \epsilon_2}(\epsilon_1 e^{-\epsilon_2 t} - \epsilon_2 e^{-\epsilon_1 t})\right]$	$1 - \dfrac{m^2}{2}\gamma^2[1 - (2 - e^{-(\lambda/\gamma)t})e^{-(\lambda/\gamma)t}]$	$\delta \sim m^3\gamma^3$
K	$\dfrac{1}{1+\gamma_c}$	$1 - \dfrac{m^2}{2}\gamma^2$	$\delta \sim \dfrac{m^3}{2}\gamma^3$
$R(t_0)$	$\dfrac{2[s_2(m\lambda - s_2)e^{-s_1 t_0} - s_1(m\lambda - s_1)e^{s_2 t_0}]}{\mu(s_1 - s_2)[2 + 2m\gamma + m^2\gamma^2]}$	$\dfrac{1}{1+\gamma_c}\exp\left(-\dfrac{1}{2+\dfrac{1}{m\gamma}}\,m\lambda t_0\right)$	$\delta \sim m^2\gamma^2$

Table 4.4.7   System of m Operating Units and One Stand-by Unit: Nonoperating Stand-by, Restricted Repair

$$s_{1,2} = \frac{\lambda}{2\gamma}[(1 + 2m\gamma) \pm \sqrt{1 + 4m\gamma}],$$

$$\epsilon_{1,2} = \frac{\lambda}{\gamma}[(1 + m\gamma) \pm \sqrt{m\gamma}],$$

$$\gamma_c = \frac{m^2\gamma^2}{1 + m\gamma}, \qquad \gamma = \frac{\lambda}{\mu}.$$

MEASURE	EXACT VALUE	APPROXIMATE VALUE	ERROR
T_1	$\dfrac{1}{m\lambda}\left(2 + \dfrac{1}{m\gamma}\right)$	$\dfrac{1}{m\lambda}\dfrac{1}{m\gamma}$	$\delta_+ = \dfrac{2}{m\lambda}$
T_2	$\dfrac{1}{m\lambda}\left(1 + \dfrac{1}{m\gamma}\right)$	$\dfrac{1}{m\lambda}\dfrac{1}{m\gamma}$	$\delta_+ = \dfrac{1}{m\lambda}$
τ	$\dfrac{1}{\mu}$	—	—
$P(t_0)$	$\dfrac{1}{s_1 - s_2}(s_1 e^{-s_2 t_0} - s_2 e^{-s_1 t_0})$	$\exp\left(-\dfrac{1}{2 + \dfrac{1}{m\gamma}}\, m\lambda t_0\right)$	$\delta \sim m^2\gamma^2$
$K(t)$	$1 - \dfrac{m^2\lambda^2}{\epsilon_1\epsilon_2}\left[1 - \dfrac{1}{\epsilon_1 - \epsilon_2}(\epsilon_1 e^{-\epsilon_2 t} - \epsilon_2 e^{-\epsilon_1 t})\right]$	$1 - m^2\gamma^2\left[1 - \left(1 + \dfrac{\lambda}{\gamma}t\right)e^{-(\lambda/\gamma)\mu}\right]$	$\delta \sim m^3\gamma^3$
K	$\dfrac{1}{1 + \gamma_c}$	$1 - m^2\gamma^2$	$\delta_+ \sim m^3\gamma^3$
$R(t_0)$	$\dfrac{s_2(m\lambda - s_2)e^{-s_1 t_0} - s_1(m\lambda - s_1)e^{-s_2 t_0}}{\mu(s_1 - s_2)(1 + m\gamma + m^2\gamma^2)}$	$\dfrac{1}{1 + \gamma_c}\exp\left(-\dfrac{1}{2 + \dfrac{1}{m\gamma}}\, m\lambda t_0\right)$	$\delta \sim m^2\gamma^2$

4.   An approximate value for the probability of failure-free operation for an initial time of $t_0 = 1$ hour is

$$\tilde{P}(t_0) = \exp\left[-\frac{m\lambda t_0}{2 + \dfrac{1}{m\gamma}}\right] = \exp\left[-\frac{10 \cdot 0.01 \cdot 1}{2 + \dfrac{1}{10 \cdot 0.01}}\right] \approx 0.992.$$

5.   An approximate value for the nonstationary readiness coefficient with unrestricted repair of the system at the moment of time $t = 1$ hour is

$$\tilde{K}(t) = 1 - \frac{m^2\gamma^2}{2}[1 - (2 - e^{-(\lambda/\gamma)t})e^{-(\lambda/\gamma)t}]$$

$$= 1 - \frac{10^2 \cdot (0.01)^2}{2}[1 - (2 - e^{-(0.01/0.01)\cdot 1})e^{-(0.01/0.01)\cdot 1}] \approx 0.998.$$

6.   An approximate value for the readiness coefficient for the case of unrestricted repair is

$$\tilde{K} = 1 - \frac{m^2\gamma^2}{2} = 1 - \frac{10^2 \cdot (0.01)^2}{2} = 0.995.$$

7.   An approximate value for the reliability coefficient of the system for a time interval of length $t_0 = 1$ hour for the case of unrestricted repair is

$$\tilde{R}(t_0) = \frac{1}{1 + \gamma_c}\exp\left(-\frac{m\lambda t_0}{2 + \dfrac{1}{m\gamma}}\right) = \frac{1}{1 + \dfrac{m^2\gamma^2}{2(1 + m\gamma)}} \cdot \exp\left(-\frac{m\lambda t_0}{2 + \dfrac{1}{m\gamma}}\right)$$

$$= \frac{1}{1 + \dfrac{10^2 \cdot (0.01)^2}{2 \cdot (1 + 10 \cdot 0.01)}}\exp\left(-\frac{10 \cdot 0.01 \cdot 1}{2 + \dfrac{1}{10 \cdot 0.01}}\right) \approx 0.9955 \cdot 0.992$$

$$\approx 0.9875.$$

Example 4.4.5   Suppose we have a redundant system consisting of two elements such that the stand-by unit is in an operating state. The units are characterized by a failure intensity of $\lambda = 0.01\,\dfrac{1}{hr}$, repair intensity $\mu = 1\,\dfrac{1}{hr}$, and

$$\gamma = \frac{\lambda}{\mu} = \frac{0.01}{1} = 0.01.$$

We want to determine the basic reliability measures of the system for unrestricted and restricted repair.

Solution.   From Tables 4.4.8 and 4.4.9 we find:

1.   Exact and approximate values of the mean operating time of the system up to failure are

$$T_1 = \frac{1}{\lambda} \cdot \frac{1 + 3\gamma}{2\gamma} = \frac{1}{0.01} \cdot \frac{1 + 3 \cdot 0.01}{2 \cdot 0.01} = 5150 \quad \text{hr},$$

$$\tilde{T}_1 = \frac{1}{\lambda} \cdot \frac{1}{2\gamma} = \frac{1}{0.01} \cdot \frac{1}{2 \cdot 0.01} = 5000 \quad \text{hr}.$$

2.   The exact value of the mean operating time between failures is

$$T_2 = \frac{1}{\lambda}\left(1 + \frac{1}{2\gamma}\right) = \frac{1}{0.01}\left(1 + \frac{1}{2 \cdot 0.01}\right) = 5100 \quad \text{hr}.$$

3.   The mean repair time for restricted repair is

$$\tau = \frac{1}{\mu} = \frac{1}{1} = 1 \quad \text{hr}.$$

4.   An approximate value for the probability of failure-free operation for an initial time of $t_0 = 10$ hours is

$$\tilde{P}(t_0) = \exp\left(-\frac{2\gamma\lambda t_0}{1 + 3\gamma}\right) = \exp\left(-\frac{2 \cdot 0.01 \cdot 0.01 \cdot 10}{1 + 3 \cdot 0.01}\right) \approx 0.998.$$

5.   An approximate value for the nonstationary readiness coefficient at the moment of time $t = 1$ hour with unrestricted repair is

$$\tilde{K}(t) = 1 - \gamma^2[1 - (2 - e^{-(\lambda/\gamma)t_0})e^{-(\lambda/\gamma)t_0}]$$
$$= 1 - (0.01)^2[1 - (2 - e^{-(0.01/0.01) \cdot 1})e^{-(0.01/0.01) \cdot 1}] \approx 0.99996.$$

6.   Exact and approximate values of the readiness coefficient with restricted repair are

$$K = \frac{1}{1 + \gamma_c} = \frac{1}{1 + \dfrac{2\gamma^2}{1 + 2\gamma}} = \frac{1}{1 + \dfrac{2 \cdot (0.01)^2}{1 + 2 \cdot 0.01}} \approx 0.999804,$$

$$\tilde{K} = 1 - 2\gamma^2 = 1 - 2 \cdot 10^{-4} = 0.9998.$$

7.   By using the results obtained in parts 4 and 6 of this example, an approximate value for the reliability coefficient for a time interval of length $t_0 = 10$ hours in the restricted-repair case is determined to be

$$\tilde{R}(t_0) = \frac{1}{1 + \gamma_c} e^{-2\gamma\lambda t_0/(1 + 3\gamma)} = 0.999804 \cdot 0.998 \approx 0.9978.$$

Table 4.4.8   Redundant System of Two Units: Operating Stand-by, Unrestricted Repair

$$s_{1,2} = \frac{\lambda}{2\gamma}\left[(1+3\gamma) \pm \sqrt{1+6\gamma+\gamma^2}\,\right],$$
$$\epsilon_{1,2} = \frac{\lambda}{2\gamma}(1+\gamma)(3\pm 1),$$
$$\gamma_c = \frac{\gamma^2}{2\gamma+1}, \qquad \gamma = \frac{\lambda}{\mu}.$$

MEASURE	EXACT VALUE	APPROXIMATE VALUE	ERROR
T_1	$\dfrac{1}{\lambda}\dfrac{1+3\gamma}{2\gamma}$	$\dfrac{1}{\lambda}\dfrac{1}{2\gamma}$	$\delta_+ = \dfrac{3}{2}\cdot\dfrac{1}{\lambda}$
T_2	$\dfrac{1}{\lambda}\dfrac{1+2\gamma}{2\gamma}$	$\dfrac{1}{\lambda}\dfrac{1}{2\gamma}$	$\delta_+ = \dfrac{1}{\lambda}$
τ	$\dfrac{1}{2\mu}$	—	—
$P(t_0)$	$\dfrac{1}{s_1-s_2}\left(s_1 e^{-s_2 t_0} - s_2 e^{-s_1 t_0}\right)$	$\exp\left(-\dfrac{2\gamma}{1+3\gamma}\lambda t_0\right)$	$\delta \sim 2\gamma^2$
$K(t)$	$1 - \dfrac{2\lambda^2}{\epsilon_1\epsilon_2}\left[1 - \dfrac{1}{\epsilon_1-\epsilon_2}\left(\epsilon_1 e^{-\epsilon_2 t} - \epsilon_2 e^{-\epsilon_1 t}\right)\right]$	$1 - \gamma^2\left[1 - \left(2 - e^{-(\lambda/\gamma)t}\right)e^{-(\lambda/\gamma)t}\right]$	$\delta \sim 8\gamma^3$
K	$\dfrac{1}{1+\gamma_c}$	$1 - \gamma^2$	$\delta_+ \sim 2\gamma^3$
$R(t_0)$	$\dfrac{s_2(\lambda-s_2)e^{-s_1 t_0} - s_1(\lambda-s_1)e^{-s_2 t_0}}{\mu(s_1-s_2)(1+\gamma)^2}$	$\dfrac{1}{1+\gamma_c}\exp\left(-\dfrac{2\gamma}{1+3\gamma}\lambda t_0\right)$	$\delta \sim 2\gamma^2$

Table 4.4.9 Redundant System of Two Units: Operating Stand-by, Restricted Repair

$$s_{1,2} = \frac{\lambda}{2\gamma}\left[(1 + 3\gamma) \pm \sqrt{1 + 6\gamma + \gamma^2}\right],$$

$$\epsilon_{1,2} = \frac{\lambda}{2\gamma}\left[(2 + 3\gamma) \pm \sqrt{4\gamma + \gamma^2}\right],$$

$$\gamma_c = \frac{2\gamma^2}{1 + 2\gamma}, \qquad \gamma = \frac{\lambda}{\mu}.$$

MEASURE	EXACT VALUE	APPROXIMATE VALUE	ERROR
T_1	$\dfrac{1}{\lambda}\dfrac{1 + 3\gamma}{2\gamma}$	$\dfrac{1}{\lambda}\dfrac{1}{2\gamma}$	$\delta_+ = \dfrac{3}{2}\dfrac{1}{\lambda}$
T_2	$\dfrac{1}{\lambda}\dfrac{1 + 2\gamma}{2\gamma}$	$\dfrac{1}{\lambda}\dfrac{1}{2\gamma}$	$\delta_+ = \dfrac{1}{\lambda}$
τ	$\dfrac{1}{\mu}$	—	—
$P(t_0)$	$\dfrac{1}{s_1 - s_2}(s_1 e^{-s_2 t_0} - s_2 e^{-s_1 t_0})$	$\exp\left(-\dfrac{2\gamma}{1 + 3\gamma}\lambda t_0\right)$	$\delta \sim 2\gamma^2$
$K(t)$	$1 - \dfrac{2\lambda^2}{\epsilon_1 \epsilon_2}\left[1 - \dfrac{1}{\epsilon_1 - \epsilon_2}(\epsilon_1 e^{-\epsilon_2 t} - \epsilon_2 e^{-\epsilon_1 t})\right]$	$1 - 2\gamma^2\left[1 - \left(1 + \dfrac{\lambda}{\gamma} - t\right)e^{-(\lambda/\gamma)t}\right]$	$\delta \sim 8\gamma^3$
K	$\dfrac{1}{1 + \gamma_c}$	$1 - 2\gamma^2$	$\delta_+ \sim 4\gamma^3$
$R(t_0)$	$\dfrac{s_2(\lambda - s_2)e^{-s_1 t_0} - s_1(\lambda - s_1)e^{-s_2 t_0}}{\mu(s_1 - s_2)(1 + 2\gamma + 2\gamma^2)}$	$\dfrac{1}{1 + \gamma_c}\exp\left(-\dfrac{2\gamma}{1 + 3\gamma}\lambda t_0\right)$	$\delta \sim 2\gamma^2$

Example 4.4.6 We are given a redundant system consisting of two structures such that the failure intensity of each structure in the operating state is $\lambda = 0.01 \dfrac{1}{\text{hr}}$, while in stand-by $\lambda_{\text{stand-by}} = \nu\lambda = 0.001 \dfrac{1}{\text{hr}}$; that is, $\nu = 0.1$. The repair intensity of a failed structure is $\mu = 1 \dfrac{1}{\text{hr}}$, and $\gamma = \lambda/\mu = 0.01/1 = 0.01$.

We want to determine the basic reliability measures of the system for unrestricted and restricted repair.

Solution. From Tables 4.4.10 and 4.4.11 we find:

1.   The exact value of the mean operating time of the system up to failure is

$$T_1 = \frac{1}{\lambda} \cdot \frac{1 + (2 + \nu)\gamma}{(1 + \nu)\gamma} = \frac{1}{0.01} \cdot \frac{1 + (2 + 0.1) \cdot 0.01}{(1 + 0.1) \cdot 0.01} = 9282 \quad \text{hr.}$$

2.   The exact value of the mean operating time of the system between failures is

$$T_2 = \frac{1}{\lambda} \cdot \frac{1 + (1 + \nu)\gamma}{(1 + \nu)\gamma} = \frac{1}{0.01} \cdot \frac{1 + (1 + 0.1) \cdot 0.01}{(1 + 0.1) \cdot 0.01} = 9191 \quad \text{hr.}$$

3.   The mean repair time in the unrestricted-repair case is

$$\tau = \frac{1}{2\mu} = \frac{1}{2 \cdot 1} = 0.5 \quad \text{hr.}$$

4.   An approximate value for the probability of failure-free operation of the system for an initial time of $t_0 = 100$ hours is

$$\tilde{P}(t_0) = \exp\left[-\frac{(1 + \nu)\gamma\lambda t_0}{1 + (2 + \nu)\gamma} \right] = \exp\left[-\frac{(1 + 0.1) \cdot 0.01 \cdot 0.01 \cdot 100}{1 + (2 + 0.1) \cdot 0.01} \right]$$
$$\approx 0.989.$$

5.   An approximate value for the nonstationary readiness coefficient at the moment of time $t = 1$ hour in the case of restricted repair is

$$\tilde{K}(t) = 1 - (1 + \nu)\gamma^2 \left[1 - \left(1 + \frac{\lambda}{\gamma} t \right) e^{-(\lambda/\gamma)t} \right]$$
$$= 1 - (1 + 0.1) \cdot (0.01)^2 \cdot \left[1 - \left(1 + \frac{0.01}{0.01} \cdot 1 \right) e^{-(0.01/0.01) \cdot 1} \right]$$
$$\approx 0.99997.$$

Table 4.4.10 Redundant System of Two Units: Partially Operating Stand-by, Unrestricted Repair

$$s_{1,2} = \frac{\lambda}{2\gamma}\{[1+(2+\nu)\gamma] \pm \sqrt{1+2(2+\nu)\gamma+\nu^2\gamma^2}\},$$

$$\epsilon_{1,2} = \frac{\lambda}{2\gamma}\{[3+(2+\nu)\gamma] \pm \sqrt{1+2(2-\nu)\gamma+\nu^2\gamma^2}\},$$

$$\gamma_c = \frac{(1+\nu)\gamma^2}{2[1+(1+\nu)\gamma]}, \quad \gamma = \frac{\lambda}{\mu}.$$

MEASURE	EXACT VALUE	APPROXIMATE VALUE	ERROR
T_1	$\dfrac{1}{\lambda}\dfrac{1+(2+\nu)\gamma}{(1+\nu)\gamma}$	$\dfrac{1}{\lambda}\dfrac{1}{(1+\nu)\gamma}$	$\delta_+ = \dfrac{2+\nu}{1+\nu}\dfrac{1}{\lambda}$
T_2	$\dfrac{1}{\lambda}\dfrac{1+(1+\nu)\gamma}{(1+\nu)\gamma}$	$\dfrac{1}{\lambda}\dfrac{1}{(1+\nu)\gamma}$	$\delta_+ = \dfrac{1}{\lambda}$
τ	$\dfrac{1}{2\mu}$	—	—
$P(t_0)$	$\dfrac{1}{s_1-s_2}(s_1 e^{-s_2 t_0} - s_2 e^{-s_1 t_0})$	$\exp\left[-\dfrac{(1+\nu)\gamma}{1+(2+\nu)\gamma}\lambda t_0\right]$	$\delta \sim (1+\nu)\gamma^2$
$K(t)$	$1 - \dfrac{(1+\nu)\lambda^2}{\epsilon_1\epsilon_2}\left[1 - \dfrac{1}{\epsilon_1-\epsilon_2}(\epsilon_1 e^{-\epsilon_2 t} - \epsilon_2 e^{-\epsilon_1 t})\right]$	$1 - \dfrac{(1+\nu)}{2}\gamma^2[1-(2-e^{-(\lambda/\gamma)t})e^{-(\lambda/\gamma)t}]$	$\delta \sim (1+\nu)^3\gamma^3$
K	$\dfrac{1}{1+\gamma_c}$	$1 - \dfrac{1+\nu}{2}\gamma^2$	$\delta_+ \sim \dfrac{(1+\nu)^2}{2}\gamma^3$
$R(t_0)$	$\dfrac{2[s_2(\lambda-s_2)e^{-s_1 t_0} - s_1(\lambda-s_1)e^{-s_2 t_0}]}{\mu(s_1-s_2)[2+2(1+\nu)\gamma+(1+\nu)\gamma^2]}$	$\dfrac{1}{1+\gamma_c}\exp\left[-\dfrac{(1+\nu)\gamma}{1+(2+\nu)\gamma}\lambda t_0\right]$	$\delta \sim (1+\nu)\gamma^2$

Table 4.4.11 Redundant System of Two Units: Partially Operating Stand-by, Restricted Repair

$$s_{1,2} = \frac{\lambda}{2\gamma}\{[1 + (2+\nu)\gamma] \pm \sqrt{1 + 2(2+\nu)\gamma + \nu^2\gamma^2}\},$$

$$\epsilon_{1,2} = \frac{\lambda}{2\gamma}\{[2 + (2+\nu)\gamma] \pm \sqrt{4\gamma + \nu^2\gamma^2}\},$$

$$\gamma_c = \frac{(1+\nu)\gamma^2}{1+(1+\nu)\gamma}, \qquad \gamma = \frac{\lambda}{\mu}.$$

MEASURE	EXACT VALUE	APPROXIMATE VALUE	ERROR
T_1	$\dfrac{1}{\lambda}\dfrac{1+(2+\nu)\gamma}{(1+\nu)\gamma}$	$\dfrac{1}{\lambda}\dfrac{1}{(1+\nu)\gamma}$	$\delta_+ = \dfrac{1}{\lambda}\dfrac{2+\nu}{1+\nu}$
T_2	$\dfrac{1}{\lambda}\dfrac{1+(1+\nu)\gamma}{(1+\nu)\gamma}$	$\dfrac{1}{\lambda}\dfrac{1}{(1+\nu)\gamma}$	$\delta_+ = \dfrac{1}{\lambda}$
τ	$\dfrac{1}{\mu}$	—	—
$P(t_0)$	$\dfrac{1}{s_1 - s_2}(s_1 e^{-s_2 t_0} - s_2 e^{-s_1 t_0})$	$\exp\left[-\dfrac{(1+\nu)\gamma}{1+(2+\nu)\gamma}\lambda t_0\right]$	$\delta \sim (1+\nu)\gamma^2$
$K(t)$	$1 - (1+\nu)\gamma^2\left[1 - \left(1+\dfrac{\lambda}{\gamma}t\right)e^{-(\lambda/\gamma)t}\right]$	$1 - (1+\nu)\gamma^2\left[1 - \left(1+\dfrac{t}{\gamma}\right)e^{-(\lambda/\gamma)t}\right]$	$\delta \sim (1+\nu)^3\gamma^3$
K	$1 - (1+\nu)\gamma^2$	$1 - (1+\nu)\gamma^2$	$\delta_+ \sim (1+\nu)^2\gamma^3$
$R(t_0)$	$\dfrac{s_2(\lambda-s_2)e^{-s_1 t_0} - s_1(\lambda-s_1)e^{-s_2 t_0}}{\mu(s_1 - s_2)[1 + (1+\nu)\gamma + (1+\nu)\gamma^2]}$	$\dfrac{1}{1+\gamma_c}\exp\left[-\dfrac{(1+\nu)\gamma}{1+(2+\nu)\gamma}\lambda t_0\right]$	$\delta \sim (1+\nu)\gamma^2$

6. An approximate value for the readiness coefficient of the system in the case of restricted repair is

$$\tilde{K} = 1 - (1 + \nu)\gamma^2 = 1 - (1 + 0.1)\cdot(0.01)^2 = 0.99989.$$

7. An approximate value for the reliability coefficient for a time interval of length $t_0 = 100$ hours in the case of restricted repair of the system is

$$\tilde{R}(t_0) = \frac{1}{1 + \gamma_c} \exp\left[-\frac{(1 + \nu)\gamma\lambda t_0}{1 + (2 + \nu)\gamma} \right]$$

$$= \frac{1}{1 + \dfrac{(1 + \nu)\gamma^2}{1 + (1 + \nu)\gamma}} \exp\left[-\frac{(1 + \nu)\gamma\lambda t_0}{1 + (2 + \nu)\gamma} \right]$$

$$= \frac{1}{1 + \dfrac{(1 + 0.1)\cdot(0.01)^2}{1 + (1 + 0.1)\cdot0.01}} \exp\left[-\frac{(1 + 0.1)\cdot0.01\cdot0.01\cdot100}{1 + (2 + 0.1)\cdot0.01} \right]$$

$$\approx 0.989.$$

Example 4.4.7 We are given a redundant system consisting of two structures such that the failure intensity of a structure in the operating state is $\lambda = 0.1\, \dfrac{1}{\text{hr}}$, and in stand-by $\lambda_{\text{stand-by}} = 0$ (nonoperating stand-by). The repair intensity of a failed structure is $\mu = 1\, \dfrac{1}{\text{hr}}$ and $\gamma = \lambda/\mu = 0.1/1 = 0.1$.

We want to determine the basic reliability measures of the system for unrestricted and restricted repair.

Solution. From Tables 4.4.12 and 4.4.13 we find:

1. The exact value of the mean operating time up to failure of the system is

$$T_1 = \frac{1}{\lambda}\left(2 + \frac{1}{\gamma}\right) = \frac{1}{0.1}\left(2 + \frac{1}{0.1}\right) = 120 \quad \text{hr.}$$

2. The exact value of the mean operating time between failures is

$$T_2 = \frac{1}{\lambda}\left(1 + \frac{1}{\gamma}\right) = \frac{1}{0.1}\left(1 + \frac{1}{0.1}\right) = 110 \quad \text{hr.}$$

3. The mean repair time for restricted repair is

$$\tau = \frac{1}{\mu} = \frac{1}{1} = 1 \quad \text{hr.}$$

Table 4.4.12   Redundant System of Two Units: Nonoperating Stand-by, Unrestricted Repair

$$s_{1,2} = \frac{\lambda}{2\gamma}[(1+2\gamma) \pm \sqrt{1+4\gamma}],$$
$$\epsilon_{1,2} = \frac{\lambda}{2\gamma}[(3+2\gamma) \pm \sqrt{1+4\gamma}],$$
$$\gamma_c = \frac{\gamma^2}{2(1+\gamma)}, \qquad \gamma = \frac{\lambda}{\mu}.$$

MEASURE	EXACT VALUE	APPROXIMATE VALUE	ERROR
T_1	$\dfrac{1}{\lambda}\left(2+\dfrac{1}{\gamma}\right)$	$\dfrac{1}{\lambda\gamma}$	$\delta_+ = \dfrac{2}{\lambda}$
T_2	$\dfrac{1}{\lambda}\left(1+\dfrac{1}{\gamma}\right)$	$\dfrac{1}{\lambda\gamma}$	$\delta_+ = \dfrac{1}{\lambda}$
τ	$\dfrac{1}{2\mu}$	—	—
$P(t_0)$	$\dfrac{1}{s_1-s_2}(s_1 e^{-s_2 t_0} - s_2 e^{s_1-t_0})$	$\exp\left(-\dfrac{1}{2+\dfrac{1}{\gamma}}\lambda t_0\right)$	$\delta \sim \gamma^2$
$K(t)$	$1 - \dfrac{\lambda^2}{\epsilon_1 \epsilon_2}\left[1 - \dfrac{1}{\epsilon_1 - \epsilon_2}(\epsilon_1 e^{-\epsilon_2 t} - \epsilon_2 e^{-\epsilon_1 t})\right]$	$1 - \dfrac{\gamma^2}{2}[1 - (2 - e^{-(\lambda/\mu)t})e^{-(\lambda/\mu)\mu t}]$	$\delta \sim \gamma^3$
K	$\dfrac{1}{1+\gamma_c}$	$1 - \dfrac{\gamma^2}{2}$	$\delta_+ \sim \dfrac{\gamma^3}{2}$
$R(t_0)$	$\dfrac{2[s_2(\lambda - s_2)e^{-s_1 t_0} - s_1(\lambda - s_1)e^{-s_2 t_0}]}{\mu(s_1 - s_2)(2+2\gamma+\gamma^2)}$	$\dfrac{1}{1+\gamma_c}\exp\left(-\dfrac{1}{2+\dfrac{1}{\gamma}}\lambda t_0\right)$	$\delta \sim \gamma^2$

Table 4.4.13 Redundant System of Two Units:
Nonoperating Stand-by, Restricted Repair

$$s_{1,2} = \frac{\lambda}{2\gamma}[(1 + 2\gamma) \pm \sqrt{1 + 4\gamma}],$$

$$\epsilon_{1,2} = \frac{\lambda}{\gamma}[(1 + \gamma) \pm \sqrt{\gamma}],$$

$$\gamma_c = \frac{\gamma^2}{1 + \gamma}, \qquad \gamma = \frac{\lambda}{\mu}.$$

MEASURE	EXACT VALUE	APPROXIMATE VALUE	ERROR
T_1	$\dfrac{1}{\lambda}\left(2 + \dfrac{1}{\gamma}\right)$	$\dfrac{1}{\lambda\gamma}$	$\delta_+ = \dfrac{2}{\lambda}$
T_2	$\dfrac{1}{\lambda}\left(1 + \dfrac{1}{\gamma}\right)$	$\dfrac{1}{\lambda\gamma}$	$\delta_+ = \dfrac{1}{\lambda}$
τ	$\dfrac{1}{\mu}$	—	—
$P(t_0)$	$\dfrac{1}{s_1 - s_2}(s_1 e^{-s_2 t_0} - s_2 e^{-s_1 t_0})$	$\exp\left(-\dfrac{1}{2 + \dfrac{1}{\gamma}}\lambda t_0\right)$	$\delta \sim \gamma^2$
$K(t)$	$1 - \dfrac{\lambda^2}{\epsilon_1 \epsilon_2}\left[1 - \dfrac{1}{\epsilon_1 - \epsilon_2}(\epsilon_1 e^{-\epsilon_2 t} - \epsilon_2 e^{-\epsilon_1 t})\right]$	$1 - \gamma^2\left[1 - \left(1 + \dfrac{\lambda}{\gamma}t\right)e^{-(\lambda/\gamma)t}\right]$	$\delta \sim \gamma^3$
K	$\dfrac{1}{1 + \gamma_c}$	$1 - \gamma^2$	$\delta_+ \sim \gamma^3$
$R(t_0)$	$\dfrac{s_2(\lambda - s_2)e^{-s_1 t_0} - s_1(\lambda - s_1)e^{-s_2 t_0}}{\mu(s_1 - s_2)(1 + \gamma + \gamma^2)}$	$\dfrac{1}{1 + \gamma_c}\exp\left(-\dfrac{1}{2 + \dfrac{1}{\gamma}}\lambda t_0\right)$	$\delta \sim \gamma^2$

4.   An approximate value for the probability of failure-free operation for an initial time of $t_0 = 10$ hours is

$$\tilde{P}(t_0) = \exp\left(-\frac{\lambda t_0}{2 + \dfrac{1}{\gamma}}\right) = \exp\left(-\frac{0.1 \cdot 10}{2 + \dfrac{1}{0.1}}\right) \approx 0.92.$$

5.   An approximate value for the nonstationary readiness coefficient at the moment of time $t = 10$ hours in the case of unrestricted repair is

$$\tilde{K}(t) = 1 - \tfrac{1}{2}\gamma^2[1 - (2 - e^{-(\lambda/\gamma)t})e^{-(\lambda/\gamma)t}]$$
$$= 1 - \tfrac{1}{2}(0.1)^2 \cdot [1 - (2 - e^{-(0.1/0.1) \cdot 10}) \cdot e^{-(0.1/0.1) \cdot 10}] \approx 0.995.$$

6.   An approximate value for the readiness coefficient in the restricted-repair case is

$$\tilde{K} = 1 - \gamma^2 = 1 - (0.1)^2 = 0.99.$$

7.   An approximate value for the reliability coefficient for a time interval of length $t_0 = 10$ hours in the case of restricted repair is

$$\tilde{R}(t_0) = \frac{1}{1 + \gamma_c} \exp\left(-\frac{\lambda t_0}{2 + \dfrac{1}{\gamma}}\right) = \frac{1}{1 + \dfrac{\gamma^2}{1 + \gamma}} \exp\left(-\frac{\lambda t_0}{2 + \dfrac{1}{\gamma}}\right)$$

$$= \frac{1}{1 + \dfrac{(0.1)^2}{1 + 0.1}} \exp\left(-\frac{0.1 \cdot 10}{2 + \dfrac{1}{0.1}}\right) \approx 0.91.$$

4.5   SYSTEMS WITH TWO STAND-BY UNITS

We examine the reliability measures of a system of m operating and two stand-by units (Figure 4.5.1). The case of a system of one operating unit and two stand-by units is treated separately (Figure 4.5.2).

A diagram of passages of a system with two stand-by units from one state to another is presented in Figure 4.5.3, where H_j is the state of the system in which j of its $N = m + 2$ units have failed, H_3 is the failed state of the system ($j = 0, 1, 2, 3$), Λ_j is the intensity of passage of the system (failure of one of its operating units) from the state with j failed units to the state with one additional failed unit—that is, $j + 1$ units ($j = 0, 1, 2$), M_j is the intensity of passage of the system (repair of one of its units) from the state with j failed units to the state with one fewer failed units—that is, $j - 1$ units ($j = 1, 2, 3$).

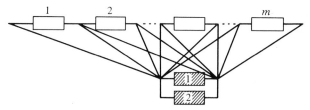

Figure 4.5.1 Block Diagram of the Model Used for Reliability Computations for a System of m Operating and Two Stand-by Units.

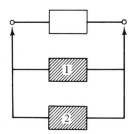

Figure 4.5.2 Block Diagram of the Model Used for Reliability Computations for a System of One Operating Unit and Two Stand-by Units.

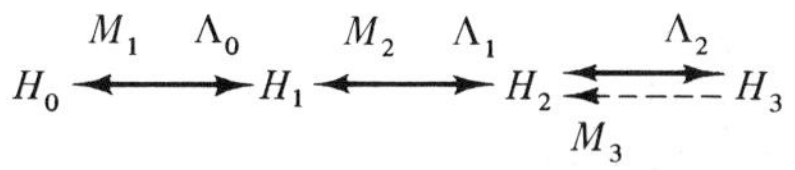

$$H_0 \xrightarrow[M_1]{\Lambda_0} H_1 \xleftarrow[M_2]{\Lambda_1} H_2 \underset{M_3}{\overset{\Lambda_2}{\rightleftarrows}} H_3$$

Figure 4.5.3 Diagram of Passages from One State to Another by a System with Two Stand-by Units.

The reliability measures of a system with two stand-by units are given for the general case and for three particular cases—namely, for operating, partially operating, and nonoperating states of the two stand-by units:

For the operating stand-by case:

$$\Lambda_0 = m\lambda + 2\lambda = (m + 2)\lambda,$$
$$\Lambda_1 = m\lambda + \lambda = (m + 1)\lambda,$$
$$\Lambda_2 = m\lambda.$$

For the partially operating stand-by case:

$$\Lambda_0 = m\lambda + 2\nu\lambda = (m + 2\nu)\lambda,$$
$$\Lambda_1 = m\lambda + \nu\lambda = (m + \nu)\lambda,$$
$$\Lambda_2 = m\lambda,$$

where ν is the load coefficient of the stand-by—that is, the failure intensity of a stand-by unit is $\nu\lambda$, $0 \leq \nu \leq 1$.

For the nonoperating stand-by case:

$$\Lambda_0 = m\lambda, \qquad \Lambda_1 = m\lambda, \qquad \Lambda_2 = m\lambda.$$

In each of these three cases we consider only two forms of repair of failed units:

(1)   unrestricted repair, where a failed unit is repaired immediately; and

(2)   restricted repair, where not more than one failed unit can undergo repair at any moment of time.

Formally,

$$M_j = \begin{cases} j\mu & \text{if repair is unrestricted,} \\ \mu & \text{if repair is restricted.} \end{cases}$$

In Tables 4.5.1–4.5.13 we give exact and approximate values of the basic reliability measures of a system with two stand-by units, and error estimates. The approximate values of the basic reliability measures are obtained subject to the condition

$$\gamma = \frac{\lambda}{\mu} \ll \frac{1}{m+2}.$$

After the tables we give examples of computation of reliability measures.

Example 4.5.1   Suppose we have a system of two operating and two stand-by units. The intensity of passage of the system from the state with all units operational to the state with one failed unit is

$$\Lambda_0 = 2\lambda.$$

With the failure of even one unit the functional load of each of the three remaining units is increased, and therefore the intensity of passage of the system to the state with two failed units is also increased and becomes

$$\Lambda_1 = 3\lambda.$$

With failure of two units the intensity of passage of the system into the failed state (a system of two operating units) will be equal to

$$\Lambda_2 = 4\lambda.$$

Repair of failed units of the system is assumed restricted—that is,

$$M_1 = M_2 = M_3 = \mu.$$

For the given system the quantities λ and μ are $\lambda = 0.1\,\dfrac{1}{\text{hr}}$ and $\mu = 1\,\dfrac{1}{\text{hr}}$.

We wish to determine the basic reliability measures for this system.

Table 4.5.1 System with Two Stand-by Units: General Case

$$\gamma_c = \frac{\Lambda_0\Lambda_1\Lambda_2}{M_3(\Lambda_0\Lambda_1 + \Lambda_0 M_2 + M_1 M_2)}$$

MEASURE	EXACT VALUE	APPROXIMATE VALUE	ERROR
T_1	$(\Lambda_0\Lambda_1 + \Lambda_0\Lambda_2 + \Lambda_2 M_1 + M_1 M_2)$ $\times (\Lambda_0\Lambda_1\Lambda_2)^{-1}$		$\delta_+ = (\Lambda_0\Lambda_1 + \Lambda_0\Lambda_2 + \Lambda_1\Lambda_2 + \Lambda_0 M_2 + \Lambda_2 M_1)(\Lambda_0\Lambda_1\Lambda_2)^{-1}$
T_2	$\dfrac{\Lambda_0\Lambda_1 + \Lambda_0 M_2 + M_1 M_2}{\Lambda_0\Lambda_1\Lambda_2}$	$\dfrac{M_1 M_2}{\Lambda_0\Lambda_1\Lambda_2}$	$\delta_+ = \dfrac{\Lambda_0\Lambda_1 + \Lambda_0 M_2}{\Lambda_0\Lambda_1\Lambda_2}$
τ	$\dfrac{1}{M_3}$	—	—
$P(t_0)$	—	$\exp\left(-\dfrac{\Lambda_0\Lambda_1\Lambda_2 t_0}{\Lambda_0\Lambda_1 + \Lambda_0\Lambda_2 + \Lambda_1\Lambda_2 + \Lambda_0 M_2 + \Lambda_2 M_1 + M_1 M_2}\right)$	$\delta \sim \dfrac{\Lambda_0\Lambda_1\Lambda_2}{M_1^2 M_2^2}(M_1 + M_2)$
K	$\dfrac{1}{1 + \gamma_c}$	$1 - \dfrac{\Lambda_0\Lambda_1\Lambda_2}{M_1 M_2 M_3}$	$\delta_+ \sim \dfrac{\Lambda_0^2\Lambda_1\Lambda_2}{M_1^2 M_2 M_3}$
$R(t_0)$	—	$\dfrac{\exp\left(-\dfrac{\Lambda_0\Lambda_1\Lambda_2 t_0}{\Lambda_0\Lambda_1 + \Lambda_0\Lambda_2 + \Lambda_1\Lambda_2 + \Lambda_0 M_2 + \Lambda_2 M_1 + M_1 M_2}\right)}{1 + \gamma_c}$	$\delta \sim \dfrac{\Lambda_0\Lambda_1\Lambda_2}{M_1^2 M_2^2}(M_1 + M_2)$

REDUNDANCY WITH REPAIR

Solution. From Table 4.5.1 we find:

1. The exact value of the mean operating time of the system up to failure is

$$
\begin{aligned}
T_1 &= \frac{\Lambda_0\Lambda_1 + \Lambda_0\Lambda_2 + \Lambda_1\Lambda_2 + \Lambda_0 M_2 + \Lambda_2 M_1 + M_1 M_2}{\Lambda_0\Lambda_1\Lambda_2} \\
&= \frac{2\lambda\cdot 3\lambda + 2\lambda\cdot 4\lambda + 3\lambda\cdot 4\lambda + 2\lambda\mu + 4\lambda\mu + \mu^2}{2\lambda\cdot 3\lambda\cdot 4\lambda} \\
&= \frac{2\cdot 0.1\cdot 3\cdot 0.1 + 2\cdot 0.1\cdot 4\cdot 0.1 + 3\cdot 0.1\cdot 4\cdot 0.1 + 2\cdot 0.1\cdot 1 + 4\cdot 0.1\cdot 1 + 1}{2\cdot 0.1\cdot 3\cdot 0.1\cdot 4\cdot 0.1} \\
&= 77.5 \quad \text{hr.}
\end{aligned}
$$

2. An approximate value for the mean operating time of the system between failures is

$$
\tilde{T}_2 = \frac{M_1 M_2}{\Lambda_0\Lambda_1\Lambda_2} = \frac{\mu^2}{2\lambda\cdot 3\lambda\cdot 4\lambda} = \frac{1}{2\cdot 0.1\cdot 3\cdot 0.1\cdot 4\cdot 0.1} = 41.7 \quad \text{hr.}
$$

The approximate value $\tilde{T}_2$ is less than the true value, and the error is

$$
\delta_+ = \frac{\Lambda_0\Lambda_1 + \Lambda_0 M_2}{\Lambda_0\Lambda_1\Lambda_2} = \frac{2\lambda\cdot 3\lambda + 2\lambda\mu}{2\lambda\cdot 3\lambda\cdot 4\lambda} = \frac{2\cdot 0.1\cdot 3\cdot 0.1 + 2\cdot 0.1\cdot 1}{2\cdot 0.1\cdot 3\cdot 0.1\cdot 4\cdot 0.1} \approx 10.8 \quad \text{hr.}
$$

3. The mean repair time of the system is

$$
\tau = \frac{1}{\mu} = \frac{1}{1} = 1 \quad \text{hr.}
$$

4. An approximate value for the probability of failure-free operation for an initial time interval of length t_0 is determined by the formula

$$
\tilde{P}(t_0) = \exp\left(-\frac{\Lambda_0\Lambda_1\Lambda_2}{\Lambda_0\Lambda_1 + \Lambda_0\Lambda_2 + \Lambda_1\Lambda_2 + \Lambda_0 M_2 + \Lambda_2 M_1 + M_1 M_2} t_0\right) = e^{-t_0/T_1}.
$$

Using the results of part 1 of this example, we obtain

$$
\tilde{P}(t_0) = e^{-t_0/77.5} = e^{-0.0129 t_0}.
$$

Thus, for example, for an initial time interval of length $t_0 = 24$ hours an approximate value for the probability of failure-free operation is

$$
\tilde{P}(t_0) = e^{-0.0129 t_0} = e^{-0.0129\cdot 24} \approx 0.738.
$$

5. The exact value of the readiness coefficient of the system is

$$
K = \frac{1}{1 + \gamma_c}.
$$

We first determine the quantity γ_c:

$$\gamma_c = \frac{\Lambda_0\Lambda_1\Lambda_2}{M_3(\Lambda_0\Lambda_1 + \Lambda_0 M_2 + M_1 M_2)} = \frac{2\lambda \cdot 3\lambda \cdot 4\lambda}{\mu(2\lambda \cdot 3\lambda + 2\lambda\mu + \mu^2)}$$

$$= \frac{2 \cdot 0.1 \cdot 3 \cdot 0.1 \cdot 4 \cdot 0.1}{1 \cdot (2 \cdot 0.1 \cdot 3 \cdot 0.1 + 2 \cdot 0.1 \cdot 1 + 1)} \approx 0.015.$$

Then,

$$K = \frac{1}{1 + \gamma_c} = \frac{1}{1 + 0.015} \approx 0.985.$$

6. An approximate value for the reliability coefficient is

$$\tilde{R}(t_0) = \frac{1}{1 + \gamma_c} e^{-t_0/T_1}.$$

Making use of the results obtained in parts 4 and 5 of this example, we obtain

$$\tilde{R}(t_0) = 0.985 e^{-0.0129 t_0}.$$

Thus, for example, for a time interval of length $t_0 = 24$ hours the reliability coefficient is approximately given by

$$\tilde{R}(t_0) = 0.985 e^{-0.0129 t_0} = 0.985 e^{-0.0129 \cdot 24} \approx 0.727.$$

Example 4.5.2 Suppose we have a system of four operating units and two stand-by units in an operating state. The failure intensity of each unit is $\lambda = 0.025 \dfrac{1}{\text{hr}}$, the repair intensity is $\mu = 0.5 \dfrac{1}{\text{hr}}$, and $\gamma = \lambda/\mu = 0.05$. We want to determine the basic reliability measures of the system for the cases of unrestricted and restricted repair.

Solution. From Tables 4.5.2 and 4.5.3 we find:

1. The exact value of the mean operating time of the system up to failure in the unrestricted-repair case is

$$T_1 = \frac{1}{\lambda} \cdot \frac{2 + (3m + 4)\gamma + (2 + 6m + 3m^2)\gamma^2}{m(m + 1)(m + 2)\gamma^2}.$$

Since in our case

$$\gamma = \frac{\lambda}{\mu} = \frac{0.025}{0.5} = 0.05,$$

it follows that

$$T_1 = \frac{1}{0.025} \cdot \frac{2 + (3 \cdot 4 + 4)0.05 + (2 + 6 \cdot 4 + 3 \cdot 4^2) \cdot (0.05)^2}{4(4 + 1)(4 + 2)(0.05)^2} = 398 \quad \text{hr}.$$

Table 4.5.2   System of m Operating and Two Stand-by
Units: Operating Stand-by, Unrestricted Repair

$$\gamma_c = \frac{m(m+1)(m+2)\gamma^3}{3[2+2(m+2)\gamma+(2+3m+m^2)\gamma^2]}, \qquad \gamma = \frac{\lambda}{\mu}.$$

MEASURE	EXACT VALUE	APPROXIMATE VALUE	ERROR
T_1	$\dfrac{1}{\lambda}\dfrac{2+(3m+4)\gamma+(2+6m+3m^2)\gamma^2}{m(m+1)(m+2)\gamma^2}$	$\dfrac{1}{\lambda}\dfrac{2}{m(m+1)(m+2)\gamma^2}$	$\delta_+ = \dfrac{1}{\lambda}\dfrac{(4+3m)+(2+6m+3m^2)\gamma}{m(m+1)(m+2)\gamma}$
T_2	$\dfrac{1}{\lambda}\dfrac{2+2(m+2)\gamma+(2+3m+m^2)\gamma^2}{m(m+1)(m+2)\gamma^2}$	$\dfrac{1}{\lambda}\dfrac{2}{m(m+1)(m+2)\gamma^2}$	$\delta_+ = \dfrac{1}{\lambda}\dfrac{2(m+2)+(2+3m+m^2)\gamma}{m(m+1)(m+2)\gamma}$
τ	$\dfrac{1}{3\mu}$	—	—
$P(t_0)$	—	$\exp\left[-\dfrac{m(m+1)(m+2)\gamma^2\lambda t_0}{2+(4+3m)\gamma+(2+6m+3m^2)\gamma^2}\right]$	$\delta \sim \tfrac{3}{4}m(m+1)(m+2)\gamma^3$
K	$\dfrac{1}{1+\gamma_s}$	$1-\dfrac{m(m+1)(m+2)}{6}\gamma^3$	$\delta_+ \sim \dfrac{m(m+1)(m+2)^2}{6}\gamma^4$
$R(t_0)$	—	$\dfrac{\exp\left[-\dfrac{m(m+1)(m+2)\gamma^2\lambda t_0}{2+(4+3m)\gamma+(2+6m+3m^2)\gamma^2}\right]}{1+\gamma_c}$	$\delta \sim \tfrac{3}{4}m(m+1)(m+2)\gamma^3$

Table 4.5.3 System of m Operating and Two Stand-by Units: Operating Stand-by, Restricted Repair

$$\gamma_c = \frac{m(m+1)(m+2)\gamma^3}{1+(m+2)\gamma+(m^2+3m+2)\gamma^2}, \qquad \gamma = \frac{\lambda}{\mu}.$$

MEASURE	EXACT VALUE	APPROXIMATE VALUE	ERROR
T_1	$\dfrac{1}{\lambda}\dfrac{1+2(m+1)\gamma+(2+6m+3m^2)\gamma^2}{m(m+1)(m+2)\gamma^2}$	$\dfrac{1}{\lambda}\dfrac{1}{m(m+1)(m+2)\gamma^2}$	$\delta_+ = \dfrac{1}{\lambda}\dfrac{2(m+1)+(2+6m+3m^2)\gamma}{m(m+1)(m+2)\gamma}$
T_2	$\dfrac{1}{\lambda}\dfrac{1+(m+2)\gamma+(2+3m+m^2)\gamma^2}{m(m+1)(m+2)\gamma^2}$	$\dfrac{1}{\lambda}\dfrac{1}{m(m+1)(m+2)\gamma^2}$	$\delta_+ = \dfrac{1}{\lambda}\dfrac{(m+2)+(2+3m+m^2)\gamma}{m(m+1)(m+2)\gamma}$
τ	$\dfrac{1}{\mu}$	—	—
$P(t_0)$	—	$\exp\left[-\dfrac{m(m+1)(m+2)\gamma^2\lambda t_0}{1+2(1+m)\gamma+(2+6m+3m^2)\gamma^2}\right]$	$\delta \sim 2m(m+1)(m+2)\gamma^3$
K	$\dfrac{1}{1+\gamma_c}$	$1-m(m+1)(m+2)\gamma^3$	$\delta_+ \sim m(m+1)(m+2)^2\gamma^4$
$R(t_0)$	—	$\exp\left[-\dfrac{m(m+1)(m+2)\gamma^2\lambda t_0}{1+2(1+m)\gamma+(2+6m+3m^2)\gamma^2}\right] \bigg/ (1+\gamma_c)$	$\delta \sim 2m(m+1)(m+2)\gamma^3$

In the restricted-repair case

$$T_1 = \frac{1}{\lambda} \cdot \frac{1 + 2(m + 1)\gamma + (2 + 6m + 3m^2)\gamma^2}{m(m + 1)(m + 2)\gamma^2}$$

$$= \frac{1}{0.025} \cdot \frac{1 + 2(4 + 1) \cdot 0.05 + (2 + 6 \cdot 4 + 3 \cdot 4^2) \cdot (0.05)^2}{4(4 + 1)(4 + 2)(0.05)^2}$$

$$\approx 225 \quad \text{hr.}$$

2.   An approximate value for the mean operating time of the system between failures in the unrestricted repair case is

$$\tilde{T}_2 = \frac{1}{\lambda} \cdot \frac{2}{m(m + 1)(m + 2)\gamma^2} = \frac{2}{0.025 \cdot 4 \cdot (4 + 1)(4 + 2) \cdot (0.05)^2}$$

$$\approx 267 \quad \text{hr.}$$

3.   The mean repair time of the system in the unrestricted-repair case is

$$\tau = \frac{1}{3\mu} = \frac{1}{3 \cdot 0.5} = 0.67 \quad \text{hr,}$$

and in the restricted-repair case it is

$$\tau = \frac{1}{\mu} = \frac{1}{0.5} = 2 \quad \text{hr.}$$

4.   An approximate value for the probability of initial failure-free operation of the system in the case of unrestricted repair is

$$\tilde{P}(t_0) = e^{-t_0/T_1}.$$

Using the result of part 1 of this example, we obtain

$$\tilde{P}(t_0) = e^{-t_0/398} = e^{-0.0025t_0}.$$

Thus, in particular, the probability of failure-free operation of the system for an initial time interval of length $t_0 = 24$ hours is approximately

$$\tilde{P}(t_0) = e^{-0.0025t_0} = e^{-0.0025 \cdot 24} \approx 0.94.$$

5.   An approximate value for the readiness coefficient of the system in the restricted-repair case is

$$\tilde{K} = 1 - m(m + 1)(m + 2)\gamma^3 = 1 - 4 \cdot (4 + 1)(4 + 2) \cdot (0.05)^3 = 0.985.$$

6.   An approximate value for the reliability coefficient of the system for the unrestricted-repair case is determined by the formula

$$\tilde{R}(t_0) = \frac{1}{1 + \gamma_c} e^{-t_0/T_1}.$$

First we determine the quantity γ_c:

$$\gamma_c = \frac{m(m+1)(m+2)\gamma^3}{3[2 + 2(m+2)\gamma + (2 + 3m + m^2)\gamma^2]}$$

$$= \frac{4\cdot(4+1)(4+2)\cdot(0.05)^3}{3[2 + 2(4+2)\cdot 0.05 + (2 + 3\cdot 4 + 4^2)\cdot(0.05)^2]} \approx 0.018.$$

Using the result of part 4 of this example, we obtain

$$\tilde{R}(t_0) = \frac{1}{1+\gamma_c}e^{-t_0/T_1} = \frac{1}{1 + 0.018}e^{-t_0/T_1} = 0.982e^{-0.0025\cdot t_0}.$$

Thus, in particular, the reliability coefficient for a time interval of length $t_0 = 24$ hours is approximately

$$\tilde{R}(t_0) = 0.982e^{-0.0025\cdot t_0} = 0.982e^{-0.0025\cdot 24} \approx 0.92.$$

Example 4.5.3 Suppose a system consists of two operating structures each with failure intensity $\lambda = 0.1\,\dfrac{1}{hr}$ and two stand-by units each with failure intensity $\lambda_{stand-by} = 0.05\,\dfrac{1}{hr}$; that is, the stand-by load coefficient is $\nu = 0.5$. Repair of a failed structure in the system is made with intensity $\mu = 0.5\,\dfrac{1}{hr}$, and $\lambda = \gamma/\mu = 0.1/0.5 = 0.2$.

We want to determine the basic reliability measures of the system for the cases of unrestricted and restricted repair.

Solution. From Tables 4.5.4 and 4.5.5 we find:

1. The exact value of the mean operating time of the system up to failure in the unrestricted-repair case is

$$T_1 = \frac{1}{\lambda}\cdot\frac{2 + (3m + 4\nu)\gamma + (2\nu^2 + 6\nu m + 3m^2)\gamma^2}{m(m+\nu)(m+2\nu)\gamma^2}$$

$$= \frac{1}{0.05}\cdot\frac{2 + (3\cdot 2 + 4\cdot 0.5)\cdot 0.2 + [2\cdot(0.5)^2 + 6\cdot 0.5\cdot 2 + 3\cdot 2^2]\cdot(0.2)^2}{2\cdot(2 + 0.5)(2 + 2\cdot 0.5)\cdot(0.2)^2}$$

$$\approx 145 \quad hr.$$

2. The exact value of the mean operating time of the system between failures in the restricted-repair case is

$$T_2 = \frac{1}{\lambda}\cdot\frac{1 + (m + 2\nu)\gamma + (m^2 + 3\nu m + 2\nu^2)\gamma^2}{m(m+\nu)(m+2\nu)\gamma^2}$$

$$= \frac{1}{0.1}\cdot\frac{1 + (2 + 2\cdot 0.5)\cdot 0.2 + [2^2 + 3\cdot 0.5\cdot 2 + 2\cdot(0.5)^2]\cdot(0.2)^2}{2\cdot(2 + 0.5)(2 + 2\cdot 0.5)\cdot(0.2)^2}$$

$$\approx 32 \quad hr.$$

Table 4.5.4 System of m Operating and Two Stand-by Units: Partially Operating Stand-by, Unrestricted Repair

$$\gamma_c = \frac{m(m+\nu)(m+2\nu)\gamma^3}{3[2+2(m+2\nu)\gamma+(m^2+3\nu m+2\nu^2)\gamma^2]}, \qquad \gamma = \frac{\lambda}{\mu}.$$

MEASURE	EXACT VALUE	APPROXIMATE VALUE	ERROR
T_1	$\dfrac{1}{\lambda}\dfrac{2+(3m+4\nu)\gamma+(2\nu^2+6\nu m+3m^2)\gamma^2}{m(m+\nu)(m+2\nu)\gamma^2}$	$\dfrac{1}{\lambda}\dfrac{2}{m(m+\nu)(m+2\nu)\gamma^2}$	$\delta_+ = \dfrac{1}{\lambda}\dfrac{(3m+4\nu)+(3m^2+6\nu m+2\nu^2)\gamma}{m(m+\nu)(m+2\nu)\gamma}$
T_2	$\dfrac{1}{\lambda}\dfrac{2+2(m+2\nu)\gamma+(m^2+3\nu m+2\nu^2)\gamma^2}{m(m+\nu)(m+2\nu)\gamma^2}$	$\dfrac{1}{\lambda}\dfrac{2}{m(m+\nu)(m+2\nu)\gamma^2}$	$\delta_+ = \dfrac{1}{\lambda}\dfrac{2(m+2\nu)+(m^2+3\nu m+2\nu^2)\gamma}{m(m+\nu)(m+2\nu)\gamma}$
τ	$\dfrac{1}{3\mu}$	—	—
$P(t_0)$	—	$\exp\left[-\dfrac{m(m+\nu)(m+2\nu)\gamma^2\lambda t_0}{2+(4\nu+3m)\gamma+(2\nu^2+6\nu m+3m^2)\gamma^2}\right]$	$\delta \sim \tfrac{3}{4}m(m+\nu)(m+2\nu)\gamma^3$
K	$\dfrac{1}{1+\gamma_c}$	$1-\dfrac{m(m+\nu)(m+2\nu)}{6}\gamma^3$	$\delta_+ \sim \dfrac{m(m+\nu)(m+2\nu)^2}{6}\gamma^4$
$R(t_0)$	—	$\dfrac{\exp\left[-\dfrac{m(m+\nu)(m+2\nu)\gamma^2\lambda t_0}{2+(4\nu+3m)\gamma+(2\nu^2+6\nu m+3m^2)\gamma^2}\right]}{1+\gamma_c}$	$\delta \sim \tfrac{3}{4}m(m+\nu)(m+2\nu)\gamma^3$

Table 4.5.5   System of m Operating and Two Stand-by Units: Partially Operating Stand-by, Restricted Repair

$$\gamma_c = \frac{m(m+\nu)(m+2\nu)\gamma^3}{1+(m+2\nu)\gamma+(m^2+3\nu m+2\nu^2)\gamma^2},$$

$$\gamma = \frac{\lambda}{\mu}.$$

MEASURE	EXACT VALUE	APPROXIMATE VALUE	ERROR
T_1	$\dfrac{1}{\lambda}\dfrac{1+2(m+\nu)\gamma+(3m^2+6\nu m+2\nu^2)\gamma^2}{m(m+\nu)(m+2\nu)\gamma^2}$	$\dfrac{1}{\lambda}\dfrac{1}{m(m+\nu)(m+2\nu)\gamma^2}$	$\delta_+ = \dfrac{1}{\lambda}\dfrac{2(m+\nu)+(3m^2+6\nu m+2\nu^2)\gamma}{m(m+\nu)(m+2\nu)\gamma}$
T_2	$\dfrac{1}{\lambda}\dfrac{1+(m+2\nu)\gamma+(m^2+3\nu m+2\nu^2)\gamma^2}{m(m+\nu)(m+2\nu)\gamma^2}$	$\dfrac{1}{\lambda}\dfrac{1}{m(m+\nu)(m+2\nu)\gamma^2}$	$\delta_+ = \dfrac{1}{\lambda}\dfrac{(m+2\nu)+(m^2+3\nu m+2\nu^2)\gamma}{m(m+\nu)(m+2\nu)\gamma}$
τ	$\dfrac{1}{\mu}$	$\dfrac{1}{\mu}$	—
$P(t_0)$	—	$\exp\left[-\dfrac{m(m+\nu)(m+2\nu)\gamma^2\lambda t_0}{1+2(\nu+m)\gamma+(2\nu^2+6\nu m+3m^2)\gamma^2}\right]$	$\delta \sim 2m(m+\nu)(m+2\nu)\gamma^3$
K	$\dfrac{1}{1+\gamma_c}$	$1-m(m+\nu)(m+2\nu)\gamma^3$	$\delta_+ \sim m(m+\nu)(m+2\nu)^2\gamma^4$
$R(t_0)$	—	$\dfrac{\exp\left[-\dfrac{m(m+\nu)(m+2\nu)\gamma^2\lambda t_0}{1+2(\nu+m)\gamma+(2\nu^2+6\nu m+3m^2)\gamma^2}\right]}{1+\gamma_c}$	$\delta \sim 2m(m+\nu)(m+2\nu)\gamma^3$

3. The mean repair time in the case of restricted repair is

$$\tau = \frac{1}{\mu} = \frac{1}{0.5} = 2 \quad \text{hr.}$$

4. An approximate value for the probability of initial failure-free operation of the system for the unrestricted-repair case is

$$\tilde{P}(t_0) = \exp\left[-\frac{m(m+\nu)(m+2\nu)}{2} \gamma^2 \lambda t_0 \right] = e^{-t_0/T_1} = e^{-t_0/145} \approx e^{-0.007t_0}.$$

Thus, for example, the probability of failure-free operation of the system for an initial time interval of length $t_0 = 10$ hours is approximately

$$\tilde{P}(t_0) = e^{-0.007t_0} = c^{-0.007 \cdot 10} \approx 0.93.$$

5. The exact value of the readiness coefficient of the system for the case of restricted repair is determined from the formula

$$K = \frac{1}{1 + \gamma_c}.$$

First we determine the quantity γ_c:

$$\gamma_c = \frac{m(m+\nu)(m+2\nu)\gamma^3}{1 + (m+2\nu)\gamma + (m^2 + 3\nu m + 2\nu^2)\gamma^2}$$

$$= \frac{2(2+0.5)(2+2 \cdot 0.5) \cdot (0.2)^3}{1 + (2+2 \cdot 0.5) \cdot 0.2 + [2^2 + 3 \cdot 0.5 \cdot 2 + 2 \cdot (0.5)^2] \cdot (0.2)^2} \approx 0.063.$$

Then

$$K = \frac{1}{1 + \gamma_c} = \frac{1}{1 + 0.063} \approx 0.941.$$

6. An approximate value for the reliability coefficient of the system for the unrestricted-repair case is

$$\tilde{R}(t_0) = \frac{1}{1 + \gamma_c} e^{-t_0/T_1}.$$

First we determine the quantity γ_c:

$$\gamma_c = \frac{m(m+\nu)(m+2\nu)\gamma^3}{3[2 + 2(m+2\nu)\gamma + (m^2 + 3\nu m + 2\nu^2)\gamma^2]}$$

$$= \frac{2(2+0.5)(2+2 \cdot 0.5) \cdot (0.2)^3}{3\{2 + 2(2+2 \cdot 0.5) \cdot 0.2 + [2^2 + 3 \cdot 0.5 \cdot 2 + 2 \cdot (0.5)^2](0.2)^2\}}$$

$$\approx 0.011.$$

Now using the result of part 4 of this example and substituting the computed value of γ_c, we obtain

$$\tilde{R}(t_0) = \frac{1}{1 + \gamma_c} e^{-t_0/T_1} \approx 0.989 e^{-0.007 t_0}.$$

Thus, for example, the reliability coefficient for a time interval of length $t_0 = 10$ hours is approximately

$$\tilde{R}(t_0) = 0.989 e^{-0.007 t_0} = 0.989 e^{-0.007 \cdot 10} \approx 0.92.$$

Example 4.5.4   Suppose we have a system consisting of $m = 4$ operating devices and two stand-by devices in a nonoperating state. The failure intensity of an operating device is $\lambda = 0.02 \, \dfrac{1}{hr}$, the repair intensity of a failed device is $\mu = 0.2 \, \dfrac{1}{hr}$, and $\gamma = \lambda/\mu = 0.02/0.2 = 0.1$.

We are to determine the basic reliability measures for the cases of unrestricted and restricted repair.

Solution.   From Tables 4.5.6 and 4.5.7 we find:

1.   The exact value of the mean operating time of the system up to failure for unrestricted repair is

$$T_1 = \frac{1}{\lambda} \cdot \frac{2 + 2m\gamma + 3m^2\gamma^2}{m^3\gamma^2}$$

$$= \frac{1}{0.02} \cdot \frac{2 + 3 \cdot 4 \cdot 0.1 + 3 \cdot 4^2(0.1)^2}{4^3(0.1)^2} \approx 287.5 \quad \text{hr.}$$

2.   An approximate value for the mean operating time of the system between failures for restricted repair is

$$\tilde{T}_2 = \frac{1}{\lambda} \cdot \frac{1}{m^3\gamma^2} = \frac{1}{0.02} \cdot \frac{1}{4^3(0.1)^2} \approx 78.1 \quad \text{hr.}$$

3.   The mean repair times for the cases of unrestricted and restricted repair are, respectively,

$$\tau = \frac{1}{3\mu} = \frac{1}{3 \cdot 0.2} = 1.7 \quad \text{hr,}$$

and

$$\tau = \frac{1}{\mu} = \frac{1}{0.2} = 5 \quad \text{hr.}$$

Table 4.5.6   System of m Operating and Two Stand-by
Units: Nonoperating Stand-by,
Unrestricted Repair

$$\gamma_c = \frac{m^3\gamma^3}{3(2 + 2m\gamma + m^2\gamma^2)},$$
$$\gamma = \frac{\lambda}{\mu}.$$

MEASURE	EXACT VALUE	APPROXIMATE VALUE	ERROR
T_1	$\dfrac{1}{\lambda}\dfrac{2 + 3m\gamma + 3m^2\gamma^2}{m^3\gamma^2}$	$\dfrac{1}{\lambda}\dfrac{2}{m^3\gamma^2}$	$\delta_+ = \dfrac{1}{\lambda}\dfrac{3(1 + m\gamma)}{m^2\gamma}$
T_2	$\dfrac{1}{\lambda}\dfrac{2 + 2m\gamma + m^2\gamma^2}{m^3\gamma^2}$	$\dfrac{1}{\lambda}\dfrac{2}{m^3\gamma^2}$	$\delta_+ = \dfrac{1}{\lambda}\dfrac{2 + m\gamma}{m^2\gamma}$
τ	$\dfrac{1}{3\mu}$	—	—
$P(t_0)$	—	$\exp\left(-\dfrac{m^3\gamma^2\lambda t_0}{2 + 3m\gamma + 3m^2\gamma^2}\right)$	$\delta \sim \frac{3}{4}m^3\gamma^3$
K	$\dfrac{1}{1 + \gamma_c}$	$1 - \dfrac{m^3}{6}\gamma^3$	$\delta_+ \sim \dfrac{m^4}{6}\gamma^4$
$R(t_0)$	—	$\dfrac{\exp\left(-\dfrac{m^3\gamma^2\lambda t_0}{2 + 3m\gamma + 3m^2\gamma^2}\right)}{1 + \gamma_c}$	$\delta \sim \frac{3}{4}m^3\gamma^3$

4.   An approximate value for the probability of initial failure-free operation for the unrestricted-repair case can be computed by making use of the result obtained in part 1 of this example:

$$\tilde{P}(t_0) = e^{-t_0/T_1} = e^{-t_0/287.5} \approx e^{-0.0035t_0}.$$

For example, the probability of failure-free operation for an initial time interval of length $t_0 = 10$ hours is approximately

$$\tilde{P}(t_0) = e^{-0.0035t_0} = e^{-0.0035\cdot 10} \approx 0.965.$$

5.   The exact value of the readiness coefficient of the system for the unrestricted-repair case is

$$K = \frac{1}{1 + \gamma_c},$$

where

$$\gamma_c = \frac{m^3\gamma^3}{3(2 + 2m\gamma + m^2\gamma^2)}.$$

First we determine γ_c:

$$\gamma_c = \frac{4^3 \cdot (0.1)^3}{3[2 + 2\cdot4\cdot0.1 + 4^2(0.1)^2]} \approx 0.0072.$$

Then

$$K = \frac{1}{1 + \gamma_c} = \frac{1}{1 + 0.0072} \approx 0.993.$$

6.   An approximate value for the reliability coefficient for the unrestricted-repair case is

$$\tilde{R}(t_0) = Ke^{-t_0/T_1}.$$

Table 4.5.7   System of m Operating and Two Stand-by
Units: Nonoperating Stand-by,
Restricted Repair

$$\gamma_c = \frac{m^3\gamma^3}{1 + m\gamma + m^2\gamma^2},$$

$$\gamma = \frac{\lambda}{\mu}.$$

MEASURE	EXACT VALUE	APPROXIMATE VALUE	ERROR
T_1	$\dfrac{1}{\lambda}\dfrac{1 + 2m\gamma + 3m^2\gamma^2}{m^3\gamma^2}$	$\dfrac{1}{\lambda}\dfrac{1}{m^3\gamma^2}$	$\delta_+ = \dfrac{1}{\lambda}\dfrac{2 + 3m\gamma}{m^2\gamma}$
T_2	$\dfrac{1}{\lambda}\dfrac{1 + m\gamma + m^2\gamma^2}{m^3\gamma^2}$	$\dfrac{1}{\lambda}\dfrac{1}{m^3\gamma^2}$	$\delta_+ = \dfrac{1}{\lambda}\dfrac{1 + m\gamma}{m^2\gamma}$
τ	$\dfrac{1}{\mu}$	—	—
$P(t_0)$	—	$\exp\left(-\dfrac{m^3\gamma^2\lambda t_0}{1 + 2m\gamma + 3m^2\gamma^2}\right)$	$\delta \sim 2m^3\gamma^3$
K	$\dfrac{1}{1 + \gamma_c}$	$1 - m^3\gamma^3$	$\delta_+ \sim m^4\gamma^4$
$R(t_0)$	—	$\dfrac{\exp\left(-\dfrac{m^3\gamma^2\lambda t_0}{1 + 2m\gamma + 3m^2\gamma^2}\right)}{1 + \gamma_c}$	$\delta \sim 2m^3\gamma^3$

Using the results obtained in parts 4 and 5 of this example, we find

$$\tilde{R}(t_0) = Ke^{-t_0/T_1} = 0.993e^{-0.0035t_0}.$$

Thus, for example, for a time interval of length $t_0 = 10$ hours the reliability coefficient is approximately

$$\tilde{R}(t_0) = 0.993e^{-0.0035t_0} = 0.993e^{-0.0035 \cdot 10} \approx 0.958.$$

Example 4.5.5 Suppose three structures are operating in parallel, where two of them are stand-bys. The failure intensity of each operating structure is $\lambda = 0.05 \dfrac{1}{hr}$, the repair intensity of a failed structure is $\mu = 0.25 \dfrac{1}{hr}$, and $\gamma = \lambda/\mu = 0.05/0.25 = 0.2$.

We wish to determine the basic reliability measures of this system for the unrestricted- and restricted-repair cases.

Solution.   From Tables 4.5.8 and 4.5.9 we find:

1.   The exact value of the mean operating time of the system up to failure for unrestricted repair is

$$T_1 = \frac{1}{\lambda} \cdot \frac{2 + 7\gamma + 11\gamma^2}{6\gamma^2} = \frac{1}{0.05} \cdot \frac{2 + 7 \cdot 0.2 + 11 \cdot (0.2)^2}{6 \cdot (0.2)^2} = 320 \quad hr.$$

2.   The exact value of the mean operating time of the system between failures for the restricted-repair case is

$$T_2 = \frac{1}{\lambda} \cdot \frac{1 + 3\gamma + 6\gamma^2}{6\gamma^2} = \frac{1}{0.05} \cdot \frac{1 + 3 \cdot 0.2 + 6 \cdot (0.2)^2}{6 \cdot (0.2)^2} \approx 153.3 \quad hr.$$

3.   The mean repair times for the cases of unrestricted and restricted repair, respectively, are

$$\tau = \frac{1}{3\mu} = \frac{1}{3 \cdot 0.25} = 1.3 \quad hr,$$

and

$$\tau = \frac{1}{\mu} = \frac{1}{0.25} = 4 \quad hr.$$

4.   An approximate value for the probability of initial failure-free operation of the system with unrestricted repair can be determined with the help of the result of part 1 of this example:

$$\tilde{P}(t_0) = e^{-6\gamma^2 \lambda t_0/(2+7\gamma+11\gamma^2)} = e^{-t_0/T_1} = e^{-t_0/320} \approx e^{-0.003t_0}.$$

Table 4.5.8 System of One Operating Unit and Two
Stand-by Units: Operating Stand-by,
Unrestricted Repair

$$\gamma_c = \frac{\gamma^3}{1 + 3\gamma + 3\gamma^2},$$

$$\gamma = \frac{\lambda}{\mu}.$$

MEASURE	EXACT VALUE	APPROXIMATE VALUE	ERROR
T_1	$\dfrac{1}{\lambda}\dfrac{2 + 7\gamma + 11\gamma^2}{6\gamma^2}$	$\dfrac{1}{\lambda}\dfrac{1}{3\gamma^2}$	$\delta_+ = \dfrac{1}{\lambda}\dfrac{7 + 11\gamma}{6\gamma}$
T_2	$\dfrac{1}{\lambda}\dfrac{1 + 3\gamma + 3\gamma^2}{3\gamma^2}$	$\dfrac{1}{\lambda}\dfrac{1}{3\gamma^2}$	$\delta_+ = \dfrac{1}{\lambda}\dfrac{1 + \gamma}{\gamma}$
τ	$\dfrac{1}{3\mu}$	—	—
$P(t_0)$	—	$\exp\left(\dfrac{-6\gamma^2 \lambda t_0}{2 + 7\gamma + 11\gamma^2}\right)$	$\delta \sim \tfrac{9}{2}\gamma^3$
K	$\dfrac{1}{1 + \gamma_c}$	$1 - \gamma^3$	$\delta_+ \sim 3\gamma^4$
$R(t_0)$	—	$\dfrac{\exp\left(-\dfrac{6\gamma^2 \lambda t_0}{2 + 7\gamma + 11\gamma^2}\right)}{1 + \gamma_c}$	$\delta \sim \tfrac{9}{2}\gamma^3$

For example, the probability of failure-free operation for an initial
time interval of length $t_0 = 100$ hours is approximately:

$$\tilde{P}(t_0) = e^{-0.003t_0} = e^{-0.003 \cdot 100} \approx 0.745.$$

5. The exact value of the readiness coefficient of the system for the
unrestricted-repair case is determined from the formula

$$K = \frac{1}{1 + \gamma_c},$$

where

$$\gamma_c = \frac{\gamma^3}{1 + 3\gamma + 3\gamma^2}.$$

Table 4.5.9   System of One Operating Unit and Two
Stand-by Units: Operating Stand-by,
Restricted Repair

$$\gamma_c = \frac{6\gamma^3}{1 + 3\gamma + 6\gamma^2},$$
$$\gamma = \frac{\lambda}{\mu}.$$

MEASURE	EXACT VALUE	APPROXIMATE VALUE	ERROR
T_1	$\dfrac{1}{\lambda}\dfrac{1 + 4\gamma + 11\gamma^2}{6\gamma^2}$	$\dfrac{1}{\lambda}\dfrac{1}{6\gamma^2}$	$\delta_+ = \dfrac{1}{\lambda}\dfrac{4 + 11\gamma}{6\gamma}$
T_2	$\dfrac{1}{\lambda}\dfrac{1 + 3\gamma + 6\gamma^2}{6\gamma^2}$	$\dfrac{1}{\lambda}\dfrac{1}{6\gamma^2}$	$\delta_+ = \dfrac{1}{\lambda}\dfrac{1 + 2\gamma}{2\gamma}$
τ	$\dfrac{1}{\mu}$	—	—
$P(t_0)$	—	$\exp\left(-\dfrac{6\gamma^2\lambda t_0}{1 + 4\gamma + 11\gamma^2}\right)$	$\delta \sim 12\gamma^3$
K	$\dfrac{1}{1 + \gamma_c}$	$1 - 6\gamma^3$	$\delta_+ \sim 18\gamma^4$
$R(t_0)$	—	$\dfrac{\exp\left(-\dfrac{6\gamma^2\lambda t_0}{1 + 4\gamma + 11\gamma^2}\right)}{1 + \gamma_c}$	$\delta \sim 12\gamma^3$

First we find the quantity γ_c:

$$\gamma_c = \frac{\gamma^3}{1 + 3\gamma + 3\gamma^2} = \frac{(0.2)^3}{1 + 3 \cdot 0.2 + 3 \cdot (0.2)^2} \approx 0.0046.$$

Then

$$K = \frac{1}{1 + \gamma_c} = \frac{1}{1 + 0.0046} \approx 0.995.$$

6.   By using the results obtained in parts 4 and 5 of this example, we
find an approximate value for the reliability coefficient of the system in
the case of unrestricted repair to be

$$\tilde{R}(t_0) = Ke^{-t_0/T_1} = 0.995e^{-0.003t_0}.$$

Thus, for example, the reliability coefficient for a time interval of length $t_0 = 24$ hours is approximately

$$\tilde{R}(t_0) = 0.995e^{-0.003 \cdot t_0} = 0.995e^{-0.003 \cdot 24} \approx 0.93.$$

Example 4.5.6 Suppose a computing complex consists of three machines, two of which are stand-bys in a partially operating state with load coefficient $\nu = 0.5$.

The failure intensity of an operating machine is $\lambda = 0.05 \dfrac{1}{\text{hr}}$, the repair intensity of a failed machine is $\mu = 0.25 \dfrac{1}{\text{hr}}$ and $\gamma = \lambda/\mu = 0.05/0.25 = 0.2$.

We want to determine the basic reliability measures of the complex for unrestricted and restricted repair.

Solution.   From Tables 4.5.10 and 4.5.11 we find:

1.   The exact value of the mean operating time of the complex up to failure for the unrestricted-repair case is

$$T_1 = \frac{1}{\lambda} \cdot \frac{2 + (3 + 4\nu)\gamma + (3 + 6\nu + 2\nu^2)\gamma^2}{(1 + \nu)(1 + 2\nu)\gamma^2}$$

$$= \frac{1}{0.05} \cdot \frac{2 + (3 + 1 \cdot 0.5) \cdot 0.2 + [3 + 6 \cdot 0.5 + 2 \cdot (0.5)^2] \cdot (0.2)^2}{(1 + 0.5)(1 + 2 \cdot 0.5) \cdot (0.2)^2}$$

$$\approx 543 \quad \text{hr}.$$

2.   An approximate value for the mean operating time of the complex between failures in the restricted repair case is

$$\tilde{T}_2 = \frac{1}{\lambda} \cdot \frac{1}{(1 + \nu)(1 + 2\nu)\gamma^2}$$

$$= \frac{1}{0.05} \cdot \frac{1}{(1 + 0.5)(1 + 2 \cdot 0.5) \cdot (0.2)^2} \approx 167 \quad \text{hr}.$$

3.   The mean repair times of the complex for the cases of unrestricted and restricted repair, respectively, are

$$\tau = \frac{1}{3\mu} = \frac{1}{3 \cdot 0.25} \approx 1.3 \quad \text{hr},$$

and

$$\tau = \frac{1}{\mu} = \frac{1}{0.25} = 4 \quad \text{hr}.$$

4.   An approximate value for the probability of initial failure-free operation of the complex in the unrestricted-repair case is found by

Table 4.5.10   System of One Operating Unit and Two Stand-by Units: Partially Operating Stand-by, Unrestricted Repair

$$\gamma_c = \frac{(1+\nu)(1+2\nu)\gamma^3}{3[2+2(1+2\nu)\gamma+(1+3\nu+2\nu^2)\gamma^2]},$$
$$\gamma = \frac{\lambda}{\mu}.$$

MEASURE	EXACT VALUE	APPROXIMATE VALUE	ERROR
T_1	$\dfrac{1}{\lambda}\dfrac{2+(3+4\nu)\gamma+(3+6\nu+2\nu^2)\gamma^2}{(1+\nu)(1+2\nu)\gamma^2}$	$\dfrac{1}{\lambda}\dfrac{2}{(1+\nu)(1+2\nu)\gamma^2}$	$\delta_+ = \dfrac{1}{\lambda}\dfrac{(3+4\nu)+(3+6\nu+2\nu^2)\gamma}{(1+\nu)(1+2\nu)\gamma}$
T_2	$\dfrac{1}{\lambda}\dfrac{2+2(1+2\nu)\gamma+(1+3\nu+2\nu^2)\gamma^2}{(1+\nu)(1+2\nu)\gamma^2}$	$\dfrac{1}{\lambda}\dfrac{2}{(1+\nu)(1+2\nu)\gamma^2}$	$\delta_+ = \dfrac{1}{\lambda}\dfrac{2(1+2\nu)+(1+3\nu+2\nu^2)\gamma}{(1+\nu)(1+2\nu)\gamma}$
τ	$\dfrac{1}{3\mu}$	—	—
$P(t_0)$		$\exp\left[-\dfrac{(1+\nu)(1+2\nu)\gamma^2\lambda t_0}{2+(3+4\nu)\gamma+(3+6\nu+2\nu^2)\gamma^2}\right]$	$\delta \sim \dfrac{3}{4}(1+\nu)(1+2\nu)\gamma^3$
K	$\dfrac{1}{1+\gamma_c}$	$1-\dfrac{(1+\nu)(1+2\nu)}{6}\gamma^3$	$\delta_+ \sim \dfrac{(1+\nu)(1+2\nu)^2}{6}\gamma^4$
$R(t_0)$		$\dfrac{\exp\left[-\dfrac{(1+\nu)(1+2\nu)\gamma^2\lambda t_0}{2+(3+4\nu)\gamma+(3+6\nu+2\nu^2)\gamma^2}\right]}{1+\gamma_c}$	$\delta \sim \dfrac{3}{4}(1+\nu)(1+2\nu)\gamma^3$

Table 4.5.11 System of One Operating Unit and Two Stand-by Units: Partially Operating Stand-by, Unrestricted Repair

$$\gamma_c = \frac{(1+\nu)(1+2\nu)\gamma^3}{1 + (1+2\nu)\gamma + (1+3\nu + 2\nu^2)\gamma^2}, \qquad \gamma = \frac{\lambda}{\mu}.$$

MEASURE	EXACT VALUE	APPROXIMATE VALUE	ERROR
T_1	$\dfrac{1}{\lambda}\dfrac{1 + 2(1+\nu)\gamma + (3+6\nu+2\nu^2)\gamma^2}{(1+\nu)(1+2\nu)\gamma^2}$	$\dfrac{1}{\lambda}\dfrac{1}{(1+\nu)(1+2\nu)\gamma^2}$	$\delta_+ = \dfrac{1}{\lambda}\dfrac{2(1+\nu)+(3+6\nu+2\nu^2)\gamma}{(1+\nu)(1+2\nu)\gamma}$
T_2	$\dfrac{1}{\lambda}\dfrac{1 + (1+2\nu)\gamma + (1+3\nu+2\nu^2)\gamma^2}{(1+\nu)(1+2\nu)\gamma^2}$	$\dfrac{1}{\lambda}\dfrac{1}{(1+\nu)(1+2\nu)\gamma^2}$	$\delta_+ = \dfrac{1}{\lambda}\dfrac{(1+2\nu)+(1+3\nu+2\nu^2)\gamma}{(1+\nu)(1+2\nu)\gamma}$
τ	$\dfrac{1}{\mu}$	—	—
$P(t_0)$	—	$\exp\left[-\dfrac{(1+\nu)(1+2\nu)\gamma^2\lambda t_0}{1+2(1+\nu)\gamma+(3+6\nu+2\nu^2)\gamma^2}\right]$	$\delta \sim 2(1+\nu)(1+2\nu)\gamma^3$
K	$\dfrac{1}{1+\gamma_c}$	$1-(1+\nu)(1+2\nu)\gamma^3$	$\delta_+ \sim (1+\nu)(1+2\nu)\gamma^4$
$R(t_0)$	—	$\dfrac{\exp\left[-\dfrac{(1+\nu)(1+2\nu)\gamma^2\lambda t_0}{1+2(1+\nu)\gamma+(3+6\nu+2\nu^2)\gamma^2}\right]}{1+\gamma_c}$	$\delta \sim 2(1+\gamma)(1+2\nu)\gamma^3$

making use of the result of part 1 of this example:

$$\tilde{P}(t_0) = e^{-t_0/T_1} = e^{-t_0/543} \approx e^{-0.0018t_0}.$$

For example, for an initial time interval of length $t_0 = 24$ hours the probability of failure-free operation of the complex is approximately

$$\tilde{P}(t_0) = e^{-0.0018t_0} = e^{-0.0018 \cdot 24} \approx 0.958.$$

5.   The exact value of the readiness coefficient of the complex for the restricted-repair case is

$$K = \frac{1}{1 + \gamma_c},$$

where

$$\gamma_c = \frac{(1 + \nu)(1 + 2\nu)\gamma^3}{1 + (1 + 2\nu)\gamma + (1 + 3\nu + 2\nu^2)\gamma^2}.$$

First we determine the quantity γ_c:

$$\gamma_c = \frac{(1 + 0.5)(1 + 2 \cdot 0.5) \cdot (0.2)^3}{1 + (1 + 2 \cdot 0.5) \cdot 0.2 + [1 + 3 \cdot 0.5 + 2 \cdot (0.5)^2] \cdot (0.2)^2} \approx 0.016.$$

Then

$$K = \frac{1}{1 + \gamma_c} = \frac{1}{1 + 0.016} \approx 0.984.$$

6.   An approximate value for the reliability coefficient of the complex in the case of unrestricted repair is

$$\tilde{R}(t_0) = \frac{1}{1 + \gamma_c} e^{-t_0/T_1},$$

where

$$\gamma_c = \frac{(1 + \nu)(1 + 2\nu)\gamma^3}{3[2 + 2(1 + 2\nu)\gamma + (1 + 3\nu + 2\nu^2)\gamma^2]}.$$

First we determine the quantity γ_c:

$$\gamma_c = \frac{(1 + 0.5)(1 + 2 \cdot 0.5) \cdot (0.2)^3}{3\{2 + 2(1 + 2 \cdot 0.5) \cdot 0.2 + [1 + 3 \cdot 0.5 + 2 \cdot (0.5)^2] \cdot (0.2)^2\}} \approx 0.0027.$$

Then, using the result of part 4 of this example, we obtain

$$\tilde{R}(t_0) = \frac{1}{1 + \gamma_c} e^{-t_0/T_1} = \frac{1}{1 + 0.0027} e^{-0.0018t_0} \approx 0.997e^{-0.0018t_0}.$$

For a time interval of length $t_0 = 72$ hours, for example, the reliability coefficient of the complex is approximately

$$\tilde{R}(t_0) = 0.997e^{-0.0018 \cdot t_0} = 0.997e^{-0.0018 \cdot 72} \approx 0.88.$$

Table 4.5.12 System of One Operating Unit and Two
Stand-by Units: Nonoperating Stand-by,
Unrestricted Repair

$$\gamma_c = \frac{\gamma^3}{3(2 + 2\gamma + \gamma^2)},$$
$$\gamma = \frac{\lambda}{\mu}.$$

MEASURE	EXACT VALUE	APPROXIMATE VALUE	ERROR
T_1	$\dfrac{1}{\lambda}\dfrac{2 + 3\gamma + 3\gamma^2}{\gamma^2}$	$\dfrac{1}{\lambda}\dfrac{2}{\gamma^2}$	$\delta_+ = \dfrac{1}{\lambda}\dfrac{3(1 + \gamma)}{\gamma}$
T_2	$\dfrac{1}{\lambda}\dfrac{2 + 2\gamma + \gamma^2}{\gamma^2}$	$\dfrac{1}{\lambda}\dfrac{2}{\gamma^2}$	$\delta_+ = \dfrac{1}{\lambda}\dfrac{2 + \gamma}{\gamma}$
τ	$\dfrac{1}{3\mu}$	—	—
$P(t_0)$	—	$\exp\left(-\dfrac{\gamma^2\lambda t_0}{2 + 3\gamma + 3\gamma^2}\right)$	$\delta \sim \frac{3}{4}\gamma^3$
K	$\dfrac{1}{1 + \gamma_c}$	$1 - \dfrac{\gamma^3}{6}$	$\delta_+ \sim \dfrac{\gamma^4}{6}$
$R(t_0)$	—	$\dfrac{\exp\left(-\dfrac{\gamma^2\lambda t_0}{2 + 3\gamma + 3\gamma^2}\right)}{1 + \gamma_c}$	$\delta \sim \frac{3}{4}\gamma^3$

Example 4.5.7 Suppose we have a system of one operating device
and two stand-by devices in a nonoperating state. The failure intensity
of the operating device is $\lambda = 0.1\,\dfrac{1}{\text{hr}}$, the repair intensity of a failed device
is $\mu = 0.25\,\dfrac{1}{\text{hr}}$ and $\gamma = \lambda/\mu = 0.1/0.25 = 0.4$.

We are to determine the basic reliability measures of the system for
the cases of unrestricted and restricted repair.

Solution. From Tables 4.5.12 and 4.5.13 we find:

1. The exact value of the mean operating time of the system up to
failure for the case of unrestricted repair is

$$T_1 = \frac{1}{\lambda}\cdot\frac{2 + 3\gamma + 3\gamma^2}{\gamma^2} = \frac{1}{0.1}\cdot\frac{2 + 3\cdot 0.4 + 3\cdot(0.4)^2}{(0.4)^2} = 162.5 \quad \text{hr.}$$

Table 4.5.13 System of One Operating Unit and Two Stand-by Units: Nonoperating Stand-by, Restricted Repair

$$\gamma_c = \frac{\gamma^3}{1 + \gamma + \gamma^2},$$
$$\gamma = \frac{\lambda}{\mu}.$$

MEASURE	EXACT VALUE	APPROXIMATE VALUE	ERROR
T_1	$\dfrac{1}{\lambda}\dfrac{1 + 2\gamma + 3\gamma^2}{\gamma^2}$	$\dfrac{1}{\lambda}\dfrac{1}{\gamma^2}$	$\delta_+ = \dfrac{1}{\lambda}\dfrac{2 + 3\gamma}{\gamma}$
T_2	$\dfrac{1}{\lambda}\dfrac{1 + \gamma + \gamma^2}{\gamma^2}$	$\dfrac{1}{\lambda}\dfrac{1}{\gamma^2}$	$\delta_+ = \dfrac{1}{\lambda}\dfrac{1 + \gamma}{\gamma}$
τ	$\dfrac{1}{\mu}$	—	—
$P(t_0)$	—	$\exp\left(-\dfrac{\gamma^2\lambda t_0}{1 + 2\gamma + 3\gamma^2}\right)$	$\delta \sim 2\gamma^3$
K	$\dfrac{1}{1 + \gamma_c}$	$1 - \gamma^3$	$\delta_+ \sim \gamma^4$
$R(t_0)$	—	$\dfrac{\exp\left(-\dfrac{\gamma^2\lambda t_0}{1 + 2\gamma + 3\gamma^2}\right)}{1 + \gamma_c}$	$\delta \sim 2\gamma^3$

2. The exact value of the mean operating time of the system between failures in the restricted-repair case is

$$T_2 = \frac{1}{\lambda}\cdot\frac{1 + \gamma + \gamma^2}{\gamma^2} = \frac{1}{0.1}\cdot\frac{1 + 0.4 + (0.4)^2}{(0.4)^2} = 97.5 \quad \text{hr.}$$

3. The mean repair time of the system for the case of unrestricted and restricted repair are, respectively,

$$\tau = \frac{1}{3\mu} = \frac{1}{3\cdot 0.25} = 1.3 \quad \text{hr,}$$

and

$$\tau = \frac{1}{\mu} = \frac{1}{0.25} = 4 \quad \text{hr.}$$

4. An approximate value for the probability of initial failure-free operation of the system in the unrestricted-repair case is found by using the result of part 1 of this example:

$$\tilde{P}(t_0) = e^{-t_0/T_1} = e^{-t_0/162.5} = e^{-0.006 \cdot t_0}.$$

For an initial time interval of length $t_0 = 24$ hours, for example, the probability of failure-free operation of the system is approximately

$$\tilde{P}(t_0) = e^{-0.006 t_0} = e^{-0.006 \cdot 24} \approx 0.87.$$

5. The exact value of the readiness coefficient of the system for the case of restricted repair is

$$K = \frac{1}{1 + \gamma_c},$$

where

$$\gamma_c = \frac{\gamma^3}{1 + \gamma + \gamma^2}.$$

We determine the quantity γ_c:

$$\gamma_c = \frac{(0.4)^3}{1 + 0.4 + (0.4)^2} \approx 0.041.$$

Then

$$K = \frac{1}{1 + \gamma_c} = \frac{1}{1 + 0.041} \approx 0.961.$$

6. An approximate value for the reliability coefficient of the system for the unrestricted-repair case is

$$\tilde{R}(t_0) = \frac{1}{1 + \gamma_c} e^{-t_0/T_1},$$

where

$$\gamma_c = \frac{\gamma^3}{3(2 + 2\gamma + \gamma^2)}.$$

First we determine the quantity γ_c:

$$\gamma_c = \frac{(0.4)^3}{3[2 + 2 \cdot 0.4 + (0.4)^2]} = 0.0072.$$

Then, using the result of part 4 of this example, we obtain

$$\tilde{R}(t_0) = \frac{1}{1 + \gamma_c} e^{-t_0/T_1} = \frac{1}{1 + 0.0072} e^{-0.006 t_0} \approx 0.993 e^{-0.006 \cdot t_0}.$$

For example, for a time interval of length $t_0 = 10$ hours the reliability coefficient of the system is approximately

$$\tilde{R}(t_0) = 0.993 e^{-0.006 t_0} = 0.993 e^{-0.006 \cdot 10} \approx 0.94.$$

4.6   SYSTEMS WITH $n - 1$ STAND-BY UNITS

We investigate the reliability measures of a system of m operating units and $n - 1$ stand-by units (Figure 4.0.1). The case of a system with one operating unit and $n - 1$ stand-by units is treated separately (Figure 4.6.1). The over-all number of units in the system is $N = m + n - 1$.

Figure 4.6.1 is a diagram of the passages of a system with $n - 1$ stand-by units from state to state, where H_j is the state of the system wherein j of its $N = m + n - 1$ units are failed ($j = 0, 1, \ldots, n$); H_n is the failed state of the system; Λ_j is the intensity of passage of the system (failure of one of its units) from the state with j failed units to the state with one additional failed unit—that is, $j + 1$ ($j = 0, 1, \ldots, n - 1$); M_j is the intensity of passage of the system (repair of one of its units) from the state with j failed units to the state with one fewer failed units —that is, $j - 1$ ($j = 1, 2, \ldots, n$).

The reliability measures for a system with $n - 1$ stand-by units are given for the general case and three particular cases—namely, for operating, partially operating, and nonoperating stand-by units:

For operating stand-by units:

$$\Lambda_j = m\lambda + [n - (j + 1)]\lambda = (N - j)\lambda, \qquad 0 \le j \le n - 1;$$

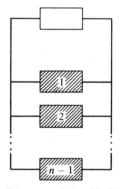

Figure 4.6.1   Block Diagram of the Model Used for Reliability Computations for a System of One Operating Unit and $n - 1$ Stand-by Units.

For partially operating stand-by units:

$$\Lambda_j = m\lambda + [n - (j + 1)]\nu\lambda, \qquad 0 \leq j \leq n - 1,$$

where ν is the load coefficient of the stand-by unit—that is, the failure intensity of the stand-by unit is $\nu\lambda$, $0 \leq \nu \leq 1$.

For nonoperating stand-by units:

$$\Lambda_j = m\lambda, \qquad 0 \leq j \leq n - 1.$$

For each of these three cases we consider only two modes of repair of failed units in the system:

(1)   unrestricted repair, where a failed unit is repaired immediately;

(2)   restricted repair, where not more than one failed unit can undergo repair at any moment of time.

Formally,

$$M_j = \begin{cases} j\mu & \text{if repair is unrestricted,} \\ \mu & \text{if repair is restricted.} \end{cases}$$

Tables 4.6.1–4.6.13 present exact and approximate values for the basic reliability measures of a system with $n - 1$ stand-by units, with error estimates. The approximate values of the basic reliability measures were obtained subject to the condition

$$\gamma = \frac{\lambda}{\mu} \ll \frac{1}{N}.$$

Next we give examples of computations of the reliability measures.

Example 4.6.1   Suppose we have a system of $m = 2$ operating units and $n - 1 = 3$ stand-by units. The system functions in such a way that when one, two, or three units fail, the load of the entire system is redistributed over the remaining operational units.

We assume that: the intensity of passage of the system (failure of one of its units) from the state with all of its units operational to the state with one failed unit is

$$\Lambda_0 = 5\lambda;$$

the intensity of passage of the system from the state with one failed unit to the state with two failed units is

$$\Lambda_1 = 8\lambda;$$

the intensity of passage of the system from the state with two failed units to the state with three failed units is

$$\Lambda_2 = 9\lambda;$$

and the intensity of passage of the system from the state with three failed units to the state with four failed units—that is, the failed state of the system—is

$$\Lambda_3 = 12\lambda.$$

In all the states of the system, it is assumed that only one failed unit can undergo repair at any one time—that is,

$$M_1 = M_2 = M_3 = M_4 = \mu.$$

The system is characterized by the following quantities:

$$\lambda = 0.01 \; \frac{1}{hr}, \qquad \mu = 0.2 \; \frac{1}{hr},$$

$$\gamma = \frac{\lambda}{\mu} = \frac{0.01}{0.2} = 0.05.$$

We wish to determine the basic reliability measures of this system.

Solution.   From Table 4.6.1 we find:

1.   The exact value of the mean operating time of the system up to failure is

$$T_1 = \sum_{s=0}^{n-1} \frac{\sum_{i=0}^{s} \Theta_i}{\Lambda_s \Theta_s} = \sum_{s=0}^{3} \frac{\sum_{i=0}^{s} \Theta_i}{\Lambda_s \Theta_s}$$

$$= \frac{\Theta_0}{\Lambda_0 \Theta_0} + \frac{1}{\Lambda_1 \Theta_1} \sum_{i=0}^{1} \Theta_i + \frac{1}{\Lambda_2 \Theta_2} \sum_{i=0}^{2} \Theta_i + \frac{1}{\Lambda_3 \Theta_3} \sum_{i=0}^{3} \Theta_i,$$

where

$$\Theta_i = \frac{\Lambda_0 \Lambda_1 \cdots \Lambda_{i-1}}{M_1 M_2 \cdots M_i}.$$

First we determine the quantity Θ_i:

$$\Theta_0 = 1,$$

$$\Theta_1 = \frac{\Lambda_0}{M_1} = \frac{5\lambda}{\mu} = 5\gamma = 5 \cdot 0.05 = 0.25,$$

$$\Theta_2 = \frac{\Lambda_0 \Lambda_1}{M_1 M_2} = \frac{5\lambda \cdot 8\lambda}{\mu^2} = 40\gamma^2 = 40 \cdot (0.05)^2 = 0.1,$$

$$\Theta_3 = \frac{\Lambda_0 \Lambda_1 \Lambda_2}{M_1 M_2 M_3} = \frac{5\lambda \cdot 8\lambda \cdot 9\lambda}{\mu^3} = 360\gamma^3 = 360 \cdot (0.05)^3 = 0.045.$$

Table 4.6.1   System with $n - 1$ Stand-by Units:
General Case

$$\Theta_i = \frac{\Lambda_0 \Lambda_1 \cdots \Lambda_{i-1}}{M_1 M_2 \cdots M_i}, \qquad \Theta_0 = 1,$$

$$\gamma_c = \frac{\Theta_n}{\displaystyle\sum_{i=0}^{n-1} \Theta_i}.$$

MEA-SURE	EXACT VALUE	APPROXIMATE VALUE	ERROR
T_1	$\displaystyle\sum_{s=0}^{n-1} \frac{\displaystyle\sum_{i=0}^{s} \Theta_i}{\Lambda_s \Theta_s}$	$\dfrac{1}{\Lambda_{n-1}\Theta_{n-1}}$	$\delta_+ \sim \dfrac{1}{\Lambda_{n-2}\Theta_{n-2}}\left(1 + \dfrac{\Lambda_0 M_{n-1}}{\Lambda_{n-1}M_1}\right)$
T_2	$\dfrac{\displaystyle\sum_{i=0}^{n-1} \Theta_i}{\Lambda_{n-1}\Theta_{n-1}}$	$\dfrac{1}{\Lambda_{n-1}\Theta_{n-1}}$	$\delta_+ \sim \dfrac{\Lambda_0}{\Lambda_{n-1}M_1\Theta_{n-1}}$
τ	$\dfrac{1}{M_n}$	—	—
$P(t_0)$	—	$\exp\left(-\dfrac{t_0}{\displaystyle\sum_{s=0}^{n-1} \frac{\displaystyle\sum_{i=0}^{s}\Theta_i}{\Lambda_s\Theta_s}}\right)$	$\delta \sim \Lambda_0 \displaystyle\prod_{i=1}^{n-1}\dfrac{\Lambda_i}{M_i}\sum_{j=1}^{n-1}\dfrac{1}{M_j}$
K	$\dfrac{1}{1+\gamma_c}$	$1 - \Theta_n$	$\delta_+ \sim \Theta_1 \Theta_n$
$R(t_0)$	—	$\dfrac{1}{1+\gamma_c}\exp\left(-\dfrac{t_0}{\displaystyle\sum_{s=0}^{n-1}\frac{\displaystyle\sum_{i=0}^{s}\Theta_i}{\Lambda_s\Theta_s}}\right)$	$\delta \sim \Lambda_0 \displaystyle\prod_{i=1}^{n-1}\dfrac{\Lambda_i}{M_i}\sum_{j=1}^{n-1}\dfrac{1}{M_j}$

Since

$$T_1 = \frac{1}{\Lambda_0} + \frac{1}{\Lambda_1 \Theta_1} (1 + \Theta_1) + \frac{1}{\Lambda_2 \Theta_2} (1 + \Theta_1 + \Theta_2)$$

$$+ \frac{1}{\Lambda_3 \Theta_3} (1 + \Theta_1 + \Theta_2 + \Theta_3)$$

$$= \frac{1}{5\lambda} + \frac{1 + \Theta_1}{8\lambda\Theta_1} + \frac{1 + \Theta_1 + \Theta_2}{9\lambda\Theta_2} + \frac{1 + \Theta_1 + \Theta_2 + \Theta_3}{12\lambda\Theta_3},$$

substituting the given numerical values we obtain

$$T_1 = \frac{1}{5 \cdot 0.01} + \frac{1 + 0.25}{8 \cdot 0.01 \cdot 0.25} + \frac{1 + 0.25 + 0.1}{9 \cdot 0.01 \cdot 0.1} + \frac{1 + 0.25 + 0.1 + 0.045}{12 \cdot 0.01 \cdot 0.045}$$

$$\approx 491 \quad \text{hr}.$$

2.   The exact value of the mean operating time of the system between failures is

$$T_2 = \frac{\sum\limits_{i=0}^{3} \Theta_i}{\Lambda_3 \Theta_3} = \frac{1 + \Theta_1 + \Theta_2 + \Theta_3}{\Lambda_3 \Theta_3}.$$

The quantities Θ_i were determined in part 1 of this example; therefore we can immediately write

$$T_2 = \frac{1 + \Theta_1 + \Theta_2 + \Theta_3}{\Lambda_3 \Theta_3} = \frac{1 + 5\gamma + 40\gamma^2 + 360\gamma^3}{12\lambda \cdot 360\gamma^3}$$

$$= \frac{1 + 5 \cdot 0.05 + 40 \cdot (0.05)^2 + 360 \cdot (0.05)^3}{12 \cdot 0.01 \cdot 360 \cdot (0.05)^3} \approx 258 \quad \text{hr}.$$

3.   The mean repair time of the system is

$$\tau = \frac{1}{M_4} = \frac{1}{\mu} = \frac{1}{0.2} = 5 \quad \text{hr}.$$

4.   By using the result of part 1 of this example, we find an approximate value for the probability of failure-free operation of the system for an initial time interval of length t_0 to be

$$\tilde{P}(t_0) = e^{-t_0/T_1} = e^{-t_0/491} \approx e^{-0.002t}.$$

Thus for an initial time interval of length $t_0 = 24$ hours, for example, the probability of failure-free operation of the system is approximately

$$\tilde{P}(t_0) = e^{-0.002t_0} = e^{-0.002 \cdot 24} = 0.95.$$

5.   An approximate value for the readiness coefficient of the system is

$$K = 1 - \Theta_n = 1 - \Theta_4,$$

where

$$\Theta_4 = \frac{\Lambda_0 \Lambda_1 \Lambda_2 \Lambda_3}{M_1 M_2 M_3 M_4}.$$

First we determine the quantity Θ_4:

$$\Theta_4 = \frac{5\lambda \cdot 8\lambda \cdot 9\lambda \cdot 12\lambda}{\mu^4} = 4320 \gamma^4.$$

Then

$$K = 1 - \Theta_4 = 1 - 4320\gamma^4 = 1 - 4320 \cdot (0.05)^4 = 0.973.$$

6.   An approximate value for the reliability coefficient is

$$\tilde{R}(t_0) = \frac{1}{1 + \gamma_c} e^{-t_0/T_1},$$

where

$$\gamma_c = \frac{\Theta_n}{\sum\limits_{i=0}^{n-1} \Theta_i} = \frac{\Theta_4}{1 + \Theta_1 + \Theta_2 + \Theta_3}.$$

Using the results obtained in parts 1, 3, 4, and 5 of this example, we find

$$\gamma_c = \frac{4320\gamma^4}{1 + 5\gamma + 40\gamma^2 + 360\gamma^3}$$

$$= \frac{4320 \cdot (0.05)^4}{1 + 5 \cdot 0.05 + 40 \cdot (0.05)^2 + 360 \cdot (0.05)^3} \approx 0.019.$$

Then

$$\tilde{R}(t_0) = \frac{1}{1 + \gamma_c} e^{-t_0/T_1} = \frac{1}{1 + 0.019} e^{-0.002 \cdot t_0} \approx 0.981 e^{-0.002 \cdot t_0}.$$

Thus, for example, the reliability coefficient of the system for a time interval of length $t_0 = 24$ hours is approximately

$$\tilde{R}(t_0) = 0.981 e^{-0.002 \cdot t_0} = 0.981 e^{-0.002 \cdot 24} \approx 0.93.$$

Example 4.6.2   Suppose we have a system of $m = 2$ operating units and $n - 1 = 4$ stand-by devices in an operating state. The failure intensity of each operating device is $\lambda = 0.05 \dfrac{1}{\text{hr}}$, the repair intensity of a failed device is $\mu = 0.5 \dfrac{1}{\text{hr}}$, and $\gamma = \lambda/\mu = 0.05/0.5 = 0.1$.

We want to determine the basic reliability measures of the system for the cases of unrestricted and restricted repair.

Solution.   From Tables 4.6.2 and 4.6.3 we find:

1.   The exact value of the mean operating time of the system up to failure for the unrestricted-repair case is

$$
T_1 = \frac{1}{N\lambda} \sum_{s=0}^{n-1} \frac{\sum_{i=0}^{s} C_N^i \gamma^i}{C_{N-1}^s \gamma^3} = \frac{1}{6\lambda} \sum_{s=0}^{4} \frac{\sum_{i=0}^{s} C_6^i \gamma^i}{C_5^s \gamma^s}
$$

$$
= \frac{1}{6\lambda} \left(1 + \frac{\sum_{i=0}^{1} C_6^i \gamma^i}{C_5^1 \gamma} + \frac{\sum_{i=0}^{2} C_6^i \gamma^i}{C_5^2 \gamma^2} + \frac{\sum_{i=0}^{3} C_6^i \gamma^i}{C_5^3 \gamma^3} + \frac{\sum_{i=0}^{4} C_6^i \gamma^i}{C_5^4 \gamma^4} \right)
$$

$$
= \frac{1}{6 \cdot 0.05} \left[1 + \frac{1 + C_6^1 \cdot 0.1}{C_5^1 \cdot 0.1} + \frac{1 + C_6^1 \cdot 0.1 + C_6^2 \cdot (0.1)^2}{C_5^2 (0.1)^2} \right.
$$

$$
+ \frac{1 + C_6^1 \cdot 0.1 + C_6^2 \cdot (0.1)^2 + C_6^3 \cdot (0.1)^3}{C_5^3 \cdot (0.1)^3}
$$

$$
\left. + \frac{1 + C_6^1 \cdot 0.1 + C_6^2 \cdot (0.1)^2 + C_6^3 \cdot (0.1)^3 + C_6^4 \cdot (0.1)^4}{C_5^4 \cdot (0.1)^4} \right]
$$

$$
\approx 12997 \quad \text{hr.}
$$

2.   An approximate value for the mean operating time of the system between failures for the restricted-repair case is

$$
\tilde{T}_2 = \frac{1}{\lambda} \cdot \frac{1}{n! C_N^n \gamma^{n-1}} = \frac{1}{\lambda} \cdot \frac{1}{5! C_6^5 \gamma^4}
$$

$$
= \frac{1}{0.05 \cdot 120 \cdot 6 \cdot (0.1)^4} \approx 278 \quad \text{hr.}
$$

3.   The mean repair times of the system for unrestricted and restricted repair are, respectively,

$$
\tau = \frac{1}{n\mu} = \frac{1}{5 \cdot 0.5} = 0.4 \quad \text{hr,}
$$

and

$$
\tau = \frac{1}{\mu} = \frac{1}{0.5} = 2 \quad \text{hr.}
$$

4.   By using the result of part 1 of this example, we find an approximate value for the probability of initial failure-free operation of the system for the case of unrestricted repair to be

$$
\tilde{P}(t_0) = e^{-t_0/T_1} = e^{-t_0/12,997} \approx e^{-0.00004 t_0}.
$$

Thus for an initial time of $t_0 = 100$ hours, for example, the probability of failure-free operation of the system is approximately

$$\tilde{P}(t_0) = e^{-0.00004t_0} = e^{-0.00004 \cdot 100} \approx 0.996.$$

Table 4.6.2   System of m Operating and $n-1$ Stand-by
Units: Operating Stand-by, Unrestricted Repair

$$N = m + (n-1), \qquad C_m^r = \frac{m!}{r!(m-r)!},$$

$$\gamma = \frac{\lambda}{\mu}, \qquad \gamma_c = \frac{C_N^m \gamma^n}{\displaystyle\sum_{i=0}^{n-1} C_N^i \gamma^i}.$$

MEA-SURE	EXACT VALUE	APPROXIMATE VALUE	ERROR
T_1	$\dfrac{1}{N\lambda} \displaystyle\sum_{s=0}^{s} \dfrac{\displaystyle\sum_{i=0}^{n-1} C_N^i \gamma^i}{C_{N-1}^s \gamma^s}$	$\dfrac{1}{m\lambda C_N^m \gamma^{n-1}}$	$\delta_+ \sim \dfrac{1 + \dfrac{N(n-1)}{m}}{(m+1)\lambda C_N^{m+1}\gamma^{n-2}}$
T_2	$\dfrac{1}{m\lambda \cdot C_N^m \gamma^{n-1}} \displaystyle\sum_{i=0}^{n-1} C_N^i \gamma^i$	$\dfrac{1}{m\lambda C_N^m \gamma^{n-1}}$	$\delta_+ \sim \dfrac{1}{\lambda C_{N-1}^{m-1}\gamma^{n-2}}$
τ	$\dfrac{1}{n\mu}$	—	—
$P(t_0)$	—	$\exp\left(-\dfrac{N\lambda t_0}{\displaystyle\sum_{s=0}^{s}\dfrac{\displaystyle\sum_{i=0}^{n-1}C_N^i\gamma^i}{C_{N-1}^s\gamma^s}}\right)$	$\delta \sim mC_N^m\gamma^n \displaystyle\sum_{j=1}^{n-1}\dfrac{1}{j}$
K	$\dfrac{1}{1+\gamma_c}$	$1 - C_N^n\gamma^n$	$\delta_+ \sim NC_N^n\gamma^{n+1}$
$R(t_0)$	—	$\dfrac{1}{1+\gamma_c}\exp\left(-\dfrac{N\lambda t_0}{\displaystyle\sum_{s=0}^{s}\dfrac{\displaystyle\sum_{i=0}^{n-1}C_N^i\gamma^i}{C_{N-1}^s\gamma^s}}\right)$	$\delta \sim mC_N^m\gamma^n \displaystyle\sum_{j=1}^{n-1}\dfrac{1}{j}$

Table 4.6.3   System of m Operating and $n − 1$ Stand-by Units: Operating Stand-by, Restricted Repair

$$N = m + (n - 1), \qquad C_m^r = \frac{m!}{r!(m-r)!}$$

$$\gamma = \frac{\lambda}{\mu}, \qquad \gamma_c = \frac{1}{\displaystyle\sum_{i=1}^{n} \dfrac{1}{i!\,C_{N-n+i}^i \gamma^i}}.$$

MEASURE	EXACT VALUE	APPROXIMATE VALUE	ERROR
T_1	$\dfrac{1}{\lambda}\displaystyle\sum_{s=0}^{n-1}\sum_{i=0}^{s}\dfrac{1}{(i+1)!\,C_{N-s+i}^{i+1}\gamma^i}$	$\dfrac{1}{\lambda}\dfrac{1}{n!\,C_N^n\gamma^{n-1}}$	$\delta_+ \sim \dfrac{1}{\lambda}\dfrac{N+m}{n!\,C_N^n\gamma^{n-2}}$
T_2	$\dfrac{1}{\lambda}\displaystyle\sum_{i=0}^{n-1}\dfrac{1}{(i+1)!\,C_{m+i}^{i+1}\gamma^i}$	$\dfrac{1}{\lambda}\dfrac{1}{n!\,C_N^n\gamma^{n-1}}$	$\delta_+ \sim \dfrac{1}{\lambda}\dfrac{1}{(n-1)!\,C_{N-1}^{n-1}\gamma^{n-2}}$
τ	$\dfrac{1}{\mu}$	—	—
$P(t_0)$	—	$\exp\left[-\dfrac{\lambda_0}{\displaystyle\sum_{s=0}^{n-1}\sum_{i=0}^{s}\dfrac{1}{(i+1)!\,C_{N-s+i}^{i+1}\gamma^i}}\right]$	$\delta \sim (n-1)n!\,C_N^m\gamma^n$
K	$\dfrac{1}{1+\gamma_c}$	$1 - n!\,C_N^n\gamma^n$	$\delta_+ \sim n!\,N C_N^m\gamma^{n+1}$
$R(t_0)$	—	$\dfrac{1}{1+\gamma_c}\exp\left[-\dfrac{\lambda_0}{\displaystyle\sum_{s=0}^{n-1}\sum_{i=0}^{s}\dfrac{1}{(i+1)!\,C_{N-s+i}^{i+1}\gamma^i}}\right]$	$\delta \sim (n-1)n!\,C_N^m\gamma^n$

5.   The exact value of the readiness coefficient of the system for the restricted-repair case is

$$K = \frac{1}{1 + \gamma_c},$$

where

$$\gamma_c = \frac{1}{\displaystyle\sum_{l=1}^{n} \frac{1}{i! C_{N-n+i}^i \gamma^i}}.$$

First we compute γ_c:

$$\gamma_c = \frac{1}{\displaystyle\sum_{i=1}^{5} \frac{1}{i! C_{1+i}^i \gamma^i}} = \frac{1}{\displaystyle\sum_{i=1}^{5} \frac{1}{(i+1)! \gamma^i}}$$

$$= \frac{1}{\dfrac{1}{2!0.1} + \dfrac{1}{3!(0.1)^2} + \dfrac{1}{4!(0.1)^3} + \dfrac{1}{5!(0.1)^4} + \dfrac{1}{6!(0.1)^5}}$$

$$\approx 0.0035.$$

Then

$$K = \frac{1}{1 + \gamma_c} = \frac{1}{1 + 0.0035} \approx 0.9965.$$

6.   An approximate value for the reliability coefficient in the case of unrestricted repair is

$$\tilde{R}(t_0) = \frac{1}{1 + \gamma_c} e^{-t_0/T_1},$$

where the quantity T_1 was determined in part 1 of this example, and

$$\gamma_c = \frac{C_N^n \gamma^n}{\displaystyle\sum_{i=0}^{n-1} C_N^i \gamma^i} = \frac{C_6^5 \gamma^5}{\displaystyle\sum_{i=0}^{4} C_6^i \gamma^i}.$$

First we determine the denominator of this fraction:

$$\sum_{i=0}^{4} C_6^i \gamma^i = 1 + C_6^1 \gamma + C_6^2 \gamma^2 + C_6^3 \gamma^3 + C_6^4 \gamma^4$$

$$= 1 + 6 \cdot 0.1 + 15 \cdot (0.1)^2 + 20 \cdot (0.1)^3 + 15 \cdot (0.1)^4 = 1.7715.$$

Then

$$\gamma_c = \frac{C_6^5 \gamma^5}{\displaystyle\sum_{i=0}^{4} C_6^i \gamma^i} = \frac{6 \cdot (0.1)^5}{1.7715} \approx 0.00003.$$

Finally,

$$\tilde{R}(t_0) = \frac{1}{1 + \gamma_c} e^{-t_0/T_1} = \frac{1}{1 + 0.00003} e^{-t_0/12,997} = 0.99997 e^{-0.00004 t_0}.$$

Thus, for example, for a time interval of length $t_0 = 100$ hours the reliability coefficient is approximately

$$\tilde{R}(t_0) = 0.99997 e^{-0.00004 t_0} = 0.99997 e^{-0.00004 \cdot 1000} \approx 0.96.$$

Example 4.6.3 Suppose a group of $m = 5$ devices has $n - 1 = 3$ stand-by devices in a partially operating state with load coefficient $\nu = 0.1$. The failure intensity of each operating device is $\lambda = 0.05 \dfrac{1}{\text{hr}}$, the repair intensity of a failed device is $\mu = 0.5 \dfrac{1}{\text{hr}}$, and $\gamma = \lambda/\mu = 0.05/0.5 = 0.1$.

We wish to determine the basic reliability coefficients of the system for the cases of unrestricted and restricted repair.

Solution. From Tables 4.6.4 and 4.6.5 we find:

1. The exact value of the mean operating time of the system up to failure in the restricted-repair case is

$$
T_1 = \frac{1}{\lambda} \cdot \sum_{s=0}^{n-1} \frac{\displaystyle\sum_{i=0}^{s} \gamma^i \prod_{l=1}^{i} [m + (n-l)\nu]}{\gamma^s \displaystyle\prod_{l=1}^{s+1} [m + (n-l)\nu]}
$$

$$
= \frac{1}{\lambda} \sum_{s=0}^{3} \frac{\displaystyle\sum_{i=0}^{s} \gamma^i \prod_{l=1}^{i} [5 + (4-l)\nu]}{\gamma^s \displaystyle\prod_{l=1}^{s+1} [5 + (4-l)\nu]}
$$

$$
= \frac{1}{\lambda} \left\{ \frac{1}{[5 + (4-1)\nu]} + \frac{\displaystyle\sum_{i=0}^{1} \gamma^i \prod_{l=1}^{i} [5 + (4-l)\nu]}{\gamma \displaystyle\prod_{l=1}^{2} [5 + (4-l)\nu]} \right.
$$

$$
+ \frac{\displaystyle\sum_{i=0}^{2} \gamma^i \prod_{l=1}^{i} [5 + (4-l)\nu]}{\gamma^2 \displaystyle\prod_{l=1}^{3} [5 + (4-l)\nu]} + \left. \frac{\displaystyle\sum_{i=0}^{3} \gamma^i \prod_{l=1}^{i} [5 + (4-l)\nu]}{\gamma^3 \displaystyle\prod_{l=1}^{4} [5 + (4-l)\nu]} \right\}.
$$

First we compute the sums involved in this expression:

$$\sum_{i=0}^{1} \gamma^i \prod_{l=1}^{i} [5 + (4 - l)\nu] = 1 + \gamma \prod_{l=1}^{1} [5 + (4 - l)\nu] = 1 + \gamma(5 + 3\nu)$$
$$= 1 + 0.1(5 + 3 \cdot 0.1) = 1.53,$$

$$\sum_{i=0}^{2} \gamma^i \prod_{l=1}^{i} [5 + (4 - l)\nu] = 1 + \gamma(5 + 3\nu) + \gamma^2(5 + 3\nu)(5 + 2\nu)$$
$$= 1 + 0.1 \cdot (5 + 3 \cdot 0.1)$$
$$+ (0.1)^2 \cdot (5 + 3 \cdot 0.1)(5 + 2 \cdot 0.1) = 1.8056,$$

$$\sum_{i=0}^{3} \gamma^i \prod_{l=1}^{i} [5 + (4 - l)\nu] = 1 + \gamma(5 + 3\nu) + \gamma^2(5 + 3\nu)(5 + 2\nu)$$
$$+ \gamma^3(5 + 3\nu)(5 + 2\nu)(5 + \nu)$$
$$= 1 + 0.1(5 + 3 \cdot 0.1)$$
$$+ (0.1)^2(5 + 3 \cdot 0.1)(5 + 2 \cdot 0.1)$$
$$+ (0.1)^3(5 + 3 \cdot 0.1)(5 + 2 \cdot 0.1) \cdot (5 + 0.1)$$
$$= 1.946156.$$

Then

$$T_1 = \frac{1}{0.05} \left\{ \frac{1}{5 + 3 \cdot 0.1} + \frac{1.53}{0.1 \cdot (5 + 3 \cdot 0.1)(5 + 2 \cdot 0.1)} \right.$$
$$+ \frac{1.8056}{(0.1)^2(5 + 3 \cdot 0.1)(5 + 2 \cdot 0.1)(5 + 0.1)}$$
$$\left. + \frac{1.946156}{(0.1)^3(5 + 3 \cdot 0.1)(5 + 2 \cdot 0.1)(5 + 0.1) \cdot 5} \right\}$$
$$\approx 96 \quad \text{hr.}$$

2.   The exact value of the mean operating time of the system between failures for the case of restricted repair is

$$T_2 = \frac{1}{\lambda} \frac{\sum_{i=0}^{n-1} \gamma^i \prod_{l=1}^{i} [m + (n - l)\nu]}{\gamma^{n-1} \prod_{l=1}^{n} [m + (n - l)\nu]}$$

$$= \frac{1}{0.05} \cdot \frac{\sum_{i=0}^{3} (0.1)^i \prod_{l=1}^{i} [5 + (4 - l) \cdot 0.1]}{(0.1)^2 \prod_{l=1}^{4} [5 + (4 - l)0.1]}.$$

Table 4.6.4 System of m Operating and $n-1$ Stand-by Units: Partially Operating Stand-by, Unrestricted Repair

$$\gamma_c = \frac{\dfrac{\gamma^n}{n!}\prod_{l=1}^{n}[m+(n-l)\nu]}{\sum_{i=0}^{n-1}\dfrac{\gamma^i}{i!}\prod_{l=1}^{i}[m+(n-l)\nu]},$$

$$\gamma = \frac{\lambda}{\mu}, \qquad \prod_{l=1}^{0}[m+(n-l)\nu] = 1.$$

MEASURE	EXACT VALUE	APPROXIMATE VALUE	ERROR
T_1	$\dfrac{1}{\lambda}\sum_{s=0}^{n-1}\dfrac{1}{s!}\sum_{i=0}^{s+1}\dfrac{\gamma^i}{i!}\prod_{l=1}^{i}[m+(n-l)\nu]\cdot\gamma^s\prod_{l=1}^{s+1}[m+(n-l)\nu]$	$\dfrac{1}{\lambda}\dfrac{(n-1)!}{\gamma^{n-1}}\dfrac{1}{\prod_{l=1}^{n}[m+(n-l)\nu]}$	$\delta_+ \sim \dfrac{1}{\lambda}\dfrac{(n-2)!}{\gamma^{n-2}}\dfrac{nm+(n-1)^2\nu}{\prod_{l=1}^{n}[m+(n-l)\nu]}$
T_2	$\dfrac{1}{\lambda}\dfrac{(n-1)!}{\gamma^{n-1}}\dfrac{\sum_{i=0}^{n-1}\dfrac{\gamma^i}{i!}\prod_{l=1}^{i}[m+(n-l)\nu]}{\prod_{l=1}^{n}[m+(n-l)\nu]}$	$\dfrac{1}{\lambda}\dfrac{(n-1)!}{\gamma^{n-1}}\dfrac{1}{\prod_{l=1}^{n}[m+(n-l)\nu]}$	$\delta_+ \sim \dfrac{1}{\lambda}\dfrac{(n-1)!}{\gamma^{n-2}}\dfrac{m+(n-1)\nu}{\prod_{l=1}^{n}[m+(n-l)\nu]}$

τ	$\dfrac{1}{n\mu}$	—	—
$P(t_0)$	—	e^{-t_0/T_1}	$\delta \sim [m+(n-1)\nu]\gamma^n \times \displaystyle\sum_{j=1}^{n-1}\frac{1}{j}\prod_{i=1}^{n-1}[m+(n-1-i)\nu]^i$
K	$\dfrac{1}{1+\gamma_c}$	$1 - \dfrac{\gamma^n}{n!}\displaystyle\prod_{l=1}^{n}[m+(n-l)\nu]$	$\delta_+ \sim \dfrac{\gamma^{n+1}}{n!}[m+(n-1)\nu]\displaystyle\prod_{l=1}^{n}[m+(n-l)\nu]$
$R(t_0)$	—	$\dfrac{1}{1+\gamma_c}\,e^{-t_0/T_1}$	$\delta \sim [m+(n-1)\nu]\gamma^n \times \displaystyle\sum_{j=1}^{n-1}\frac{1}{j}\prod_{i=1}^{n-1}[m+(n-1-i)\nu]^i$

Table 4.6.5 System of m Operating and $n-1$ Stand-by Units: Partially Operating Stand-by, Restricted Repair

$$\gamma_c = \frac{\gamma^n \prod_{l=1}^{n}[m+(n-l)\nu]}{\sum_{i=0}^{n-1}\gamma^i \prod_{l=1}^{i}[m+(n-l)\nu]},$$

$$\gamma = \frac{\lambda}{\mu}, \qquad \prod_{l=1}^{0}[m+(n-l)\nu]=1.$$

MEASURE	EXACT VALUE	APPROXIMATE VALUE	ERROR
T_1	$\dfrac{1}{\lambda}\displaystyle\sum_{s=0}^{n-1}\dfrac{\displaystyle\sum_{i=0}^{s}\gamma^i \prod_{l=1}^{i}[m+(n-l)\nu]}{\gamma^s \prod_{l=1}^{s+1}[m+(n-l)\nu]}$	$\dfrac{1}{\lambda}\dfrac{1}{\gamma^{n-1}\prod_{l=1}^{n}[m+(n-l)\nu]}$	$\delta_+ \sim \dfrac{1}{\lambda}\dfrac{2m+(n-1)\nu}{\gamma^n \prod_{l=1}^{n}[m+(n-l)\nu]}$
T_2	$\dfrac{1}{\lambda}\dfrac{\displaystyle\sum_{i=0}^{n-1}\gamma^i \prod_{l=1}^{i}[m+(n-l)\nu]}{\gamma^{n-1}\prod_{l=1}^{n}[m+(n-l)\nu]}$	$\dfrac{1}{\lambda}\dfrac{1}{\gamma^{n-1}\prod_{l=1}[m+(n-l)\nu]}$	$\delta_+ \sim \dfrac{1}{\lambda}\dfrac{m+(n-1)\nu}{\gamma^{n-2}\prod_{l=1}^{n}[m+(n-l)\nu]}$

τ	$\dfrac{1}{\mu}$	—	—
$P(t_0)$	—	e^{-t_0/T_1}	$\delta \sim (n-1)\gamma^n \displaystyle\prod_{i=0}^{n-1}[m+(n-1-i)\nu]$
K	$\dfrac{1}{1+\gamma_c}$	$1 - \gamma^n \displaystyle\prod_{l=1}^{n}[m+(n-l)\nu]$	$\delta_+ \sim \gamma^{n+1}[m+(n-1)\nu]\displaystyle\prod_{l=1}^{n}[m+(n-l)\nu]$
$R(t_0)$	—	$\dfrac{1}{1+\gamma_c}\,e^{-t_0/T_1}$	$\delta \sim (n-1)\gamma^n \displaystyle\prod_{i=0}^{n-1}[m+(n-1-i)\nu]$

First we compute the numerator of this fraction:

$$\sum_{i=0}^{3} (0.1)^i \prod_{l=1}^{i} [5 + (4 - l)0.1] = 1 + 0.1[5 + (4 - 1)0.1]$$

$$+ (0.1)^2 \prod_{l=1}^{2} [5 + (4 - l)0.1]$$

$$+ (0.1)^3 \prod_{l=1}^{3} [5 + (4 - l)0.1]$$

$$\approx 1.846.$$

In the denominator we obtain

$$\prod_{l=1}^{4} [5 + (4 - l) \cdot 0.1] = (5 + 0.3)(5 + 0.2)(5 + 0.1) \cdot 5 = 702.78.$$

Finally, we obtain

$$T_2 = \frac{1}{0.05} \cdot \frac{1.846}{(0.1)^3 \cdot 702.78} = 52.7 \quad \text{hr.}$$

3. The mean repair time of the system for the unrestricted-repair case is

$$\tau = \frac{1}{n\mu} = \frac{1}{4 \cdot 0.5} = 0.5 \quad \text{hr.}$$

4. By using the result obtained in part 1 of this example, we find an approximate value for the probability of initial failure-free operation of the system for the case of restricted repair to be

$$\tilde{P}(t_0) = e^{-t_0/T_1} = e^{-t_0/96} = e^{-0.0104t_0}.$$

Thus for an initial time of $t_0 = 24$ hours, for example, the probability of failure-free operation is approximately

$$\tilde{P}(t_0) = e^{-0.0101t_0} = e^{-0.0101 \cdot 24} \approx 0.78.$$

5. The exact value of the readiness coefficient of the system for the restricted-repair case is

$$K = \frac{1}{1 + \gamma_c} = \frac{T_2}{T_2 + \tau}.$$

The quantity T_2 for the case of restricted repair was computed in part 2 of this example to be

$$T_2 = 52.7 \quad \text{hr.}$$

The mean repair time of the system in the restricted-repair case is

$$\tau = \frac{1}{\mu} = \frac{1}{0.5} = 2 \quad \text{hr.}$$

Then

$$K = \frac{T_2}{T_2 + \tau} = \frac{52.7}{52.7 + 2} \approx 0.963.$$

An approximate value of the readiness coefficient is

$$\tilde{K} = 1 - \gamma^n \prod_{l=1}^{n} [m + (n - l)\nu]$$

$$= 1 - (0.1)^4 \prod_{l=1}^{4} [5 + (4 - l)0.1]$$

$$\approx 0.930.$$

6.   An approximate value for the reliability coefficient of the system in the case of restricted repair is

$$\tilde{R}(t_0) = Ke^{-t_0/T_1}.$$

The quantity K was determined in part 5 of this example to be

$$K = 0.963.$$

The quantity T_1 for the restricted-repair case was computed in part 1 of this example to be

$$T_1 = 96 \quad \text{hr.}$$

Then

$$\tilde{R}(t_0) = 0.963e^{-t_0/96} = 0.963e^{-0.0104t_0}.$$

For example, for a time interval of length $t_0 = 24$ hours the reliability coefficient of the system is approximately

$$\tilde{R}(t_0) = 0.963e^{-0.0104t_0} = 0.963e^{-0.0104 \cdot 24} \approx 0.75.$$

Example 4.6.4   Suppose we have a system of $m = 3$ operating devices and $n - 1 = 3$ stand-by devices in a nonoperating state. The failure intensity of each operating device is $\lambda = 0.1 \dfrac{1}{\text{hr}}$, the repair intensity of a failed device is $\mu = 1 \dfrac{1}{\text{hr}}$, and $\gamma = \lambda/\mu = 0.1$.

We wish to determine the basic reliability measures of the system for the unrestricted- and restricted-repair cases.

Solution. From Tables 4.6.6 and 4.6.7 we find:

1. The exact value of the mean operating time of the system up to failure in the case of unrestricted repair is

$$
\begin{aligned}
T_1 &= \frac{1}{m\lambda} \sum_{i=0}^{n-1} C_n^{i+1} \frac{i!}{(m\gamma)^i} \\
&= \frac{1}{3\lambda} \sum_{i=0}^{3} C_4^{i+1} \frac{i!}{(3\gamma)^i} \\
&= \frac{1}{3\lambda} \left[4 + 6 \cdot \frac{1}{3\gamma} + 4 \cdot \frac{2!}{(3\gamma)^2} + \frac{3!}{(3\gamma)^3} \right] \\
&= \frac{1}{3 \cdot 0.1} \left(4 + \frac{2}{0.1} + \frac{8}{9 \cdot 0.01} + \frac{2}{9 \cdot 0.001} \right) \quad \text{hr} \\
&\approx 1117 \quad \text{hr.}
\end{aligned}
$$

2. The exact value of the mean operating time of the system between failures for the restricted-repair case is

$$
T_2 = \frac{1}{m\lambda} \cdot \frac{1 - (m\gamma)^n}{(m\gamma)^{n-1} \cdot (1 - m\gamma)} = \frac{1}{3 \cdot 0.1} \cdot \frac{1 - (3 \cdot 0.1)^4}{(3 \cdot 0.1)^3 \cdot (1 - 3 \cdot 0.1)} \approx 175 \quad \text{hr.}
$$

3. The mean repair time of the system for restricted repair is

$$
\tau = \frac{1}{\mu} = \frac{1}{1} = 1 \quad \text{hr.}
$$

4. By using the result obtained in part 1 of this example, we find an approximate value for the probability of initial failure-free operation of the system in the case of unrestricted repair to be

$$
\tilde{P}(t_0) = e^{-t_0/T_1} = e^{-t_0/1117} = e^{-0.0009 t_0}.
$$

For an initial time of $t_0 = 100$ hours, for example, the probability of failure-free operation of the system is approximately

$$
\tilde{P}(t_0) = e^{-0.0009 t_0} = e^{-0.0009 \cdot 1000} \approx 0.41.
$$

5. The exact value of the readiness coefficient of the system in the case of restricted repair is

$$
K = \frac{1}{1 + \gamma_c} = \frac{T_2}{T_2 + \tau}.
$$

Making use of the results obtained in parts 2 and 3 of this example, we find

$$
K = \frac{T_2}{T_2 + \tau} = \frac{175}{175 + 1} \approx 0.994.
$$

Table 4.6.6 System of m Operating and $n-1$ Stand-by Units: Nonoperating Stand-by, Unrestricted Repair

$$\gamma_c = \frac{1}{\sum_{i=1}^{n} C_n^i \dfrac{i!}{(m\gamma)^i}}, \qquad \gamma = \frac{\lambda}{\mu}, \qquad C_m^r = \frac{m!}{r!(m-r)!}.$$

MEASURE	EXACT VALUE	APPROXIMATE VALUE	ERROR
T_1	$\dfrac{1}{m\lambda} \displaystyle\sum_{i=0}^{n-1} C_n^{i+1} \dfrac{i!}{(m\gamma)^i}$	$\dfrac{1}{m\lambda} \dfrac{(n-1)!}{(m\gamma)^{n-1}}$	$\delta_+ \sim \dfrac{1}{m\lambda} \dfrac{n!}{(n-1)(m\gamma)^{n-2}}$
T_2	$\dfrac{1}{m\lambda} \displaystyle\sum_{i=0}^{n-1} C_{n-1}^{i} \dfrac{i!}{(m\gamma)^i}$	$\dfrac{1}{m\lambda} \dfrac{(n-1)!}{(m\gamma)^{n-1}}$	$\delta_+ \sim \dfrac{1}{m\lambda} \dfrac{(n-1)!}{(m\gamma)^{n-2}}$
τ	$\dfrac{1}{n\mu}$	—	—
$P(t_0)$	—	$\exp\left[-\dfrac{m\lambda t_0}{\displaystyle\sum_{i=0}^{n-1} C_n^{i+1} \dfrac{i!}{(m\gamma)^i}}\right]$	$\delta \sim \dfrac{(m\gamma)^n}{(n-1)!} \displaystyle\sum_{j=1}^{n-1} \dfrac{1}{j}$
K	$\dfrac{1}{1+\gamma_c}$	$1 - \dfrac{(m\gamma)^n}{n!}$	$\delta_+ \sim \dfrac{(m\gamma)^{n+1}}{n!}$
$R(t_0)$	—	$\dfrac{1}{1+\gamma_c} \exp\left[-\dfrac{m\lambda t_0}{\displaystyle\sum_{i=0}^{n-1} C_n^{i+1} \dfrac{i!}{(m\gamma)^i}}\right]$	$\delta \sim \dfrac{(m\gamma)^n}{(n-1)!} \displaystyle\sum_{j=1}^{n-1} \dfrac{1}{j}$

Table 4.6.7 System of m Operating and $n - 1$ Stand-by Units: Nonoperating Stand-by, Restricted Repair

$$\gamma_c = \frac{(m\gamma)^n(1-m\gamma)}{1-(m\gamma)^n},$$
$$\gamma = \frac{\lambda}{\mu}.$$

MEASURE	EXACT VALUE	APPROXIMATE VALUE	ERROR
T_1	$\dfrac{1}{m\lambda}\dfrac{1-(m\gamma)^n[1+n(1-m\gamma)]}{(m\gamma)^{n-1}(1-m\gamma)^2}$	$\dfrac{1}{m\lambda}\dfrac{1}{(m\gamma)^{n-1}}$	$\delta_+ \sim \dfrac{2}{m\lambda(m\gamma)^{n-2}}$
T_2	$\dfrac{1}{m\lambda}\dfrac{1-(m\gamma)^n}{(m\gamma)^{n-1}(1-m\gamma)}$	$\dfrac{1}{m\lambda}\dfrac{1}{(m\gamma)^{n-1}}$	$\delta_+ \sim \dfrac{1}{m\lambda(m\gamma)^{n-2}}$
τ	$\dfrac{1}{\mu}$	—	—
$P(t_0)$	—	$\exp\left\{-\dfrac{(m\gamma)^{n-1}(1-m\gamma)^2 m\lambda t_0}{1-(m\gamma)^n[1+n(1-m\gamma)]}\right\}$	$\delta \sim (n-1)(m\gamma)^n$
K	$\dfrac{1}{1+\gamma_c}$	$1-(m\gamma)^n$	$\delta_+ \sim (m\gamma)^{n+1}$
$R(t_0)$	—	$\dfrac{\exp\left\{-\dfrac{(m\gamma)^{n-1}(1-m\gamma)^2 m\lambda t_0}{1-(m\gamma)^n[1+n(1-m\gamma)]}\right\}}{1+\gamma_c}$	$\delta \sim (n-1)(m\gamma)^n$

6. An approximate value for the reliability coefficient for the unrestricted-repair case is

$$R(t_0) = \frac{1}{1 + \gamma_c}\, e^{-t_0/T_1},$$

where T_1 was determined in part 1 of this example and

$$\gamma_c = \frac{1}{\displaystyle\sum_{i=1}^{n} C_n^i \frac{i!}{(m\gamma)^i}} = \frac{1}{\displaystyle\sum_{i=1}^{4} C_4^i \frac{i!}{(3\gamma)^i}}.$$

First we find the denominator of this fraction:

$$\sum_{i=1}^{4} C_4^i \frac{i!}{(3\gamma)^i} = C_4^1 \frac{1}{3\gamma} + C_4^2 \frac{2!}{(3\gamma)^2} + C_4^3 \frac{3!}{(3\gamma)^3} + C_4^4 \frac{4!}{(3\gamma)^4}$$

$$= 4 \cdot \frac{1}{3 \cdot 0.1} + \frac{6 \cdot 2}{9 \cdot (0.1)^2} + \frac{4 \cdot 6}{27 \cdot (0.1)^3} + \frac{24}{81 \cdot (0.1)^4}$$

$$\approx 4000.$$

Then

$$\gamma_c = \frac{1}{4000} = 0.00025.$$

In this case,

$$\tilde{R}(t_0) = \frac{1}{1 + \gamma_c}\, e^{-t_0/T_1} = \frac{1}{1 + 0.00025} \cdot e^{-t_0/1117} = 0.99975 \cdot e^{-0.0004 t_0}.$$

Thus, for example, the reliability coefficient for a time interval of length $t_0 = 1000$ hours is approximately

$$\tilde{R}(t_0) = 0.99975 e^{-0.0004 t_0} = 0.99975 e^{-0.0004 \cdot 1000} \approx 0.41.$$

Example 4.6.5 Suppose we have a system of one operating device and $n - 1 = 3$ stand-by devices in an operating state. The failure intensity of each operating device is $\lambda = 0.05\, \dfrac{1}{\text{hr}}$, the repair intensity of a failed device is $\mu = 0.5\, \dfrac{1}{\text{hr}}$, and $\gamma = \lambda/\mu = 0.05/0.5 = 0.1$.

We wish to determine the basic reliability measures of the system for the cases of unrestricted and restricted repair.

Solution.   From Tables 4.6.8 and 4.6.9 we find:

1.   The exact value of the mean operating time of the system up to failure in the restricted-repair case is

$$
T_1 = \frac{1}{\lambda} \sum_{s=0}^{n-1} \sum_{i=0}^{s} \frac{1}{(i+1)!\, C_{n-s+i}^{i+1} \gamma^i} = \frac{1}{\lambda} \sum_{s=0}^{3} \sum_{i=0}^{s} \frac{1}{(i+1)!\, C_{4-s+i}^{i+1} \gamma^i}
$$

$$
= \frac{1}{\lambda} \left\{ \frac{1}{4} + \sum_{i=0}^{1} \frac{1}{(i+1)!\, C_{3+i}^{i+1}\gamma^i} + \sum_{i=0}^{2} \frac{1}{(i+1)!\, C_{2+i}^{i+1}\gamma^i} \right.
$$

$$
\left. + \sum_{i=0}^{3} \frac{1}{(i+1)!\, C_{1+i}^{i+1}\gamma^i} \right\}
$$

$$
= \frac{1}{\lambda} \left\{ \frac{1}{4} + \sum_{i=0}^{1} \frac{2}{(3+i)!\,\gamma^i} + \sum_{i=0}^{2} \frac{1}{(2+i)!\,\gamma^i} + \sum_{i=0}^{3} \frac{1}{(i+1)!\,\gamma^i} \right\}
$$

$$
= \frac{1}{\lambda} \left\{ \frac{1}{4} + \left(\frac{2}{3!} + \frac{2}{4!\gamma} \right) + \left(\frac{1}{2!} + \frac{1}{3!\gamma} + \frac{1}{4!\gamma^2} \right) \right.
$$

$$
\left. + \left(1 + \frac{1}{2!\gamma} + \frac{1}{3!\gamma^2} + \frac{1}{4!\gamma^3} \right) \right\}
$$

$$
= \frac{1}{\lambda} \left\{ \frac{1}{4} + \frac{1}{3} + \frac{1}{12\gamma} + \frac{1}{2} + \frac{1}{6\gamma} + \frac{1}{24\gamma^2} + 1 + \frac{1}{2\gamma} + \frac{1}{6\gamma^2} + \frac{1}{24\gamma^3} \right\}
$$

$$
= \frac{1}{\lambda} \left[\frac{25}{12} + \frac{3}{4\cdot\gamma} + \frac{5}{24\gamma^2} + \frac{1}{24\gamma^3} \right]
$$

$$
= \frac{1}{0.05} \left[\frac{25}{12} + \frac{3}{4\cdot 0.1} + \frac{5}{24\cdot(0.1)^2} + \frac{1}{24\cdot(0.1)^3} \right] \approx 1442 \quad \text{hr.}
$$

2.   The exact value of the mean operating time of the system between failures for the case of restricted repair is

$$
T_2 = \frac{1}{\lambda} \sum_{i=0}^{n-1} \frac{1}{(i+1)!\,\gamma^i} = \frac{1}{0.05} \sum_{i=0}^{3} \frac{1}{(i+1)!\,(0.1)^i} = 1287 \quad \text{hr.}
$$

3.   The mean repair times of the system for the unrestricted- and restricted-repair cases are, respectively,

$$
\tau = \frac{1}{n\mu} = \frac{1}{4\cdot 0.5} = 0.5 \quad \text{hr,}
$$

Table 4.6.8 System of One Operating Unit and $n - 1$ Stand-by Units: Operating Stand-by, Unrestricted Repair

$$\gamma = \frac{\lambda}{\mu}, \qquad C_m^r = \frac{m!}{r!(m-r)!},$$

$$\gamma_c = \frac{1}{\displaystyle\sum_{i=1}^{n} \frac{C_n^i}{\gamma^i}}.$$

MEA-SURE	EXACT VALUE	APPROXIMATE VALUE	ERROR
T_1	$\dfrac{1}{n\lambda} \displaystyle\sum_{s=0}^{n-1} \dfrac{\displaystyle\sum_{i=0}^{s} C_n^i \gamma^i}{C_{n-1}^s \gamma^s}$	$\dfrac{1}{n\lambda\gamma^{n-1}}$	$\delta_+ \sim \dfrac{1 + n(n-1)}{\lambda(n-1)n\gamma^{n-2}}$
T_2	$\dfrac{1}{n\lambda\gamma^{n+1}} \displaystyle\sum_{i=0}^{n-1} C_n^i \gamma^i$	$\dfrac{1}{n\lambda\gamma^{n-1}}$	$\delta_+ \sim \dfrac{1}{\lambda\gamma^{n-2}}$
τ	$\dfrac{1}{n\mu}$	—	—
$P(t_0)$	—	$\exp\left(- \dfrac{n\lambda t_0}{\displaystyle\sum_{s=0}^{n-1} \dfrac{\displaystyle\sum_{i=0}^{s} C_n^i \gamma^i}{C_{n-1}^s \gamma^s}} \right)$	$\delta \sim n\gamma^n \displaystyle\sum_{j=1}^{n-1} \dfrac{1}{j}$
K	$\dfrac{1}{1 + \gamma_c}$	$1 - \gamma^n$	$\delta_+ \sim n\gamma^{n+1}$
$R(t_0)$	—	$\dfrac{1}{1 + \gamma_c} \exp\left(- \dfrac{n\lambda t_0}{\displaystyle\sum_{s=0}^{n-1} \dfrac{\displaystyle\sum_{i=0}^{s} C_n^i \gamma^i}{C_{n-1}^s \gamma^s}} \right)$	$\delta \sim n\gamma^n \displaystyle\sum_{j=1}^{n-1} \dfrac{1}{j}$

Table 4.6.9   System of One Operating Unit and $n-1$ Stand-by Units: Operating Stand-by, Restricted Repair

$$C_m^r = \frac{m!}{r!(m-r)!}, \quad \gamma = \frac{\lambda}{\mu}, \quad \gamma_c = \frac{1}{\sum\limits_{i=1}^{n} \dfrac{1}{i!\,\gamma^i}}.$$

MEASURE	EXACT VALUE	APPROXIMATE VALUE	ERROR
T_1	$\dfrac{1}{\lambda} \sum\limits_{s=0}^{n-1} \sum\limits_{i=0}^{s} \dfrac{1}{(i+1)!\,C_{n-s+i}^{i+1}\,\gamma^i}$	$\dfrac{1}{\lambda}\,\dfrac{1}{n!\,\gamma^{n-1}}$	$\delta_+ \sim \dfrac{1}{\lambda}\,\dfrac{n+1}{n!\,\gamma^{n-2}}$
T_2	$\dfrac{1}{\lambda} \sum\limits_{i=0}^{n-1} \dfrac{1}{(i+1)!\,\gamma^i}$	$\dfrac{1}{\lambda}\,\dfrac{1}{n!\,\gamma^{n-1}}$	$\delta_+ \sim \dfrac{1}{\lambda}\,\dfrac{1}{(n-1)!\,\gamma^{n-2}}$
τ	$\dfrac{1}{\mu}$	—	—
$P(t_0)$	—	$\exp\left[-\dfrac{\lambda t_0}{\sum\limits_{s=0}^{n-1} \sum\limits_{i=0}^{s} \dfrac{1}{(i+1)!\,C_{n-s+i}^{i+1}\,\gamma^i}} \right]$	$\delta \sim \dfrac{1}{\lambda}(n-1)\,n!\,\gamma^n$
K	$\dfrac{1}{1+\gamma_c}$	$1 - n!\,\gamma^n$	$\delta_+ \sim n \cdot n!\,\gamma^{n+1}$
$R(t_0)$	—	$\dfrac{1}{1+\gamma_c} \exp\left[-\dfrac{\lambda t_0}{\sum\limits_{s=0}^{n-1} \sum\limits_{i=0}^{s} \dfrac{1}{(i+1)!\,C_{n-s+i}^{i+1}\,\gamma^i}} \right]$	$\delta \sim (n-1)\,n!\,\gamma^n$

and

$$\tau = \frac{1}{\mu} = \frac{1}{0.5} = 2 \quad \text{hr.}$$

4. By using the result obtained in part 1 of this example, we find an approximate value for the probability of initial failure-free operation of the system for the case of restricted repair to be

$$\tilde{P}(t_0) = e^{-t_0/T_1} = e^{t_0/1442} = e^{-0.0007t_0}.$$

For example, for an initial time of $t_0 = 24$ hours the probability of failure-free operation of the system is approximately

$$\tilde{P}(t_0) = e^{-0.0007t_0} = e^{-0.0007 \cdot 24} \approx 0.983.$$

5. The exact value of the readiness coefficient of the system for the case of restricted repair is

$$K = \frac{1}{1 + \gamma_c},$$

where

$$\gamma_c = \frac{1}{\displaystyle\sum_{i=1}^{n} \frac{1}{i!\gamma^i}}.$$

First we compute the quantity γ_c:

$$\gamma_c = \frac{1}{\displaystyle\sum_{i=1}^{n} \frac{1}{i!\gamma^i}} = \frac{1}{\displaystyle\sum_{i=1}^{4} \frac{1}{i!(0.1)^i}} \approx 0.0016.$$

Then

$$K = \frac{1}{1 + \gamma_c} = \frac{1}{1 + 0.0016} \approx 0.9984.$$

6. An approximate value for the reliability coefficient in the restricted-repair case is

$$\tilde{R}(t_0) = \frac{1}{1 - \gamma_c} e^{-t_0/T_1} = K e^{-t_0/T_1}.$$

Using the results obtained in parts 4 and 5 of this example and substituting the appropriate numerical values, we obtain

$$\tilde{R}(t_0) = 0.9984 e^{-0.0007t_0}.$$

Thus, for example, for a time interval of length $t_0 = 24$ hours the reliability coefficient is approximately

$$\tilde{R}(t_0) = 0.9984e^{-0.0007 \cdot t_0} = 0.9984e^{-0.0007 \cdot 24} \approx 0.98.$$

Example 4.6.6   Suppose a device has a group of $n - 1 = 3$ stand-by devices in a partially operating state with load coefficient $\nu = 0.5$. The failure intensity of the operating device is $\lambda = 0.1 \dfrac{1}{\text{hr}}$, the repair intensity of a failed device is $\mu = 0.25 \dfrac{1}{\text{hr}}$, and $\gamma = \lambda/\mu = 0.1/0.25 = 0.4$.

We are to determine the basic reliability measures of the system for the cases of unrestricted and restricted repair.

Solution.   From Tables 4.6.10 and 4.6.11 we find:

1.   The exact value of the mean operating time of the system up to failure for the unrestricted-repair case is

$$T_1 = \frac{1}{\lambda} \sum_{s=0}^{n-1} \frac{s! \displaystyle\sum_{i=0}^{s} \frac{\gamma^i}{i!} \prod_{l=1}^{i} [1 + (n - l)\nu]}{\gamma^s \displaystyle\prod_{l=1}^{s+1} [1 + (n - l)\nu]}$$

$$= \frac{1}{\lambda} \sum_{s=0}^{3} \frac{s! \displaystyle\sum_{i=0}^{s} \frac{\gamma^i}{i!} \prod_{l=1}^{i} [1 + (4 - l)\nu]}{\gamma^s \displaystyle\prod_{l=1}^{s+1} [1 + (4 - l)\nu]}$$

$$= \frac{1}{\lambda} \left\{ \frac{1}{1 + 3\nu} + \frac{\displaystyle\sum_{i=0}^{1} \frac{\gamma^i}{i!} \prod_{l=1}^{i} [1 + (4 - l)\nu]}{\gamma \displaystyle\prod_{l=1}^{2} [1 + (4 - l)\nu]} \right.$$

$$\left. + \frac{2! \displaystyle\sum_{i=0}^{2} \frac{\gamma^i}{i!} \prod_{l=1}^{i} [1 + (4 - l)\nu]}{\gamma^2 \displaystyle\prod_{l=1}^{3} [1 + (4 - l)\nu]} + \frac{3! \displaystyle\sum_{i=0}^{3} \frac{\gamma^i}{i!} \prod_{l=1}^{i} [1 + (4 - l)\nu]}{\gamma^3 \displaystyle\prod_{l=1}^{4} [1 + (4 - l)\nu]} \right\}.$$

We successively compute the sums in the numerators:

$$\sum_{i=0}^{1} \frac{\gamma^i}{i!} \prod_{l=1}^{i} [1 + (4 - l)\nu] = 1 + \gamma(1 + 3\nu) = 1 + 0.4(1 + 3 \cdot 0.5) = 2,$$

$$\sum_{i=0}^{2} \frac{\gamma^i}{i!} \prod_{l=1}^{i} [1 + (4 - l)\nu] = 1 + \gamma(1 + 3\nu) + \frac{\gamma^2}{2}(1 + 3\nu)(1 + 2\nu)$$

$$= 1 + 0.4(1 + 3 \cdot 0.5)$$

$$+ \frac{(0.4)^2}{2}(1 + 3 \cdot 0.5)(1 + 2 \cdot 0.5)$$

$$= 2.4,$$

$$\sum_{i=0}^{3} \frac{\gamma^i}{i!} \prod_{l=1}^{i} [1 + (4 - l)\nu] = 1 + \gamma(1 + 3\nu) + \frac{\gamma^2}{2}(1 + 3\nu)(1 + 2\nu)$$

$$+ \frac{\gamma^3}{6}(1 + 3\nu)(1 + 2\nu)(1 + \nu)$$

$$= 1 + 0.4(1 + 3 \cdot 0.5)$$

$$+ \frac{(0.4)^2}{2} \cdot (1 + 3 \cdot 0.5)(1 + 2 \cdot 0.5)$$

$$+ \frac{(0.4)^3}{6}(1 + 3 \cdot 0.5)(1 + 2 \cdot 0.5)(1 + 0.5)$$

$$= 2.48.$$

Further, the products in the denominators are

$$\prod_{l=1}^{2} [1 + (4 - l)\nu] = (1 + 3\nu)(1 + 2\nu) = (1 + 3 \cdot 0.5)(1 + 2 \cdot 0.5) = 5,$$

$$\prod_{l=1}^{3} [1 + (4 - l)\nu] = (1 + 3\nu)(1 + 2\nu)(1 + \nu)$$

$$= (1 + 3 \cdot 0.5)(1 + 2 \cdot 0.5)(1 + 0.5) = 7.5,$$

$$\prod_{l=1}^{4} [1 + (4 - l)\nu] = (1 + 3\nu)(1 + 2\nu)(1 + \nu)1 = 7.5.$$

Then

$$T_1 = \frac{1}{0.1}\left[\frac{1}{1 + 3 \cdot 0.5} + \frac{2}{0.4 \cdot 5} + \frac{2 \cdot 2.4}{(0.4)^2 \cdot 7.5} + \frac{6 \cdot 2.48}{(0.4)^3 \cdot 7.5}\right] \approx 364 \quad \text{hr.}$$

Table 4.6.10 System of One Operating Unit and $n-1$ Stand-by Units: Partially Operating Stand-by, Unrestricted Repair

$$\gamma_c = \frac{\dfrac{\gamma^n}{n!}\displaystyle\prod_{l=1}^{n}[1+(n-l)\nu]}{\displaystyle\sum_{i=0}^{n-1}\frac{\gamma^i}{i!}\prod_{l=1}^{i}[1+(n-l)\nu]},$$

$$\gamma = \frac{\lambda}{\mu}, \qquad \prod_{l=1}^{0}[1+(n-l)\nu] = 1.$$

MEASURE	EXACT VALUE	APPROXIMATE VALUE	ERROR
T_1	$\dfrac{1}{\lambda}\displaystyle\sum_{s=0}^{n-1}s!\,\dfrac{\displaystyle\sum_{i=0}^{l}\frac{\gamma^i}{i!}\prod_{l=1}^{l}[1+(n-l)\nu]}{\gamma^s\displaystyle\prod_{l=1}^{s+1}[1+(n-l)\nu]}$	$\dfrac{1}{\lambda}\dfrac{(n-1)!}{\gamma^{n-1}}\dfrac{1}{\displaystyle\prod_{l=1}^{n}[1+(n-l)\nu]}$	$\delta_+ \sim \dfrac{1}{\lambda}\dfrac{(n-2)!}{\gamma^{n-2}}\dfrac{n+(n-1)^2\nu}{\displaystyle\prod_{l=1}^{n}[1+(n-l)\nu]}$

T_2	$\dfrac{1}{\lambda}\dfrac{(n-1)!}{\gamma^{n-1}}\displaystyle\sum_{i=0}^{n-1}\dfrac{\gamma^i}{i!}\dfrac{\prod_{l=1}^{i}[1+(n-l)\nu]}{\prod_{l=1}^{n}[1+(n-l)\nu]}$	$\dfrac{1}{\lambda}\dfrac{(n-1)!}{\gamma^{n-1}}\dfrac{1}{\prod_{l=1}^{n}[1+(n-l)\nu]}$	$\delta_+ \sim \dfrac{1}{\lambda}\dfrac{(n-1)!}{\gamma^{n-2}}\dfrac{1+(n-1)2\nu}{\prod_{l=1}^{n}[1+(n-l)\nu]}$
τ	$\dfrac{1}{n\mu}$	—	—
$P(t_0)$	—	e^{-t_0/T_1}	$\delta \sim [1+(n-1)\nu]\gamma^n \displaystyle\sum_{j=1}^{n-1}\dfrac{1}{j}\prod_{i=1}^{i}\dfrac{[1+(n-1-i)\nu]}{[1+(n-1)\nu]}$
K	$\dfrac{1}{1+\gamma_c}$	$1-\dfrac{\gamma^n}{n!}\prod_{l=1}^{n}[1+(n-l)\nu]$	$\delta_+ \sim \dfrac{\gamma^{n+1}}{n!}[1+(n-1)\nu]\prod_{l=1}^{n}[1+(n-l)\nu]$
$R(t_0)$	—	$\dfrac{1}{1+\gamma_c}e^{-t_0/T_1}$	$\delta \sim [1+(n-1)\nu]\gamma^n \displaystyle\sum_{j=1}^{n-1}\dfrac{1}{j}\prod_{i=1}^{i}\dfrac{[1+(n-1-i)\nu]}{[1+(n-1)\nu]}$

Table 4.6.11   System of One Operating Unit and $n-1$ Stand-by Units: Partially Operating Stand-by, Restricted Repair

$$\gamma_c = \frac{\gamma^n \displaystyle\prod_{l=1}^{n}[1+(n-l)\nu]}{\displaystyle\sum_{i=0}^{n-1}\gamma^i \prod_{l=1}^{i}[1+(n-l)\nu]},$$

$$\gamma = \frac{\lambda}{\mu}, \qquad \prod_{l=1}^{0}[1+(n-l)\nu]=1.$$

MEASURE	EXACT VALUE	APPROXIMATE VALUE	ERROR
T_1	$\dfrac{1}{\lambda}\displaystyle\sum_{s=0}^{n-1}\dfrac{\displaystyle\sum_{i=0}^{i}\gamma^i \prod_{l=1}^{i}[1+(n-l)\nu]}{\gamma^s \displaystyle\prod_{l=1}^{s+1}[1+(n-l)\nu]}$	$\dfrac{1}{\lambda}\,\dfrac{1}{\gamma^{n-1}\displaystyle\prod_{l=1}^{n}[1+(n-l)\nu]}$	$\delta_+ \sim \dfrac{1}{\lambda}\,\dfrac{2+(n-1)\nu}{\gamma^{n-2}\displaystyle\prod_{l=1}^{n}[1+(n-l)\nu]}$

T_2	$\dfrac{1}{\lambda}\dfrac{\sum_{i=0}^{n-1}\gamma^i\prod_{l=1}^{i}[1+(n-l)\nu]}{\gamma^{n-1}\prod_{l=1}^{n}[1+(n-l)\nu]}$	$\dfrac{1}{\lambda}\dfrac{1}{\gamma^{n-1}\prod_{l=1}^{n}[1+(n-l)\nu]}$	$\delta_+\sim\dfrac{1}{\lambda}\dfrac{1+(n-1)\nu}{\gamma^{n-2}\prod_{l=1}^{n}[1+(n-l)\nu]}$
τ	$\dfrac{1}{\mu}$	—	—
$P(t_0)$	—	e^{-t_0/T_1}	$\delta\sim(n-1)\gamma^n\prod_{i=0}^{n-1}[1+(n-1)\nu]\prod_{l=1}^{n}[1+(n-1-i)\nu]$
K	$\dfrac{1}{1+\gamma_c}$	$1-\gamma^n\prod_{l=1}^{n}[1+(n-l)\nu]$	$\kappa_{\cdot\cdot}\sim\gamma^{n+1}[1+(n-1)\nu]\prod_{l=1}^{n}[1+(n-1-i)\nu]$
$R(t_0)$	—	$\dfrac{1}{1+\gamma_c}e^{-t_0/T_1}$	$\delta_+\sim(n-1)\gamma^n\prod_{i=0}^{n-1}[1+(n-1-i)\nu]$

2. The exact value of the mean operating time of the system between failures for the case of restricted repair is

$$T_2 = \frac{1}{\lambda} \cdot \frac{\sum_{i=0}^{n-1} \gamma^i \prod_{l=1}^{i} [1 + (n - l)\nu]}{\gamma^{n-1} \prod_{l=1}^{n} [1 + (n - l)\nu]}$$

$$= \frac{1}{0.1} \cdot \frac{\sum_{i=0}^{3} (0.4)^i \prod_{l=1}^{i} [1 + (4 - l)0.5]}{(0.4)^3 \prod_{l=1}^{4} [1 + (4 - l) \cdot 0.5]} \quad \text{hr.}$$

First we compute the numerator of this expression:

$$\sum_{l=0}^{3} (0.4)^i \prod_{l=1}^{i} [1 + (4 - l) \cdot 0.5]$$
$$= 1 + 0.4[1 + (4 - 1) \cdot 0.5] - (0.4)^2 \cdot [1 + (4 - 1) \cdot 0.5][1 + (4 - 2) \cdot 0.5]$$
$$+ (0.4)^3[1 + (4 - 1) \cdot 0.5][1 + (4 - 2) \cdot 0.5][1 + (4 - 3) \cdot 0.5]$$
$$= 3.28.$$

The quantity

$$\prod_{l=1}^{4} [1 + (4 - l) \cdot 0.5] = 7.5$$

was computed in the previous part of this example. Thus,

$$T_2 = \frac{1}{0.1} \cdot \frac{3.28}{(0.4)^3 \cdot 7.5} \approx 68.3 \quad \text{hr.}$$

3. The mean repair times of the system for the unrestricted- and restricted-repair cases, respectively, are

$$\tau = \frac{1}{n\mu} = \frac{1}{4 \cdot 0.25} = 1 \quad \text{hr,}$$

and

$$\tau = \frac{1}{\mu} = \frac{1}{0.25} = 4 \quad \text{hr.}$$

4.   By using the result obtained in part 1 of this example, we find an approximate expression for the probability of initial failure-free operation of the system in the case of unrestricted repair to be

$$\tilde{P}(t_0) = e^{-t_0/T_1} = e^{-t_0/364} = e^{-0.0027 t_0}.$$

For example, the probability of failure-free operation of the system for the initial $t_0 = 720$ hours is approximately

$$\tilde{P}(t_0) = e^{-0.0027 t_0} = e^{-0.0027 \cdot 720} \approx 0.14.$$

5.   The exact value of the readiness coefficient of the system for the case of restricted repair is

$$K = \frac{1}{1 + \gamma_c} = \frac{T_2}{T_2 + \tau}.$$

By using the results obtained in parts 2 and 3 of this example, we find

$$K = \frac{T_2}{T_2 + \tau} = \frac{68.3}{68.3 + 4} \approx 0.945.$$

6.   An approximate expression for the reliability coefficient of the system in the unrestricted-repair case is

$$\tilde{R}(t_0) = \frac{1}{1 + \gamma_c} e^{-t_0/T_1},$$

where $T_1 = 364$ hours (computed in part 1 of this example) and

$$\gamma_c = \frac{\dfrac{\gamma^n}{n!} \displaystyle\prod_{l=1}^{n} [1 + (n - l)\nu]}{\displaystyle\sum_{i=0}^{n-1} \frac{\gamma^i}{i!} \prod_{l=1}^{i} [1 + (n - l)\nu]} = \frac{\dfrac{\gamma^4}{4!} \displaystyle\prod_{l=1}^{4} [1 + (4 - l)\nu]}{\displaystyle\sum_{i=0}^{3} \frac{\gamma^i}{i!} \prod_{l=1}^{i} [1 + (4 - l)\nu]}.$$

But in part 1 we showed that

$$\prod_{l=1}^{4} [1 + (4 - l)\nu] = 7.5,$$

$$\sum_{i=0}^{3} \frac{\gamma^i}{i!} \prod_{l=1}^{i} [1 + (4 - l)\nu] = 2.48.$$

Then

$$\gamma_c = \frac{\dfrac{(0.4)^4}{4!} \cdot 7.5}{2.48} \approx 0.0032.$$

Thus,

$$\tilde{R}(t_0) = \frac{1}{1 + \gamma_c} e^{-t_0/T_1} = \frac{1}{1 + 0.0032} e^{-t_0/364} \approx 0.997 e^{-0.0027 t_0}.$$

For a time interval of length $t_0 = 24$ hours the reliability coefficient is approximately given by

$$\tilde{R}(t_0) = 0.997 e^{-0.0027 t_0} = 0.997 e^{-0.0027 \cdot 24} \approx 0.94.$$

Example 4.6.7 Suppose we have a system of one operating device and $n - 1 = 4$ stand-by devices in a nonoperating state. The failure intensity of the operating device is $\lambda = 0.2 \dfrac{1}{\text{hr}}$, the repair intensity of a failed device is $\mu = 0.5 \dfrac{1}{\text{hr}}$, and $\gamma = \lambda/\mu = 0.2/0.5 = 0.4$.

We wish to determine the basic reliability measures of the system for the cases of unrestricted and restricted repair.

Solution. From Tables 4.6.12 and 4.6.13 we find:

1. The exact value of the mean operating time of the system up to failure in the case of unrestricted repair is

$$T_1 = \frac{1}{\lambda} \sum_{i=0}^{n-1} C_n^{i+1} \frac{i!}{\gamma^i} = \frac{1}{\lambda} \sum_{i=0}^{4} C_5^{i+1} \frac{i!}{\gamma^i}$$

$$= \frac{1}{0.2} \cdot \left[5 + 10 \frac{1}{0.4} + 10 \cdot \frac{2!}{(0.4)^2} + 5 \frac{3!}{(0.4)^3} + \frac{4!}{(0.4)^4} \right] \quad \text{hr}$$

$$\approx 7806 \quad \text{hr.}$$

2. The exact value of the mean operating time of the system between failures in the restricted-repair case is

$$T_2 = \frac{1}{\lambda} \cdot \frac{1 - \gamma^n}{\gamma^{n-1}(1 - \gamma)} = \frac{1}{0.2} \cdot \frac{1 - (0.4)^5}{(0.4)^4 \cdot (1 - 0.4)} \approx 322 \quad \text{hr.}$$

3. The mean repair times for the cases of unrestricted and restricted repair are, respectively,

$$\tau = \frac{1}{n\mu} = \frac{1}{4 \cdot 0.5} = 0.5 \quad \text{hr,}$$

Table 4.6.12   System of One Operating Unit and $n - 1$
Stand-by Units: Nonoperating Stand-by,
Unrestricted Repair

$$\gamma_c = \frac{1}{\displaystyle\sum_{i=1}^{n} C_n^i \frac{i!}{\gamma^i}},$$

$$\gamma = \frac{\lambda}{\mu}, \qquad C_m^r = \frac{m!}{r!(m-r)!}$$

MEASURE	EXACT VALUE	APPROXIMATE VALUE	ERROR
T_1	$\dfrac{1}{\lambda}\displaystyle\sum_{i=0}^{n-1} C_n^{i+1}\dfrac{i!}{\gamma^i}$	$\dfrac{1}{\lambda}\dfrac{(n-1)!}{\gamma^{n-1}}$	$\delta_+ \sim \dfrac{1}{\lambda}\dfrac{n!}{(n-1)\gamma^{n-2}}$
T_2	$\dfrac{1}{\lambda}\displaystyle\sum_{i=0}^{n-1} C_{n-1}^{i}\dfrac{i!}{\gamma^i}$	$\dfrac{1}{\lambda}\dfrac{(n-1)!}{\gamma^{n-1}}$	$\delta_+ \sim \dfrac{1}{\lambda}\dfrac{(n-1)!}{\gamma^{n-2}}$
τ	$\dfrac{1}{n\mu}$	—	—
$P(t_0)$	—	$\exp\left(-\dfrac{\lambda t_0}{\displaystyle\sum_{i=0}^{n-1} C_n^{i+1}\dfrac{i!}{\gamma^i}}\right)$	$\delta \sim \dfrac{\gamma^n}{(n-1)!}\displaystyle\sum_{j=1}^{n-1}\dfrac{1}{j}$
K	$\dfrac{1}{1+\gamma_c}$	$1 - \dfrac{\gamma^n}{n!}$	$\delta_+ \sim \dfrac{\gamma^{n+1}}{n!}$
$R_0(t)$	—	$\dfrac{1}{1+\gamma_c}\exp\left(-\dfrac{\lambda t_0}{\displaystyle\sum_{i=0}^{n-1} C_n^{i+1}\dfrac{i!}{\gamma^i}}\right)$	$\delta \sim \dfrac{\gamma^n}{(n-1)!}\displaystyle\sum_{j=1}^{n-1}\dfrac{1}{j}$

Table 4.6.13   System of One Operating Unit and $n-1$ Stand-by Units: Nonoperating Stand-by, Restricted Repair

$$\gamma_c = \frac{\gamma^n(1-\gamma)}{1-\gamma^n}, \quad \gamma = \frac{\lambda}{\mu}.$$

MEASURE	EXACT VALUE	APPROXIMATE VALUE	ERROR
T_1	$\dfrac{1}{\lambda}\dfrac{1-\gamma^n[1+n(1-\gamma)]}{\gamma^{n-1}(1-\gamma)^2}$	$\dfrac{1}{\lambda}\dfrac{1}{\gamma^{n-1}}$	$\delta_+ \sim \dfrac{2}{\lambda\gamma^{n-2}}$
T_2	$\dfrac{1}{\lambda}\dfrac{1-\gamma^n}{\gamma^{n-1}(1-\gamma)}$	$\dfrac{1}{\lambda}\dfrac{1}{\gamma^{n-1}}$	$\delta_+ \sim \dfrac{1}{\lambda\gamma^{n-2}}$
τ	$\dfrac{1}{\mu}$	—	—
$P(t_0)$	—	$\exp\left\{-\dfrac{(1-\gamma)^2\gamma^{n-1}\lambda t_0}{1-\gamma^n[1+n(1-\gamma)]}\right\}$	$\delta \sim (n-1)\gamma^n$
K	$\dfrac{1}{1+\gamma_c}$	$1-\gamma^n$	$\delta_+ \sim \gamma^{n+1}$
$R(t_0)$	—	$\dfrac{1}{1+\gamma_c}\exp\left\{-\dfrac{(1-\gamma)^2\gamma^{n-1}\lambda t_0}{1-\gamma^n[1+n(1-\gamma)]}\right\}$	$\delta \sim (n-1)\gamma^n$

and

$$\tau = \frac{1}{\mu} = \frac{1}{0.5} = 2 \quad \text{hr.}$$

4.   An approximate value for the probability of initial failure-free operation of the system in the case of unrestricted repair is found by using the result obtained in part 1 of this example:

$$\tilde{P}(t_0) = e^{-t_0/T_1} = e^{-t_0/7806} = e^{-0.00013t_0}.$$

For an initial time of $t_0 = 100$ hours, for example, the probability of failure-free operation of the system is

$$\tilde{P}(t_0) = e^{-0.00013t_0} = e^{-0.00013 \cdot 100} \approx 0.987.$$

5.   The exact value of the readiness coefficient of the system in the case of restricted repair is

$$K = \frac{1}{1 + \gamma_c},$$

where

$$\gamma_c = \frac{\gamma^n(1 - \gamma)}{1 - \gamma^n}.$$

First we compute the quantity γ_c:

$$\gamma_c = \frac{(0.4)^5 \cdot (1 - 0.4)}{1 - (0.4)^5} \approx 0.0062.$$

Then

$$K = \frac{1}{1 + \gamma_c} = \frac{1}{1 + 0.0062} \approx 0.994.$$

6.   An approximate value for the reliability coefficient of the system for the unrestricted-repair case is

$$\tilde{R}(t_0) = \frac{1}{1 + \gamma_c} e^{-t_0/T_1}.$$

Making use of the result obtained in part 4 of this example, we find

$$\tilde{R}(t_0) = \frac{1}{1 + \gamma_c} e^{-0.00013t_0}.$$

For this we determine

$$\gamma_c = \cfrac{1}{\displaystyle\sum_{i=1}^{n} C_n^i \frac{i!}{\gamma^i}} = \cfrac{1}{\displaystyle\sum_{i=1}^{5} C_5^i \frac{i!}{\gamma^i}}$$

$$= \cfrac{1}{C_5^1 \dfrac{1}{\gamma} + C_5 \dfrac{2!}{\gamma^2} + C_5^3 \dfrac{3!}{\gamma^3} + C_5^4 \dfrac{4!}{\gamma^4} + C_5^5 \dfrac{5!}{\gamma^5}}$$

$$= \cfrac{1}{\dfrac{5}{\gamma} + \dfrac{20}{\gamma^2} + \dfrac{60}{\gamma^3} + \dfrac{120}{\gamma^4} + \dfrac{120}{\gamma^5}}$$

$$= \cfrac{1}{\dfrac{5}{0.4} + \dfrac{20}{(0.4)^2} + \dfrac{60}{(0.4)^3} + \dfrac{120}{(0.4)^4} + \dfrac{120}{(0.4)^5}} \approx 0.00005.$$

Then

$$\tilde{R}(t_0) = 0.99995 \cdot e^{-0.00013 t_0}.$$

Thus, for example, the reliability coefficient for a time interval of length $t_0 = 100$ hours is approximately given by

$$\tilde{R}(t_0) = 0.99995 e^{-0.00013 t_0} = 0.99995 e^{-0.00013 \cdot 100} \approx 0.987.$$

BIBLIOGRAPHY

1. KHINCHIN, A. Y., *Mathematical Methods in the Theory of Queuing* (Statistical monographs and courses, no. 7). Translated by D. M. Andrews and M. H. Quenouille. London: Hafner Publishing Co., Inc., 1960.
2. SOLOV'EV, A. D., "Limit Theorems in Birth and Death Processes," *Theory of Probability and Its Applications* (Moscow), no. 6, 1968.
3. SMITH, W. L., "Renewal Theory and Its Ramifications," *Royal Statistical Society, series B.* (London), vol. 20, no. 2, 1958.
4. GNEDENKO, B. V., BELYAEV, Y. K., AND SOLOV'EV, A. D., *Mathematical Methods in Reliability Theory.* New York: Academic Press, Inc., 1968.
5. SOLOV'EV, A. D., "Reliability of Systems with Repair," in the collection *Cybernetics in the Service of Communism.* Moscow: Energiya, 1964.
6. LEVIN, B. R., AND EPISHIN, Y. G., "Theory of Redundancy of Systems with Repair." Supplement to the book by I. Bazovsky, *Reliability: Theory and Practice.* Moscow: Mir, 1965.
7. VASIL'EV, B. V., KOZLOV, B. A., AND TKACHENKO, L. G., *Reliability and Efficiency of Radioelectronic Structures,* chap. 2. Moscow: Soviet Radio, 1964.

5

SPECIAL STAND-BY PROBLEMS

In this chapter we calculate the reliability measures for some special stand-by systems, assuming exponential distribution of operating time and repair time of each of the units of the system.

Detailed descriptions of the models for the systems under consideration are given in the appropriate sections.

5.1 REPAIRABLE STAND-BY SYSTEMS OF n UNITS WITH DIFFERENT RELIABILITY MEASURES[1]

We consider a stand-by system consisting of n units with different reliability measures characterized by the failure intensities and repair intensities $\lambda_1, \mu_1; \lambda_2, \mu_2; \ldots; \lambda_i, \mu_i; \ldots, \lambda_n, \mu_n$, respectively. The stand-by system operates continuously, and control of the operationality of the units is assumed complete and faultless. We also make the assumption that the switching structure in the system is characterized by high reliability and very short switching time.

The system is assumed to have failed at some moment of time t if at that moment all n of its units are nonoperational and are undergoing repair (Figure 5.1.1).

In Table 5.1.1 we give exact and approximate formulas for the basic reliability measures for the case of unrestricted repair of a system; in Table 5.1.2 we give lower bounds for measures in the case of restricted repair. Note that, as before, unrestricted repair is the case wherein any

---

[1] In this chapter, units with different reliability measures will be called different units, for brevity.

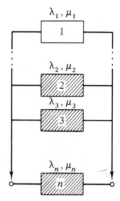

Figure 5.1.1 Block Diagram of the Model Used for Reliability Computations for a Redundant System of n Units with Different Reliability Measures.

number of failed units (from 1 to n) in the system can undergo repair at the same time. Restricted repair is the case wherein at any one time only one unit (of any number of units that are faulty at a given moment) in the system can undergo repair.

The approximate formulas given are valid under the condition

$$\gamma_m = \gamma_{\max} = \frac{\lambda_{\max}}{\mu_{\min}} \ll \frac{1}{n}, \tag{5.1.1}$$

where

$$\lambda_{\max} = \max \{\lambda_1, \lambda_2, \ldots, \lambda_i, \ldots, \lambda_n\},$$
$$\mu_{\min} = \min \{\mu_1, \mu_2, \ldots, \mu_i, \ldots, \mu_n\}.$$

In this section we also use the following notation:

$$\gamma_i = \frac{\lambda_i}{\mu_i}.$$

Lower bounds for reliability measures of a stand-by system consisting of n different units [also including the case of dependent different units— for example, when the failure and repair intensities of the jth unit equal to $\lambda_j^{(i,k,l,\ldots)}$, $\mu_j^{(i,k,l,\ldots)}$ depend on which units $(i, k, l, \ldots)$ have already failed (for more detail see Section 5.2)] are the precise reliability measures of a system consisting of n identical units with failure and repair intensities (see Chapter 4) equal to

$$\lambda = \lambda_{\max}, \qquad \mu = \mu_{\min}.$$

Upper bounds are obtained by setting

$$\lambda = \lambda_{\min}, \qquad \mu = \mu_{\max}.$$

Table 5.1.1   Stand-by System of n Different Units:
Unrestricted Repair

MEASURE	EXACT VALUE	APPROXIMATE VALUE	ERROR
T_1	—	$\dfrac{1}{\sum\limits_{i=1}^{n} \mu_i \prod\limits_{j=1}^{n} \gamma_j}$	$\delta_+ \sim \dfrac{\sum\limits_{i=1}^{n} \gamma_i}{\sum\limits_{i=1}^{n} \mu_i \prod\limits_{j=1}^{n} \gamma_j}$
T_2	$\dfrac{1}{\sum\limits_{i=1}^{n} \mu_i} \left[\dfrac{\prod\limits_{j=1}^{n}(1+\gamma_j)}{\prod\limits_{j=1}^{n} \gamma_j} - 1 \right]$	$\dfrac{1}{\sum\limits_{i=1}^{n} \mu_i \prod\limits_{j=1}^{n} \gamma_j}$	$\delta_+ \sim \dfrac{\sum\limits_{i=1}^{n} \gamma_i}{\sum\limits_{i=1}^{n} \mu_i \prod\limits_{j=1}^{n} \gamma_j}$
τ	$\dfrac{1}{\sum\limits_{i=1}^{n} \mu_i}$	—	—
K	$1 - \prod\limits_{i=1}^{n} \dfrac{\gamma_i}{1+\gamma_i}$	$1 - \prod\limits_{i=1}^{n} \gamma_i$	$\delta_+ \sim \sum\limits_{i=1}^{n} \gamma_i \prod\limits_{j=1}^{n} \gamma_j$

Thus, in particular, the lower bounds of the probability of failure-free operation and the coefficient of reliability of the system in the case of unrestricted repair have the form:

$$\tilde{P}(t_0) = \exp\left[- \frac{\lambda_{\max} t_0}{\sum\limits_{s=0}^{n-1} \dfrac{\sum\limits_{i=0}^{s} C_n^i \gamma_m^i}{(n-s)C_n^s \gamma_m^s}} \right]. \qquad (5.1.2)$$

$$\tilde{R}(t_0) = \left(1 - \prod\limits_{j=1}^{n} \frac{\gamma_j}{1+\gamma_j}\right) \exp\left[- \frac{\lambda_{\max} t_0}{\sum\limits_{s=0}^{n-1} \dfrac{\sum\limits_{i=0}^{s} C_n^i \gamma_m^i}{(n-s)C_n^s \gamma_m^s}} \right]. \qquad (5.1.3)$$

Table 5.1.2   Stand-by System of n Different Units
Restricted Repair

MEASURE	LOWER BOUND
$\tilde{T}_1$	$\dfrac{1}{\lambda_{max}} \displaystyle\sum_{s=0}^{n-1} (n-s-1)! \sum_{i=0}^{s} \dfrac{\gamma_m^{i-s}}{(n-i)!}$
$\tilde{T}_2$	$\dfrac{1}{\lambda_{max}} \displaystyle\sum_{i=1}^{n} \dfrac{1}{i!\gamma_m^{i-1}}$
$\tilde{P}(t_0)$	$\exp\left[-\dfrac{\lambda_{max} t_0}{\displaystyle\sum_{s=0}^{n-1} (n-s-1)! \sum_{i=0}^{s} \dfrac{\gamma_m^{i-s}}{(n-i)!}} \right]$
$\tilde{K}$	$\left(1 + \dfrac{1}{\displaystyle\sum_{i=1}^{n} \dfrac{1}{i!\gamma_m^{i}}} \right)^{-1}$
$\tilde{R}(t_0)$	$\left(1 + \dfrac{1}{\displaystyle\sum_{i=1}^{n} \dfrac{1}{i!\gamma_m^{i}}} \right)^{-1} \exp\left[-\dfrac{\lambda_{max} t_0}{\displaystyle\sum_{s=0}^{n-1} (n-s-1)! \sum_{i=0}^{s} \dfrac{\gamma_m^{i-s}}{(n-i)!}} \right]$

An upper bound for the mean repair time of the system in the case of restricted repair is

$$\tilde{\tau} = \frac{1}{\mu_{min}}. \tag{5.1.4}$$

Example 5.1.1   Consider a system of three different units with failure and repair intensities equal to

$$\lambda_1 = \lambda, \qquad \lambda_2 = 1.1\lambda, \qquad \lambda_3 = 1.2\lambda,$$
$$\mu_1 = \mu, \qquad \mu_2 = 0.9\mu, \qquad \mu_3 = 0.8\mu,$$

where

$$\lambda = 0.1\frac{1}{hr}, \qquad \mu = 1\frac{1}{hr}.$$

We wish to determine the basic reliability measures of this system for the case of unrestricted repair.

Solution.   We use Table 5.1.1.

First we compute the quantities γ_i:

$$\gamma_1 = \frac{\lambda_1}{\mu_1} = \frac{\lambda}{\mu} = \gamma = \frac{0.1}{1} = 0.10,$$

$$\gamma_2 = \frac{\lambda_2}{\mu_2} = \frac{1.1\lambda}{0.9\mu} = \tfrac{11}{9}\gamma = \tfrac{11}{9}\cdot 0.1 \approx 0.12,$$

$$\gamma_3 = \frac{\lambda_3}{\mu_3} = \frac{1.2\lambda}{0.8\mu} = \tfrac{12}{8}\gamma = \tfrac{3}{2}\cdot 0.1 = 0.15,$$

$$\gamma_m = \gamma_{max} = \frac{\lambda_{max}}{\mu_{min}} = \frac{\lambda_3}{\mu_3} = \gamma_3 = 0.15.$$

1.   An approximate value of the mean operating time of the system up to failure is

$$\tilde{T}_1 = \frac{1}{\displaystyle\sum_{i=1}^{n} \mu_i \cdot \prod_{j=1}^{n} \gamma_j} = \frac{1}{\gamma_1\gamma_2\gamma_3(\mu_1 + \mu_2 + \mu_3)}$$

$$= \frac{1}{0.10\cdot 0.12\cdot 0.15(1 + 0.9 + 0.8)} \approx 206 \quad \text{hr.}$$

2.   The exact value of the mean operating time of the system between failures is

$$T_2 = \frac{1}{\displaystyle\sum_{i=1}^{n} \mu_i}\left[\frac{\displaystyle\prod_{j=1}^{n}(1 + \gamma_j)}{\displaystyle\prod_{j=1}^{n}\gamma_j} - 1\right]$$

$$= \frac{1}{\mu_1 + \mu_2 + \mu_3}\left[\frac{(1 + \gamma_1)(1 + \gamma_2)(1 + \gamma_3)}{\gamma_1\gamma_2\gamma_3} - 1\right]$$

$$= \frac{1}{1 + 0.9 + 0.8}\left[\frac{(1 + 0.1)(1 + 0.12)(1 + 0.15)}{0.1\cdot 0.12\cdot 0.15} - 1\right] \approx 291 \quad \text{hr.}$$

3.   The exact value of the mean repair time is

$$\tau = \frac{1}{\displaystyle\sum_{i=1}^{n} \mu_i} = \frac{1}{\mu_1 + \mu_2 + \mu_3} = \frac{1}{1 + 0.9 + 0.8} \approx 0.37 \quad \text{hr.}$$

4. A lower bound for the probability of initial failure-free operation of the system for $t_0 = 24$ hours is

$$\tilde{P}(t_0) = \exp\left[-\frac{\lambda_{\max}t_0}{\displaystyle\sum_{s=0}^{n-1}\frac{\displaystyle\sum_{i=0}^{s}C_n^i\gamma_m^i}{(n-s)C_n^s\gamma_m^i}}\right].$$

In our case,

$$\sum_{s=0}^{n-1}\frac{\displaystyle\sum_{i=0}^{s}C_n^i\gamma_m^i}{(n-s)C_n^s\gamma_m^s} = \sum_{s=0}^{2}\frac{\displaystyle\sum_{i=0}^{s}C_3^i\gamma_m^i}{(3-s)C_3^s\gamma_m^s}$$

$$= \frac{1}{3} + \frac{\displaystyle\sum_{i=0}^{1}C_3^i\gamma_m^i}{2C_3^1\gamma_m} + \frac{\displaystyle\sum_{i=0}^{2}C_3^i\gamma_m^i}{C_3^2\gamma_m^2}$$

$$= \frac{1}{3} + \frac{1+C_3^1\gamma_m}{2C_3^1\gamma_m} + \frac{1+C_3^1\gamma_m+C_3^2\gamma_m^2}{C_3^2\gamma_m^2}$$

$$= \frac{1}{3} + \frac{1+3\gamma_m}{2\cdot3\gamma_m} + \frac{1+3\gamma_m+3\gamma_m^2}{3\gamma_m^2}$$

$$= \frac{1}{3} + \frac{1+3\cdot0.15}{6\cdot0.15} + \frac{1+3\cdot0.15+3\cdot(0.15)^2}{3\cdot(0.15)^2} \approx 24.42.$$

Since by hypothesis

$$\lambda_{\max} = \lambda_3 = 1.2\lambda = 1.2\cdot0.1 = 0.12 \ \frac{1}{\text{hr}},$$

then

$$\tilde{P}(t_0 = 24 \text{ hr}) = e^{-0.12\cdot24/24.42} \approx e^{-0.12} \approx 0.89.$$

5. The exact value of the readiness coefficient of the system can be found by using the appropriate formula from Table 5.1.1. However, it may be easier and more direct to find this value by using the results obtained in parts 2 and 3 of this example:

$$K = \frac{T_2}{T_2+\tau} = \frac{291}{291+0.37} \approx 0.9987.$$

6. A lower bound for the reliability coefficient for a time period of $t_0 = 24$ hours is found by using the results of parts 4 and 5 of this example:

$$\tilde{R}(t_0 = 24 \text{ hr}) = K\cdot\tilde{P}(t_0 = 24 \text{ hr}) = 0.9987\cdot0.89 \approx 0.89.$$

Example 5.1.2   Consider a redundant system of units characterized by failure and repair intensities:

$$\lambda_1 = \lambda, \quad \lambda_2 = 1.2\lambda, \quad \mu_1 = \mu, \quad \mu_2 = 0.8\mu,$$

where

$$\lambda = 0.05 \; \frac{1}{\text{hr}}, \quad \mu = 2 \; \frac{1}{\text{hr}}.$$

We wish to determine lower bounds for the basic reliability measures of the system in the case of restricted repair.

Solution.   We use Table 5.1.2.

1.   A lower bound for the mean operating time of the redundant system up to failure is

$$\tilde{T}_1 = \frac{1}{\lambda_{\max}} \sum_{s=0}^{n-1} (n - s - 1)! \sum_{i=0}^{s} \frac{\gamma_m^{i-s}}{(n-i)!}.$$

In our case,

$$\lambda_{\max} = \lambda_2 = 1.2\lambda = 1.2 \cdot 0.05 = 0.06 \; \frac{1}{\text{hr}},$$

$$n = 2,$$

$$\gamma_m = \gamma_{\max} = \frac{\lambda_{\max}}{\mu_{\min}} = \frac{\lambda_2}{\mu_2} = \frac{1.2\lambda}{0.8\mu} = \frac{3}{2} \cdot \frac{0.05}{2} = 0.0375.$$

Then,

$$\tilde{T}_1 = \frac{1}{0.06} \sum_{s=0}^{1} (1 - s)! \sum_{i=0}^{s} \frac{(0.0375)^{i-s}}{(2-i)!}$$

$$= \frac{1}{0.06} \left[\frac{1}{2} + \sum_{i=0}^{1} \frac{(0.0375)^{i-1}}{(2-i)!} \right] = \frac{1}{0.06} \left[\frac{1}{2} + \frac{(0.0375)^{-1}}{2} + 1 \right]$$

$$\approx 247 \quad \text{hr.}$$

2.   A lower bound for the mean operating time of the system between failures is

$$\tilde{T}_2 = \frac{1}{\lambda_{\max}} \sum_{i=0}^{n-1} \frac{\gamma_m^{i-n+1}}{(n-i)!}.$$

Using the relationships given in part 1 of this example, we obtain

$$\tilde{T}_2 = \frac{1}{\lambda_{\max}} \sum_{i=0}^{1} \frac{\gamma_m^{i-1}}{(2-i)!} = \frac{2}{0.06} \left[\frac{(0.0375)^{-1}}{2} + 1 \right] \approx 238 \quad \text{hr.}$$

3. An upper bound for the mean repair time of the system [Formula (5.1.4)] is

$$\bar{\tau} = \frac{1}{\mu_{\min}} = \frac{1}{\mu_2} = \frac{1}{0.8\mu} = \frac{1}{0.8 \cdot 2} = 0.625 \quad \text{hr.}$$

4. A lower bound for the probability of failure-free operation of the system for $t_0 = 24$ hours is found by using the result obtained in part 1 of this example. It is

$$\tilde{P}(t_0 = 24 \text{ hr}) = e^{-t_0/\tilde{T}_1} = e^{-24/247} \approx 0.91 \quad \text{hr.}$$

5. A lower bound for the readiness coefficient of the system is

$$K = \cfrac{1}{1 + \cfrac{1}{\displaystyle\sum_{i=1}^{n} \frac{1}{i!\gamma_m^i}}} = \cfrac{1}{1 + \cfrac{1}{\displaystyle\sum_{i=1}^{2} \frac{1}{i!\gamma_m^i}}}.$$

First we compute the following quantity:

$$\sum_{i=1}^{2} \frac{1}{i!\gamma_m^i} = \frac{1}{\gamma_m} + \frac{1}{2\gamma_m^2} = \frac{1}{0.0375} + \frac{1}{2 \cdot (0.0375)^2} \approx 382.$$

Then

$$\tilde{K} = \frac{1}{1 + \frac{1}{382}} \approx 0.997.$$

6. By using the results obtained in parts 4 and 5 of this example, we find an approximate value (lower bound) for the reliability coefficient for the time interval of $t_0 = 24$ hours to be

$$\tilde{R}(t_0 = 24 \text{ hr}) = \tilde{K}\tilde{P}(t_0 = 24 \text{ hr}) = 0.997 \cdot 0.91 \approx 0.91.$$

5.2 REPAIRABLE REDUNDANT SYSTEMS OF DEPENDENT UNITS WITH DIFFERENT RELIABILITY PARAMETERS

In order to increase the reliability of more complicated or particularly important structures, besides ordinary stand-by units one also frequently makes use of so-called functional stand-by units, which are structures not necessarily of analogous design to the basic (operating) structure and often are based on completely different principles. Thus, for example, an electronic structure may have as stand-by a mechanical or optical system, and so on.

Such structures are characterized by very strongly differing failure and repair intensities, and almost always depend on load (input parameters), which, in the case of failure of one of the structures, is redistributed among the structures that are still operational. (Problems of the priority of functioning of structures in such stand-by systems will not be discussed here.)

In this section we obtain reliability measures for only two cases of a system with functional stand-by—namely, a redundant system consisting of two units, and a system of three units where failure of any two units leads to failure of the system (a system of elementary majority unit type).

5.2.1   Redundant systems of two units where the probabilities of failure and repair of both units depend on the state of the system

For the given system (Figure 5.2.1) we let

$\lambda_1^{(2)}(\lambda_2^{(1)})$ denote the intensity of failures of the first (second) unit given that the second (first) has already failed, and

$\mu_1^{(2)}(\mu_2^{(1)})$ denote the intensity of repair of the first (second) unit given that the second (first) unit is already undergoing repair.

The diagram of transitions of a redundant system with two units from one state into another is shown in Figure 5.2.2, where H_0 is the state of the system in which no units have failed, H_1 (H_2) is the state of the system in which the first (second) unit has failed, and H_{12} is the state of failure of the system wherein both units have failed.

In Table 5.2.1 we give exact and approximate formulas for the basic reliability measures of a redundant system consisting of two units for the case of unrestricted repair—that is, repair where all failed units in the system can undergo repair simultaneously and independently.

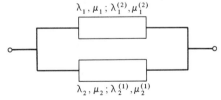

$$\lambda_1, \mu_1 ; \lambda_1^{(2)}, \mu_1^{(2)}$$

$$\lambda_2, \mu_2 ; \lambda_2^{(1)}, \mu_2^{(1)}$$

Figure 5.2.1 Block Diagram of the Model Used for Reliability Computations for a Redundant System Consisting of Two Units with Different Reliability Measures.

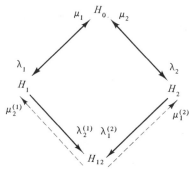

Figure 5.2.2 Diagram of Crossings from One State to Another by a Redundant System Consisting of Two Dependent Units with Different Reliability Measures.

The approximate formulas given in Table 5.2.1 hold under the condition

$$\gamma_m = \lambda_{\max} = \frac{\lambda_{\max}}{\mu_{\min}} \ll \tfrac{1}{2}, \tag{5.2.1}$$

where

$$\lambda_{\max} = \max\{\lambda_1, \lambda_2, \lambda_1^{(2)}, \lambda_2^{(1)}\},$$
$$\mu_{\min} = \min\{\mu_1, \mu_2, \mu_1^{(2)}, \mu_2^{(1)}\}.$$

Lower bounds for the probability of initial failure-free operation and the reliability coefficient of the system are

$$\tilde{P}(t_0) = e^{-2\gamma_m\lambda_{\max}t_0/(1+3\gamma_m)}, \tag{5.2.2}$$

$$\tilde{R}(t_0) = \left(1 + \frac{\alpha_1\lambda_1 + \alpha_2\lambda_2}{\mu_1^{(2)}\psi_1 + \mu_2^{(1)}\psi_2}\right)^{-1} \cdot e^{-2\gamma_m\lambda_{\max}t_0/(1+3\gamma_m)}, \tag{5.2.3}$$

where the values of the quantities α_1, α_2, ψ_1, and ψ_2 are given in Table 5.2.1.

Example 5.2.1 Consider a system consisting of two different structures that function as stand-bys to each other. The failure and repair intensities of the structures are, respectively,

$$\lambda_1 = \lambda, \qquad \lambda_2 = 1.6\lambda, \qquad \mu_1 = \mu, \qquad \mu_2 = 0.9\mu,$$

Table 5.2.1   Redundant System of Two Different
Dependent Units: Unrestricted Repair

$$\alpha_1 = \frac{\lambda_2^{(1)}}{\lambda_2^{(1)} + \mu_1}, \quad \alpha_2 = \frac{\lambda_1^{(2)}}{\lambda_1^{(2)} + \mu_2}, \quad \beta_1 = \frac{\lambda_1}{\lambda_2^{(1)}}, \quad \beta_2 = \frac{\lambda_2}{\lambda_1^{(2)}},$$

$$\psi_1 = (1 - \alpha_2)(1 + \alpha_1\beta_1) + \alpha_2\beta_2\left(1 + \alpha_1\frac{\lambda_1}{\lambda_2}\right),$$

$$\psi_2 = (1 - \alpha_1)(1 + \alpha_2\beta_2) + \alpha_1\beta_1\left(1 + \alpha_2\frac{\lambda_2}{\lambda_1}\right).$$

MEASURE	EXACT VALUE	APPROXIMATE VALUE	ERROR
T_1	$\dfrac{1 + \alpha_1\beta_1 + \alpha_2\beta_2}{\alpha_1\lambda_1 + \alpha_2\lambda_2}$	$\dfrac{\mu_1\mu_2}{\lambda_2\lambda_1^{(2)}\mu_1 + \lambda_1\lambda_2^{(1)}\mu_2}$	$\delta_+ \sim \dfrac{3}{2\lambda_{\max}}$
T_2	$\dfrac{\mu_1^{(2)}\psi_1 + \mu_2^{(1)}\psi_2}{(\mu_1^{(2)} + \mu_2^{(1)})(\alpha_1\lambda_1 + \alpha_2\lambda_2)}$	$\dfrac{\mu_1\mu_2}{\lambda_2\lambda_1^{(2)}\mu_1 + \lambda_1\lambda_2^{(1)}\mu_2}$	$\delta_+ \sim \dfrac{1}{\lambda_{\max}}$
τ	$\dfrac{1}{\mu_1^{(2)} + \mu_2^{(1)}}$	—	—
K	$\left(1 + \dfrac{\alpha_1\lambda_1 + \alpha_2\lambda_2}{\mu_1^{(2)}\psi_1 + \mu_2^{(1)}\psi_2}\right)^{-1}$	$\left(1 + \dfrac{\lambda_2\lambda_1^{(2)}\mu_1 + \lambda_1\lambda_2^{(1)}\mu_2}{\mu_1\mu_2(\mu_1^{(2)} + \mu_2^{(1)})}\right)^{-1}$	$\delta_+ \sim 2\gamma_m^3$

where

$$\lambda = 0.01\frac{1}{\text{hr}}, \quad \mu = 1\frac{1}{\text{hr}},$$

$$\gamma = \frac{\lambda}{\mu} = \frac{0.01}{1} = 0.01.$$

When either of the structures fails, the functions of the defective structure are transferred to the operational one. The failure intensities are redistributed according to the conditional intensities:

$$\lambda_1^{(2)} = 2.2\lambda, \quad \lambda_2^{(1)} = 1.9\lambda.$$

In the case of repair of the structures comprising the system, repair depends in large measure on whether or not the other structure is undergoing repair. The following redistribution of the repair intensities is adopted:

$$\mu_1^{(2)} = 0.7\mu, \quad \mu_2^{(1)} = 0.8\mu.$$

We wish to determine the basic reliability measures of the system.

Solution. We make use of Table 5.2.1.

First we determine the values of the quantities α_1, α_2, β_1, and β_2:

$$\alpha_1 = \frac{\lambda_2^{(1)}}{\lambda_2^{(1)} + \mu_1} = \frac{1.9\lambda}{1.9\lambda + \mu} = \frac{1.9\gamma}{1.9\gamma + 1} = \frac{1.9 \cdot 0.01}{1.9 \cdot 0.01 + 1} \approx 0.0186,$$

$$\alpha_2 = \frac{\lambda_1^{(2)}}{\lambda_1^{(2)} + \mu_2} = \frac{2.2\lambda}{2.2\lambda + 0.9\mu} = \frac{2.2\gamma}{2.2\gamma + 0.9} = \frac{2.2 \cdot 0.01}{2.2 \cdot 0.01 + 0.9} \approx 0.0239,$$

$$\beta_1 = \frac{\lambda_1}{\lambda_2^{(1)}} = \frac{\lambda}{1.9\lambda} \approx 0.526,$$

$$\beta_2 = \frac{\lambda_2}{\lambda_1^{(2)}} = \frac{1.6\lambda}{2.2\lambda} = 0.727.$$

1. The mean operating time of the system up to failure is

$$T_1 = \frac{1 + \alpha_1\beta_1 + \alpha_2\beta_2}{\alpha_1\lambda_1 + \alpha_2\lambda_2} = \frac{1 + 0.0186 \cdot 0.526 + 0.0239 \cdot 0.727}{0.0186 \cdot 0.01 + 0.0239 \cdot 1.6 \cdot 0.01} \approx 1807 \quad \text{hr.}$$

2. An approximate value for the mean operating time of the system up to failure is

$$\tilde{T}_1 = \frac{\mu_1\mu_2}{\lambda_2\lambda_1^{(2)}\mu_1 + \lambda_1\lambda_2^{(1)}\mu_2} = \frac{\mu \cdot 0.9\mu}{1.6\lambda \cdot 2.2\lambda \cdot \mu + \lambda \cdot 1.9\lambda \cdot 0.9\mu}$$

$$= \frac{1}{\lambda} \cdot \frac{0.9}{1.6 \cdot 2.2\gamma + 1.9 \cdot 0.9 \cdot \gamma}$$

$$= \frac{1}{0.01} \cdot \frac{0.9}{1.6 \cdot 2.2 \cdot 0.01 + 1.9 \cdot 0.9 \cdot 0.01} \approx 1721 \quad \text{hr.}$$

The relative error of the approximate estimate is

$$\delta_+ = \frac{1807 - 1721}{1807} \approx 0.042$$

—that is, about 4 per cent.

3. The mean time between failures of the system is

$$T_2 = \frac{\mu_1^{(2)}\psi_1 + \mu_2^{(1)}\psi_2}{(\mu_1^{(2)} + \mu_2^{(1)})(\alpha_1\lambda_1 + \alpha_2\lambda_2)}.$$

We first compute the quantities ψ_1 and ψ_2:

$$\psi_1 = (1 - \alpha_2)(1 + \alpha_1\beta_1) + \alpha_2\beta_2\left(1 + \alpha_1\frac{\lambda_1}{\lambda_2}\right)$$

$$= (1 - 0.0239)(1 + 0.0186 \cdot 0.526) + 0.0239 \cdot 0.727 \cdot \left(1 + 0.0186 \cdot \frac{1}{1.6}\right)$$

$$\approx 1.003.$$

$$\psi_2 = (1 - \alpha_1)(1 + \alpha_2\beta_2) + \alpha_1\beta_1\left(1 + \alpha_2\frac{\lambda_2}{\lambda_1}\right)$$
$$= (1 - 0.0186)(1 + 0.0239\cdot0.727) + 0.0186\cdot0.526(1 + 0.0239\cdot1.6)$$
$$\approx 1.008.$$

Then

$$T_2 = \frac{0.7\mu\cdot1.003 + 0.8\mu\cdot1.008}{(0.7\mu + 0.8\mu)(0.0186\lambda + 0.0239\cdot1.6\lambda)}$$
$$= \frac{0.7\cdot1.003 + 0.8\cdot1.008}{(0.7 + 0.8)(0.0186\cdot0.01 + 0.0239\cdot1.6\cdot0.01)} \approx 1752 \quad \text{hr.}$$

4.   The mean repair time of the system is

$$\tau = \frac{1}{\mu_1^{(2)} + \mu_2^{(1)}} = \frac{1}{0.7\mu + 0.8\mu} = \frac{1}{1.5\mu} = \frac{1}{1.5} \approx 0.67 \quad \text{hr.}$$

5.   An approximate value (lower bound) for the probability of failure-free operation for the initial $t_0 = 100$ hours is computed by Formula (5.2.2) and is

$$\tilde{P}(t_0) = e^{-2\gamma_m\lambda_{\max}t_0/(1+3\gamma_m)}.$$

In our case

$$\gamma_m = \gamma_{\max} = \frac{\lambda_{\max}}{\mu_{\min}} = \frac{\lambda_1^{(2)}}{\mu_1^{(2)}} = \frac{2.2\lambda}{0.7\mu} = \frac{22}{7}\gamma = \frac{22}{7}\cdot0.01 \approx 0.031,$$

$$\lambda_{\max} = \lambda_1^{(2)} = 2.2\lambda = 2.2\cdot0.01 = 0.022 \quad \frac{1}{\text{hr}}\cdot$$

Then

$$\tilde{P}(t_0 = 100 \text{ hr}) = e^{-2\cdot0.031\cdot0.022\cdot100/(1+3\cdot0.031)} \approx e^{-0.125} \approx 0.88.$$

6.   The exact value of the readiness coefficient of the system can be determined from the formula

$$K = \left(1 + \frac{\alpha_1\lambda_1 + \alpha_2\lambda_2}{\mu_1^{(2)}\psi_1 + \mu_2^{(1)}\psi_2}\right)^{-1}.$$

However, we can find the exact value of the readiness coefficient more easily by making use of the results obtained in parts 3 and 4 of this example. It is

$$K = \frac{T_2}{T_2 + \tau} = \frac{1752}{1752 + 0.67} \approx 0.9996.$$

7.   By using Formula (5.2.3) and the results obtained in part 6 of this example, we find an approximate value (lower bound) for the reliability coefficient of the system for a time interval of length $t_0 = 100$ hours to be

$$\tilde{R}(t_0 = 100 \text{ hr}) = K\tilde{P}(t_0 = 100 \text{ hr}) = 0.9996 \cdot 0.88 \approx 0.88.$$

5.2.2 A system consisting of three different dependent units, where intensities of failure and repair depend on the state of the system

For the given system (Figure 5.2.3) we use a double-index notation for the intensities of failure and repair. For example:

$\lambda_3^{(1)}$ is the failure intensity of the third unit given that the first unit has already failed;

$\mu_1^{(2)}$ is the repair intensity of the first unit given that the second unit is undergoing repair, and so on (see Section 5.2.1).

By failure of the system we mean failure of any two of its units, where there is no separate utilization of the remaining operational (third) unit.

A diagram of passages by the system from one state into another is given in Figure 5.2.4, where H_0 is the state of the system with no failed units, H_2 is the state of the system where the second unit has failed, and H_{13} is one of the three states of failure of the system where the failed units are the first and third, and so on.

The mean operating time up to failure of the given system is given by the expression

$$T_1 = \frac{b_1 b_2 b_3 + \lambda_1 b_2 b_3 + \lambda_2 b_1 b_3 + \lambda_3 b_1 b_2}{\lambda_1 \lambda^{(1)} b_2 b_3 + \lambda_2 \lambda^{(2)} b_1 b_3 + \lambda_3 \lambda^{(3)} b_1 b_2}, \tag{5.2.4}$$

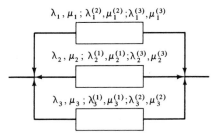

Figure 5.2.3 Block Diagram of the Model Used for Reliability Computations for a System of Three Units with Different Reliability Measures.

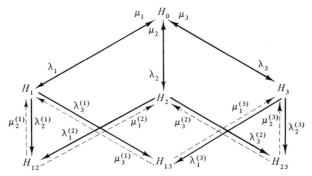

Figure 5.2.4   Diagram of Passages from One State to Another by a System of Three Dependent Units with Different Reliability Measures Where Failure of Any Two Units Leads to Failure of the System.

where

$$\lambda^{(1)} = \lambda_2^{(1)} + \lambda_3^{(1)}, \qquad b_1 = \lambda^{(1)} + \mu_1,$$
$$\lambda^{(2)} = \lambda_1^{(2)} + \lambda_3^{(2)}, \qquad b_2 = \lambda^{(2)} + \mu_2, \qquad (5.2.5)$$
$$\lambda^{(3)} = \lambda_1^{(3)} + \lambda_2^{(3)}; \qquad b_3 = \lambda^{(3)} + \mu_3.$$

If the condition

$$\gamma_m = \gamma_{\max} = \frac{\lambda_{\max}}{\mu_{\min}} \ll \tfrac{1}{3}, \qquad (5.2.6)$$

is satisfied, where

$$\lambda_{\max} = \max \{\lambda_1, \lambda_2, \lambda_3, \lambda_1^{(2)}, \lambda_1^{(3)}, \lambda_2^{(1)}, \lambda_2^{(3)}, \lambda_3^{(1)}, \lambda_3^{(2)}\},$$
$$\mu_{\min} = \min \{\mu_1, \mu_2, \mu_3, \mu_1^{(2)}, \mu_1^{(3)}, \mu_2^{(1)}, \mu_2^{(3)}, \mu_3^{(1)}, \mu_3^{(2)}\},$$

then we can obtain the following approximate formulas:

$$\tilde{T}_1 = \frac{\mu_1\mu_2\mu_3}{\lambda_1\lambda^{(1)}\mu_2\mu_3 + \lambda_2\lambda^{(2)}\mu_1\mu_3 + \lambda_3\lambda^{(3)}\mu_1\mu_2}, \qquad (5.2.7)$$

$$\tilde{T}_2 = \frac{\mu_1\mu_2\mu_3}{\lambda_1\lambda^{(1)}\mu_2\mu_3 + \lambda_2\lambda^{(2)}\mu_1\mu_3 + \lambda_3\lambda^{(3)}\mu_1\mu_2}. \qquad (5.2.8)$$

The absolute errors of Formulas (5.2.7) and (5.2.8) have the following orders of magnitude, respectively:

$$\delta_+ \sim \frac{5}{6\lambda_{\max}}, \qquad (5.2.9)$$

$$\delta_+ \sim \frac{1}{2\lambda_{\max}}. \qquad (5.2.10)$$

Table 5.2.2 Repairable System of Three Different Dependent Units, Where Failure of Any Two Is Equivalent to Failure of the System: Unrestricted Repair

MEASURE	LOWER BOUND
$\tilde{P}(t_0)$	$e^{-6\gamma_m \lambda_{\max} t_0/(1+5\gamma_m)}$
$\tilde{K}$	$\dfrac{1 + 3\gamma_m}{1 + 3\gamma_m + 3\gamma_m^2}$
$\tilde{R}(t_0)$	$\dfrac{1 + 3\gamma_m}{1 + 3\gamma_m + 3\gamma_m^2}\, e^{-6\gamma_m \lambda_{\max} t_0/(1+5\gamma_m)}$

An upper bound for the mean repair time of the system is

$$\tilde{\tau} = \frac{1}{\mu_{\min}}. \tag{5.2.11}$$

In Table 5.2.2 lower bounds are given for the basic reliability measures of the system in the case of unrestricted repair—that is, repair where all failed units in the system can undergo repair simultaneously and independently.

Example 5.2.2 Consider a system of three devices with failure and repair intensities, respectively, equal to

$$\lambda_1 = \lambda = 0.01 \ \frac{1}{\text{hr}},$$

$$\mu_1 = \mu = 2 \ \frac{1}{\text{hr}}, \qquad \gamma = \frac{\lambda}{\mu} = \frac{0.01}{2} = 0.005,$$

$$\lambda_2 = 1.1\lambda, \qquad \mu_2 = 0.9\mu,$$

$$\lambda_3 = 1.2\lambda, \qquad \mu_3 = 0.8\mu.$$

This system functions with two operational devices, and failure of only one of the three devices does not lead to breakdown, while failure of any two does. In the case of failure of one of the devices, owing to a load redistribution, there is a change in the failure intensities of the remaining operational devices. We assume these intensities to have the following values:

$$\lambda_2^{(1)} = 1.2\lambda, \qquad \lambda_3^{(1)} = 1.3\lambda,$$

$$\lambda_1^{(2)} = 1.5\lambda, \qquad \lambda_3^{(2)} = 1.4\lambda,$$

$$\lambda_1^{(3)} = 1.9\lambda, \qquad \lambda_2^{(3)} = 1.3\lambda.$$

We shall also assume that the intensities of repair of each of the three devices depend in an essential way on whether one device is undergoing repair. The values of these intensities are taken to be:

$$\mu_2^{(1)} = 0.8\mu, \qquad \mu_3^{(1)} = 0.7\mu,$$
$$\mu_1^{(2)} = 0.5\mu, \qquad \mu_3^{(2)} = 0.6\mu,$$
$$\mu_1^{(3)} = 0.1\mu, \qquad \mu_2^{(3)} = 0.7\mu.$$

Obviously

$$\gamma_m = \gamma_{\max} = \frac{\lambda_{\max}}{\mu_{\min}} = \frac{\lambda_1^{(3)}}{\mu_1^{(3)}} = \frac{1.9\lambda}{0.1\mu} = 19\gamma = 19 \cdot 0.005 = 0.095.$$

We wish to determine the basic reliability measures of the system.
Solution.   First we use Formulas (5.2.5) to find the quantities

$$\lambda^{(1)} = \lambda_1^{(2)} + \lambda_3^{(1)} = 2.5\lambda = 2.5 \cdot 0.01 = 0.025 \;\; \frac{1}{\mathrm{hr}},$$

$$\lambda^{(2)} = \lambda_1^{(2)} + \lambda_3^{(2)} = 2.9\lambda = 2.9 \cdot 0.01 = 0.029 \;\; \frac{1}{\mathrm{hr}},$$

$$\lambda^{(3)} = \lambda_1^{(3)} + \lambda_2^{(3)} = 3.2\lambda = 3.2 \cdot 0.01 = 0.032 \;\; \frac{1}{\mathrm{hr}},$$

$$b_1 = \mu_1 + \lambda_2^{(1)} + \lambda_3^{(1)} = \mu + 2.5\lambda = (2 + 2.5 \cdot 0.01) = 2.025 \;\; \frac{1}{\mathrm{hr}},$$

$$b_2 = \mu_2 + \lambda_1^{(2)} + \lambda_3^{(2)} = 0.9\mu + 2.9\lambda = (0.9 \cdot 2 + 2.9 \cdot 0.01) = 1.829 \;\; \frac{1}{\mathrm{hr}},$$

$$b_3 = \mu_3 + \lambda_1^{(3)} + \lambda_2^{(3)} = 0.8\mu + 3.2\lambda = (0.8 \cdot 2 + 3.2 \cdot 0.01) = 1.632 \;\; \frac{1}{\mathrm{hr}}.$$

1.   The mean operating time of the system up to failure is determined by Formula (5.2.4) to be

$$T_1 = \frac{b_1 b_2 b_3 + \lambda_1 b_2 b_3 + \lambda_2 b_1 b_3 + \lambda_3 b_1 b_2}{\lambda_1 \lambda^{(1)} b_2 b_3 + \lambda_2 \lambda^{(2)} b_1 b_3 + \lambda_1 \lambda^{(3)} b_1 b_2}$$

$$= \frac{\left[\begin{array}{l} 2.025 \cdot 1.829 \cdot 1.632 + 0.01 \cdot 1.829 \cdot 1.632 + 0.011 \cdot 2.025 \cdot 1.632 \\ \qquad\qquad\qquad\qquad\qquad\qquad\qquad + 0.012 \cdot 2.025 \cdot 1.829 \end{array} \right]}{\left[\begin{array}{l} 0.01 \cdot 0.025 \cdot 1.829 \cdot 1.632 + 0.011 \cdot 0.029 \cdot 2.025 \cdot 1.632 \\ \qquad\qquad\qquad\qquad\qquad\qquad\qquad + 0.012 \cdot 0.032 \cdot 2.025 \cdot 1.829 \end{array} \right]}$$

$$\approx 1925 \;\; \mathrm{hr}.$$

2.   An approximate value for the mean operating time of the system between failures is determined from Formula (5.2.8) to be

$$
\begin{aligned}
\tilde{T}_2 &= \frac{\mu_1\mu_2\mu_3}{\lambda_1\lambda^{(1)}\mu_2\mu_3 + \lambda_2\lambda^{(2)}\mu_1\mu_3 + \lambda_3\lambda^{(3)}\mu_1\mu_2} \\
&= \frac{\mu\cdot 0.9\mu\cdot 0.8\mu}{\lambda\lambda^{(1)}\cdot 0.9\mu\cdot 0.8\mu + 1.1\lambda\cdot\lambda^{(2)}\mu\cdot 0.8\mu + 1.2\lambda\cdot\lambda^{(3)}\mu\cdot 0.9\mu} \\
&= \frac{0.72}{(0.72\lambda^{(1)} + 0.88\lambda^{(2)} + 1.08\lambda^{(3)})\gamma} \\
&= \frac{0.72}{(0.72\cdot 0.025 + 0.88\cdot 0.029 - 1.08\cdot 0.032)\cdot 0.005} \\
&\approx 1846 \quad \text{hr.}
\end{aligned}
$$

3.   An upper bound for the mean repair time of the system is obtained by use of Formula (5.2.11) to be

$$
\tilde{\tau} = \frac{1}{\mu_{\min}} = \frac{1}{\mu^{(3)}} = \frac{1}{0.1\mu} = \frac{1}{0.1\cdot 2} = 5 \quad \text{hr.}
$$

4.   A lower bound for the probability of initial failure-free operation of the system for a time of $t_0 = 24$ hours is

$$
\tilde{P}(t_0 = 24 \text{ hr}) = e^{-6\gamma_m\Lambda_{\max}t_0/(1+5\gamma_m)} = e^{-6\cdot 0.095\cdot 1.9\cdot 0.01\cdot 24/(1+5\cdot 0.095)}
$$
$$
\approx e^{-0.18} \approx 0.84.
$$

5.   A lower bound for the readiness coefficient of the system is

$$
\tilde{K} = \frac{1 + 3\gamma_m}{1 + 3\gamma + 3\gamma_m^2} = \frac{1 + 3\cdot 0.095}{1 + 3\ 0.095 + 3\cdot (0.095)^2} \approx 0.98.
$$

6.   We find a lower bound for the reliability coefficient for a time interval of length $t_0 = 24$ hours by making use of the results obtained in parts 4 and 5 of this example. It is

$$
\tilde{R}(t_0 = 24 \text{ hr}) = \tilde{K}\tilde{P}(t_0 = 24 \text{ hr}) = 0.98\cdot 0.84 \approx 0.82.
$$

5.3   REPAIRABLE REDUNDANT SYSTEMS OF TWO UNITS WITH PARTIAL CONTROL OF THE STAND-BY UNIT

In this section we investigate the effect of the completeness of the control of stand-by units on the reliability measure of a redundant system consisting of two repairable units with repair intensity μ, in which only a $(1 - \alpha)$th part of the stand-by unit is controlled, and the αth part is not controlled. We assume that the operating unit is completely controlled.

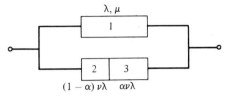

Figure 5.3.1   Block Diagram of the Model
Used for Reliability Compu-
tations for a Redundant System Consisting of Two Units
with Partial Control.

The intensities of failures of the controlled and uncontrolled parts of
the stand-by unit are $(1 - \alpha)\nu\lambda$ and $\alpha\nu\lambda$, respectively, where ν is the load
coefficient of the stand-by, $0 \le \nu \le 1$, and as in the block diagram (Figure
5.3.1) the stand-by unit is composite (units 2 and 3 in Figure 5.3.1).

The diagram of passages of the system from one state to another is
given in Figure 5.3.2—where, for example, H_0 is the state of the system
with no failed units, H_2 is the state of the system where the controlled
part of the stand-by unit (unit 2) has failed, and H_{23} is the state of the

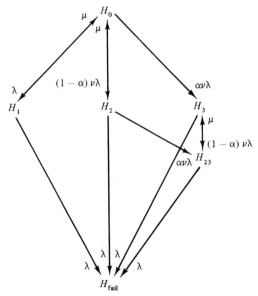

Figure 5.3.2   Diagram of Passages from One State
to Another by a Doubled System
with Partial Control of the Stand-by
Unit.

system in which the controlled and uncontrolled parts of the stand-by unit have failed.

All the approximate expressions given in this section hold subject to the condition

$$\gamma = \frac{\lambda}{\mu} \ll \frac{1}{2}. \qquad (5.3.1)$$

1.   The mean operating time of the given system up to failure is determined from the formula

$$T_1 = \frac{1}{\lambda} \cdot \frac{(1 + \gamma + \alpha\nu\gamma)\gamma + (1 + \alpha\nu)(1 + \gamma)(1 + \gamma + \nu\gamma)}{\alpha\nu + \gamma(1 + \alpha\nu)(1 + \nu)(1 + \gamma)}. \qquad (5.3.2)$$

2.   A lower bound for the mean operating time of the system up to failure has the form

$$\tilde{T}_1 = \frac{1}{\lambda} \frac{1 + \alpha\nu}{\alpha\nu + \gamma + \nu\gamma}. \qquad (5.3.3)$$

3.   A lower bound for the mean operating time of the system between failures is determined by the formula

$$\tilde{T}_2 = \tilde{T}_1 = \frac{1}{\lambda} \frac{1 + \alpha\nu}{\alpha\nu + \gamma + \nu\gamma}. \qquad (5.3.4)$$

Special cases

A.   If the stand-by unit is in an operating state ($\nu = 1$), then the corresponding reliability measure of the system is

$$T_1 = \frac{1}{\lambda} \cdot \frac{(1 + \gamma + \alpha\gamma)\gamma + (1 + \alpha)(1 + \gamma)(1 + 2\gamma)}{\alpha + 2\gamma(1 + \alpha)(1 + \gamma)}, \qquad (5.3.5)$$

$$\tilde{T}_1 = \tilde{T}_2 = \frac{1}{\lambda} \frac{1 + \alpha}{\alpha + 2\gamma}. \qquad (5.3.6)$$

B.   If the stand-by unit is in an operating state and is entirely uncontrolled—that is, $\nu = \alpha = 1$—then

$$T_1 = \frac{1}{\lambda} \frac{2 + 3\gamma}{1 + 2\gamma}, \qquad (5.3.7)$$

$$\tilde{T}_1 = \tilde{T}_2 = \frac{2}{\lambda}. \qquad (5.3.8)$$

The last expression becomes a better approximation under the condition

$$\gamma = \frac{\lambda}{\mu} \to 0$$

—that is, for very intense repair.

Moreover, for a system in which the basic unit has $n - 1$ stand-by units in an operating state, where $\gamma \to 0$ (that is as $\mu \to \infty$ and for a fixed value of λ), the mean operating time up to failure is

$$T_1 = \frac{n}{\lambda} \qquad (5.3.9)$$

—that is, it is equal to the mean operating time up to failure of a system in which the basic unit has $n - 1$ stand-by units in a nonoperating state and there is no repair.

Remark. It can be shown that if the control of a stand-by unit can be effected only by putting it into operation, then this control is advisable even if

$$\lambda_{\text{storage}} \geq \lambda\gamma.$$

C.   The case of nonoperating stand-by is not considered, since if a stand-by unit does not fail, then its control does not make sense.

Example 5.3.1   Consider a redundant system of two units where the operationality of the basic unit is completely controlled and that of the stand-by unit partially controlled. Specifically we assume that only half of the stand-by unit is controlled $(1 - \alpha = 0.5)$. The stand-by unit is in a partially operating state characterized by the coefficient $\nu = 0.6$.

The intensity of failures of the operating unit is $\lambda = 0.005 \, \dfrac{1}{\text{hr}}$, the intensity of repair in the system is $\mu = 0.5 \, \dfrac{1}{\text{hr}}$, and

$$\gamma = \frac{\lambda}{\mu} = \frac{0.005}{0.5} = 0.01.$$

We wish to determine the mean operating time of the system up to failure.

Solution.   The mean operating time of the system up to failure is determined from Formula (5.3.2) to be

$$T_1 = \frac{1}{\lambda} \frac{(1 + \gamma + \alpha\nu\gamma)\gamma + (1 + \alpha\nu)(1 + \gamma)(1 + \gamma + \nu\gamma)}{\alpha\nu + \gamma(1 + \alpha\nu)(1 + \nu)(1 + \gamma)}$$

$$= \frac{1}{0.005} \frac{\left[\begin{array}{c}(1 + 0.01 + 0.5 \cdot 0.6 \cdot 0.01) \cdot 0.01 \\ + (1 + 0.5 \cdot 0.6)(1 + 0.01)(1 + 0.01 + 0.6 \cdot 0.01)\end{array}\right]}{0.5 \cdot 0.6 + 0.01(1 + 0.5 \cdot 0.6)(1 + 0.6)(1 + 0.01)}$$

$$\approx 845 \quad \text{hr}.$$

Example 5.3.2   Consider a redundant system consisting of 2 units with an uncontrolled stand-by unit in an operating state. The parameters of the system are

$$\lambda = 0.01\,\frac{1}{\mathrm{hr}}, \qquad \mu = 1\,\frac{1}{\mathrm{hr}},$$

$$\gamma = \frac{\lambda}{\mu} = \frac{0.01}{1} = 0.01.$$

1.   We wish to determine the advantage (gain in mean time) that would be realized in allowing the repair of the given system, by determining the quantities

$$T_1|_{\gamma=0.01} \quad \text{and} \quad T|_{\gamma \to \infty}.$$

Obviously $\gamma \to \infty$ corresponds to the case of an unrepairable ($\mu = 0$) stand-by unit.

2.   Determination of $T_1|_{\gamma=0.01}$ in the case of a completely controlled stand-by unit.

Solution.   1.   Making use of Formula (5.3.7), we obtain

$$T_1|_{\gamma=0.01} = \frac{1}{\lambda}\frac{2+3\gamma}{1+2\gamma} = \frac{1}{0.01}\frac{2+3\cdot0.01}{1+2\cdot0.01} \approx 199 \quad \mathrm{hr},$$

$$T_1|_{\gamma\to\infty} = \frac{1}{\lambda}\frac{3}{2} = \frac{1}{0.01}\frac{3}{2} = 150 \quad \mathrm{hr}.$$

2.   The case of a completely controlled stand-by corresponds to $\alpha = 0$, whence, via Formula (5.3.2), we obtain

$$T_1|_{\substack{\gamma=0.01 \\ \alpha=0}} = \frac{1}{\lambda}\frac{1+3\gamma}{2\gamma} = \frac{1}{0.01}\frac{1+3\cdot0.01}{2\cdot0.01} = 5150 \quad \mathrm{hr}.$$

5.4   RELIABILITY OF INDIVIDUAL SUBSYSTEMS IN A REPAIRABLE STAND-BY SYSTEM

In this section we investigate a repairable stand-by system consisting of N units, m of which are designed to operate in m equivalent but different "operating subsystems," with the remaining $N - m$ units used as a universal component stand-by pool. We determine the reliability measures of one of the m (arbitrary but fixed) individual operating subsystems under the condition that the stand-by units can be used to replace any failed operating unit (the so-called "channel" reliability problem).

In practice one often encounters the use of universal component stand-by units to raise system reliability. In Chapter 4 we determined reliability measures for a system of m units in series connection with a universal component stand-by of $N - m$ units in the case of the usual failure criterion for the system [failure of any $(N - m) + 1$ units is equivalent to failure of the system].

Here we investigate a different problem—namely, the problem of determining the reliability of only one arbitrary fixed "subsystem" out of m, given that any of the stand-by units can be used to replace any failed unit of the given type in any subsystem—that is, we determine the effect of using universal component stand-by units on a given unit of a fixed but arbitrary subsystem (Figure 5.4.1). (It is assumed that the given subsystem is in a failed state whenever this particular unit is in a failed state. An important special case is that in which the unit is the entire subsystem.)

Such a problem is most often encountered in designing the connection or control of several channels (hence the term "channel" reliability). In this case we have m isolated connection channels possibly scattered territorially. If we want to use another $N - m$ channels as stand-by in order to raise the reliability of the entire network and of each of the m channels, then it is of interest to estimate the reliability not only of the entire system of N channels, but also of each of the m channels separately. This is the problem of estimating "channel" reliability.

*

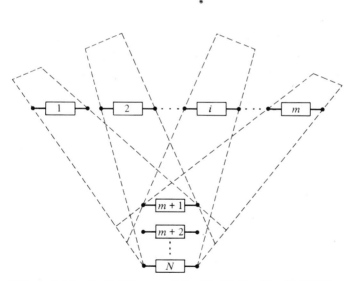

Figure 5.4.1   Block Diagram of the Model Used for Reliability Computations for an Individual Subsystem in a Redundant System.

5.4.1   Systems consisting of N identical units

Each of the N units of the system is designed to operate in one of m operating locations (subsystems consisting of one unit), and the $N - m$ "free" units are on operating stand-by for the m operating units. The units are characterized by a failure intensity of λ and repair intensity of μ. As always, we consider only two types of repair in the system: unrestricted repair, where all failed units can undergo repair simultaneously and independently; and restricted repair, where only one failed unit in the system can undergo repair at any given time.

Figures 5.4.2 and 5.4.3 show the respective diagrams of passages of a system from one state to another for these two cases. Here H_j is the state of the system with any j units failed $(0 \leq j \leq N - m - 1)$; H_i is the state of the system with i failed units, with an operational unit in the given operating location $(N - m \leq i \leq N - 1)$; $H_{i,1}$ is the state of the system with $i + 1$ failed units with no operational unit in the given location $(N - m \leq i \leq N - 1)$.

Tables 5.4.1 and 5.4.2 list exact and approximate values for the basic reliability measures of an individual subsystem for the cases of unrestricted and restricted repair, respectively. The approximate formulas are subject to the condition

$$\gamma = \frac{\lambda}{\mu} \leq \frac{1}{N}. \tag{5.4.1}$$

Remark. The formulas for the reliability measures of a subsystem in the unrestricted-repair case given in Table 5.4.1 and in Section 5.4.2 will also hold for arbitrary distribution laws of individual units operating and repair times if we assume the following:

(1)   The operating times between successive failures for the αth unit $(\alpha = 1, 2, \ldots, N - 1, N)$,

$$\theta_{\mathrm{I}}^{(\alpha)}, \; \theta_{\mathrm{II}}^{(\alpha)}, \; \theta_{\mathrm{III}}^{(\alpha)}, \; \ldots, \; \theta_{M-1}^{(\alpha)}, \; \theta_{M}^{(\alpha)}, \; \ldots$$

[where $\theta_M^{(\alpha)}$ is the operating time of the αth unit between the $(M - 1)$st and Mth failures], are independent, identically distributed random variables with mean value T_α $(\alpha = 1, 2, \ldots, N - 1, N)$.

(2)   The repair times of the αth unit,

$$\xi_{\mathrm{I}}^{(\alpha)}, \; \xi_{\mathrm{II}}^{(\alpha)}, \; \xi_{\mathrm{III}}^{(\alpha)}, \; \ldots, \; \xi_{M-1}^{(\alpha)}, \; \xi_{M}^{(\alpha)}, \; \ldots$$

(where $\xi_M^{(\alpha)}$ is the repair time of the αth unit after its Mth failure), are also independent, identically distributed random variables with mean value τ_α $(\alpha = 1, 2, \ldots, N - 1, N)$.

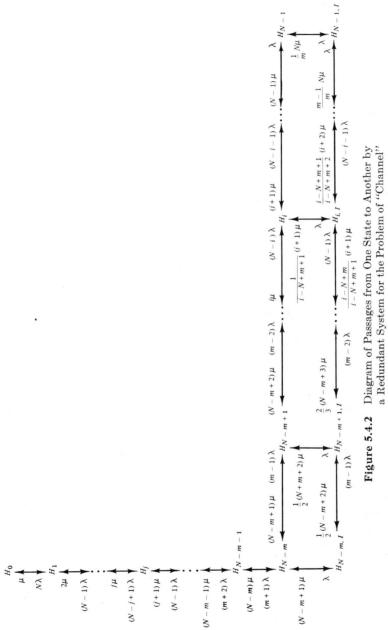

Figure 5.4.2   Diagram of Passages from One State to Another by a Redundant System for the Problem of "Channel" Reliability in the Case of Unrestricted Repair.

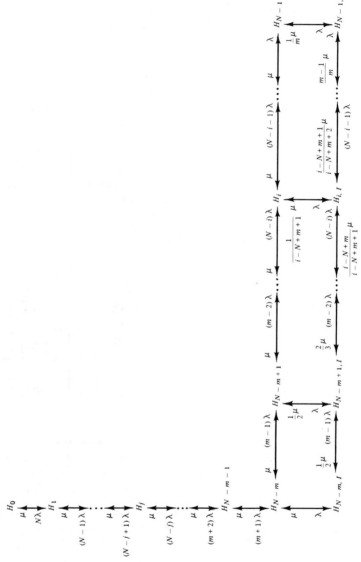

Figure 5.4.3   Diagram of Passages from One State to Another by a Redundant System for the "Channel" Reliability Problem in the Restricted Repair Case.

(3)   For any pair of units with indices α and β ($\alpha, \beta = 1, 2, \ldots, N - 1,$ N), the variables

$$\theta_I^{(\alpha)}, \; \theta_{II}^{(\alpha)}, \; \theta_{III}^{(\alpha)}, \; \ldots, \; \theta_{M-1}^{(\alpha)}, \; \theta_M^{(\alpha)}, \; \ldots$$

are statistically independent of

$$\xi_I^{(\alpha)}, \; \xi_{II}^{(\alpha)}, \; \xi_{III}^{(\alpha)}, \; \ldots, \; \xi_{M-1}^{(\alpha)}, \; \xi_M^{(\alpha)}, \; \ldots$$

and of

$$\theta_I^{(\beta)}, \; \theta_{II}^{(\beta)}, \; \theta_{III}^{(\beta)}, \; \ldots, \; \theta_{M-1}^{(\beta)} \; \theta_M^{(\beta)}, \; \ldots.$$

Furthermore, $\xi_I^{(\beta)}, \; \xi_{II}^{(\beta)}, \; \ldots, \; \xi_{M-1}^{(\beta)}, \; \xi_M^{(\beta)}$ are independent of the θ variables corresponding to the unit with index β.

(4)   Besides satisfying conditions (1), (2), and (3), all these random variables have distributions that are not "lattice"—that is, there do not exist numbers a, h such that all possible values of the random variables can be represented in the form $a + kh$, where the parameter k can take on any integral values.

These four assumptions do not impose very stringent restrictions on the distribution laws of the random variables involved in the process that describes the behavior of our system. Thus, in particular, this process is not necessarily Markovian.

In the computations involving the appropriate "exponential" formulas we take

$$\gamma_\alpha = \frac{\tau_\alpha}{T_\alpha}, \qquad \lambda_\alpha = \frac{1}{T_\alpha}.$$

Example 5.4.1   Suppose we have a system consisting of $m = 2$ operating and $N - m = 2$ stand-by units (for a total of $N = 4$ units in the system). All four units are identical and have intensities of failure and repair $\lambda = 0.05 \, \dfrac{1}{\text{hr}}$, $\mu = 1 \, \dfrac{1}{\text{hr}}$, and

$$\gamma = \frac{\lambda}{\mu} = \frac{0.05}{1} = 0.05.$$

We want to determine the basic reliability measures of one separate subsystem for the cases of unrestricted and restricted repair.

Solution.   We use Tables 5.4.1 and 5.4.2.

1.   An approximate value for the mean operating time of the subsystem up to failure for restricted repair is

$$\tilde{T}_1 = \frac{m!}{\lambda N! \gamma^{N-m}} [1 + (N - m + 2)\gamma]$$

$$= \frac{2![1 + (4 - 2 + 2) \cdot 0.05]}{0.05 \cdot 4! (0.05)^{4-2}} = 800 \quad \text{hr.}$$

Table 5.4.1   Reliability Measures of an Individual Subsystem in a Repairable Redundant System: Unrestricted Repair

MEASURE	EXACT VALUE	APPROXIMATE VALUE	ERROR
T_2	$\displaystyle\sum_{i=m}^{N} \frac{C_N^i}{\gamma^i} + \frac{N}{m\gamma}\sum_{i=0}^{m-2}\frac{C_{N-1}^i}{\gamma^i}$	$\displaystyle\frac{1}{\lambda\gamma^{N-m}C_N^m} \times \left[1 + \frac{N^2 - (m-1)(N+1)}{N-m+1}\,\gamma\right]$	$\displaystyle\delta_+ \sim \frac{(m-1)\gamma}{12\lambda},\ \text{if } N-m=1;$ $\displaystyle\delta_+ \sim \frac{N(N-m)\{2N + (N-m+3)[N(N-m)-(m-1)]\}}{2C_N^m\lambda\gamma^{N-m-2}(N-m+1)^2(N-m+2)}$ $\text{if } N-m>1.$
τ	$\displaystyle\frac{\displaystyle\sum_{i=0}^{m}\frac{C_N^i}{\gamma^i}}{\displaystyle\frac{\lambda N}{m\gamma}\sum_{i=0}^{m-1}\frac{C_{N-1}^i}{\gamma^i}} - \frac{1}{\lambda}$	$\displaystyle\frac{\gamma}{\lambda(N-m+1)} \times \left[1 + \frac{(m-1)(N-m)}{(N-m+1)(N-m+2)}\,\gamma\right]$	$\displaystyle\delta \sim \frac{(m-1)(N-m)[N(m-3)-(m-1)^2]\gamma^3}{\lambda(N-m+1)^3(N-m+2)(N-m+3)}$
K	$\displaystyle\frac{\displaystyle\sum_{i=m}^{N}\frac{C_N^i}{\gamma^i} + \frac{N}{m\gamma}\sum_{i=0}^{m-2}\frac{C_{N-1}^i}{\gamma^i}}{\displaystyle N\sum_{i=0}^{m}\frac{C_N^i}{\gamma^i}}$	$\displaystyle 1 - \frac{C_N^{m-1}}{m}\gamma^{N-m+1}$	$\displaystyle\delta_+ \sim C_N^{m-1}\frac{2 + (N-m)(N+2)}{m(N-m+2)}\gamma^{N-m+2}$

2. The mean operating time of the subsystem between failures for unrestricted repair is

$$
T_2 = \frac{\displaystyle\sum_{i=m}^{N} \frac{C_N^i}{\lambda^i} + \frac{N}{m\gamma} \sum_{i=0}^{m-2} \frac{C_{N-1}^i}{\gamma^i}}{\displaystyle\frac{\lambda N}{m\gamma} \sum_{i=0}^{m-1} \frac{C_{N-1}^i}{\gamma^i}} = \frac{\dfrac{C_4^2}{\gamma^2} + \dfrac{C_4^3}{\gamma^3} + \dfrac{C_4^4}{\gamma^4} + \dfrac{4}{2\gamma}}{\dfrac{4\lambda}{2\gamma}\left(1 + \dfrac{C_3^1}{\gamma}\right)}
$$

$$
= \frac{\dfrac{6}{(0.05)^2} + \dfrac{4}{(0.05)^3} + \dfrac{1}{(0.05)^4} + \dfrac{2}{0.05}}{\dfrac{2\cdot 0.05}{0.05}\left(1 + \dfrac{3}{0.05}\right)} \approx 1594 \quad \text{hr.}
$$

3. The mean repair time of the subsystem for the case of restricted repair is

$$
\tau = \frac{\displaystyle\sum_{i=0}^{m} \frac{1}{\gamma^j i!}}{\displaystyle\frac{\lambda}{m\gamma} \sum_{i=0}^{m-1} \frac{1}{\gamma^i i!}} - \frac{1}{\lambda} = \frac{1 + \dfrac{1}{\gamma} + \dfrac{1}{2\gamma^2}}{\dfrac{\lambda}{2\gamma}\left(1 + \dfrac{1}{\gamma}\right)} - \frac{1}{\lambda}
$$

$$
= \frac{1 + \dfrac{1}{0.05} + \dfrac{1}{2\cdot (0.05)^2}}{\dfrac{0.05}{2\cdot 0.05}\left(1 + \dfrac{1}{0.05}\right)} - \frac{1}{0.05} \approx 1.05 \quad \text{hr.}
$$

4. An approximate value for the readiness coefficient of the subsystem in the unrestricted-repair case is

$$
\tilde{K} = 1 - \frac{C_N^{m-1}}{m}\gamma^{N-m+1} = 1 - \frac{C_4^1}{2}\gamma^3 = 1 - 2\cdot (0.05)^3 = 0.99975.
$$

5.4.2 Systems consisting of N different units

Suppose the units of a system are characterized by failure and repair intensity $\lambda_1, \mu_1; \lambda_2, \mu_2; \ldots; \lambda_j, \mu_j; \ldots; \lambda_N, \mu_N$, respectively.

As before, these units are designed to operate in m operating locations (subsystems), and the remaining $N - m$ units, not in operating locations at the given moment, are stand-bys.

Table 5.4.2 Reliability Measures of an Individual Subsystem in a Repairable Redundant System: Restricted Repair

MEASURE	EXACT VALUE	APPROXIMATE VALUE	ERROR
T_1	—	$\dfrac{m!}{\lambda \cdot N! \gamma^{N-m}}[1 + (N - m + 2)\gamma]$	$\delta_- \sim \dfrac{2\gamma(-2m^2 + 5m - 3)}{(m + 1)\lambda}$, if $N - m = 1$; $\delta \sim \dfrac{-3m^2 + 17m + 3}{(m + 1)(m + 2)\lambda}$, if $N - m = 2$; $\delta \sim \dfrac{m![(N + 1)(N - m) - 4m^2 + 12m - 5]}{N! \lambda \gamma^{N-m-2}}$, if $N - m > 2$.
T_2	$\dfrac{\displaystyle\sum_{i=m}^{N} \frac{1}{\gamma^i \cdot i!} + \frac{1}{m\gamma}\sum_{i=0}^{m-2} \frac{1}{\gamma^i \cdot i!}}{\dfrac{\lambda}{m\gamma}\displaystyle\sum_{i=0}^{m-1}\frac{1}{\gamma^i \cdot i!}} - \dfrac{1}{\lambda}$	$\dfrac{m!}{\lambda \cdot N! \gamma^{N-m}}[1 + (N - m + 1)\gamma]$	$\delta_+ \sim \dfrac{(m - 1)\gamma}{(m + 1)\lambda}$, if $N - m = 1$; $\delta_+ \sim \dfrac{m! \cdot (N - 1)(N - m + 1)}{\lambda \cdot N! \gamma^{N-m-2}}$, if $N - m > 1$.
τ	$\dfrac{\displaystyle\sum_{i=0}^{m}\frac{1}{\gamma^i \cdot i!}}{\dfrac{\lambda}{m\gamma}\displaystyle\sum_{i=0}^{m-1}\frac{1}{\gamma^i \cdot i!}} - \dfrac{1}{\lambda}$	$\dfrac{\gamma}{\lambda}[1 + (m - 1)\gamma]$	$\delta \sim \dfrac{1}{\lambda} \cdot (m - 1)(m - 3)\gamma^3$

$$K \quad \frac{\displaystyle\sum_{i=m}^{N} \frac{1}{\gamma^i \cdot i!} + \frac{1}{m\gamma} \sum_{i=0}^{m-2} \frac{1}{\gamma^i \cdot i!}}{\displaystyle\sum_{i=0}^{N} \frac{1}{\gamma^i i!}}$$

$$1 - \frac{N!}{m!}\gamma^{N-m+1}$$

$$\delta_+ \sim \frac{N!}{m!}(N - 2m + 2)\gamma^{N-m+2}$$

Obviously the reliability characteristics of each subsystem (each location) will be identical for "equivalent" operating locations and for different units. The basic reliability measures of (any) one subsystem in the unrestricted repair case are determined by the formulas:

$$T_2 = \frac{K}{\Lambda}, \tag{5.4.2}$$

$$\tau = \frac{1 - K}{\Lambda}, \tag{5.4.3}$$

where

$$K = 1 - \frac{Q}{m} \sum_{l=0}^{m-1} \left[(m-l) \sum_{s=1}^{N-(l-1)} \sum_{r>s}^{N-(l-2)} \cdots \right.$$
$$\left. \sum_{m>n}^{N-2} \sum_{j>m}^{N-1} \sum_{i>j}^{N} \frac{1}{\gamma_s \gamma_r \cdots \gamma_m \gamma_j \gamma_i} \right]; \tag{5.4.4}$$

$$\Lambda = \frac{Q}{m} \sum_{l=1}^{m} \left[\sum_{s=1}^{N-(l-1)} \sum_{r>s}^{N-(l-2)} \cdots \right.$$
$$\left. \sum_{m>n}^{N-2} \sum_{j>m}^{N-1} \sum_{i>j}^{N} \frac{\lambda_s + \lambda_r + \cdots + \lambda_m - \lambda_j - \lambda_i}{\gamma_s \gamma_r \cdots \gamma_m \gamma_j \gamma_i} \right]; \tag{5.4.5}$$

$$Q = \prod_{\alpha=1}^{N} \frac{\lambda_\alpha}{\lambda_\alpha + \mu_\alpha} = \prod_{\alpha=1}^{N} \frac{\gamma_\alpha}{1 + \gamma_\alpha}, \qquad \gamma_\alpha = \frac{\lambda_\alpha}{\mu_\alpha}. \tag{5.4.6}$$

Example 5.4.2 Suppose we have a system of $N = 5$ different units. Of these, $m = 4$ are operating in the system and one is a stand-by. The failure and repair intensities of the units of the system are

$$\lambda_1, \mu_1; \lambda_2, \mu_2; \lambda_3, \mu_3; \lambda_4, \mu_4; \lambda_5, \mu_5.$$

We want to find formulas for the readiness coefficient of the subsystem and the quantity Λ for unrestricted repair of the system.

Solution. 1. By Formula (5.4.4) we have

$$K = 1 - \frac{Q}{4} \left[4 + 3 \sum_{i=1}^{5} \frac{1}{\gamma_i} + 2 \sum_{j=1}^{4} \sum_{i>j}^{5} \frac{1}{\gamma_j \gamma_i} + \sum_{m=1}^{3} \sum_{j>m}^{4} \sum_{i>j}^{5} \frac{1}{\gamma_m \gamma_j \gamma_i} \right],$$

where, according to Formula (5.4.6),

$$Q = \prod_{\alpha=1}^{5} \frac{\gamma_\alpha}{1 + \gamma_\alpha}, \qquad \gamma_\alpha = \frac{\lambda_\alpha}{\mu_\alpha}.$$

2. The quantity Λ determined by Formula (5.4.5) is equal to

$$\Lambda = \frac{Q}{4}\left[\sum_{i=1}^{5} \frac{\lambda_i}{\gamma_i} + \sum_{j=1}^{4}\sum_{i>j}^{5} \frac{\lambda_j + \lambda_i}{\gamma_j\gamma_i} + \sum_{m=1}^{3}\sum_{j>m}^{4}\sum_{i>j}^{5} \frac{\lambda_m + \lambda_j + \lambda_i}{\gamma_m\gamma_j\gamma_i} \right.$$
$$\left. + \sum_{n=1}^{2}\sum_{m>n}^{3}\sum_{j>m}^{4}\sum_{i>j}^{5} \frac{\lambda_n + \lambda_m + \lambda_j + \lambda_i}{\gamma_n\gamma_m\gamma_j\gamma_i} \right].$$

3. Knowing the quantities K and Λ and using Formulas (5.4.2) and (5.4.3), we can easily compute the mean operating time of the subsystem T_2 between failures and its mean repair time τ.

Remark. The multiple sums involved in the expression for the quantities K and Λ can be written in the usual order—for example,

$$\sum_{j=1}^{4}\sum_{i>j}^{5} \frac{\lambda_j + \lambda_i}{\gamma_j\gamma_i} = \sum_{i>1}^{5} \frac{\lambda_1 + \lambda_i}{\gamma_1\gamma_i} + \sum_{i>2}^{5} \frac{\lambda_2 + \lambda_i}{\gamma_2\gamma_i} + \sum_{i>3}^{5} \frac{\lambda_3 + \lambda_i}{\gamma_3\gamma_i} + \sum_{i>4}^{5} \frac{\lambda_4 + \lambda_i}{\gamma_4\gamma_i}$$
$$= \left(\frac{\lambda_1 + \lambda_2}{\gamma_1\gamma_2} + \frac{\lambda_1 + \lambda_3}{\gamma_1\gamma_3} + \frac{\lambda_1 + \lambda_4}{\gamma_1\gamma_4} + \frac{\lambda_1 + \lambda_5}{\gamma_4\gamma_5} \right)$$
$$+ \left(\frac{\lambda_2 + \lambda_3}{\gamma_2\gamma_3} + \frac{\lambda_2 + \lambda_4}{\gamma_2\gamma_4} + \frac{\lambda_2 + \lambda_5}{\gamma_2\gamma_5} \right)$$
$$+ \left(\frac{\lambda_3 + \lambda_4}{\gamma_3\gamma_4} + \frac{\lambda_3 + \lambda_5}{\gamma_3\gamma_5} \right) + \frac{\lambda_4 + \lambda_5}{\gamma_4\gamma_5}.$$

5.5 REPAIRABLE SYSTEMS WITH UNREPAIRABLE STAND-BY UNITS

We consider a system of n identical units of which one is operating and the remaining $n - 1$ are nonoperating stand-bys. The system functions in such a way that after failure of the operating unit it is replaced (repair of the system) by one of the stand-by units, and no repair of the failed unit is ever made. The failed unit takes no further part in the operation of the system. The system functions until the last (nth) unit fails. This stand-by system has the advantage of allowing the immediate return

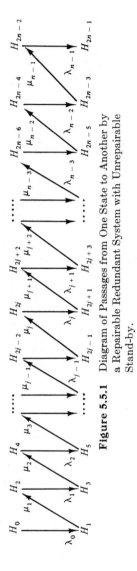

Figure 5.5.1 Diagram of Passages from One State to Another by a Repairable Redundant System with Unrepairable Stand-by.

of the system to operation after a failure. This is a characteristic model for systems of one-time action.

Having started operation, the system alternates between operational and failed states. Obviously, while it is functioning, the system can be in only a finite number of states: $H_0, H_1, H_2, \ldots, H_{2n-2}, H_{2n-1}$. We say that the system is in an operational state if it is in one of the states $H_0, H_2, \ldots, H_{2j}, \ldots, H_{2n-2}$ and that it is in a failed state when it is in one of the states $H_1, H_3, \ldots, H_{2n+1}, \ldots, H_{2n-1}$. Such an indexing of the states is characterized as follows: H_{2j} is the operational state entered by the system after the jth failure and the corresponding jth replacement (jth repair), H_{2j+1} is the failed state entered by the system after the jth replacement and the $(j + 1)$st failure. A diagram of the transitions of this system from state to state is shown in Figure 5.5.1. In this diagram all lower (odd) states are failed states of the system, and the upper states are the operational ones. If we use the notation

$$2n - 1 = N, \qquad \lambda_j = \omega_{2j},$$
$$\mu_j = \omega_{2j-1}, \qquad 0 \leq j \leq n - 1,$$

then the diagram of transitions can be presented in a different form (Figure 5.5.2). This diagram coincides perfectly with the passage diagram of a system for a so-called "pure death process," a system considered in Chapter 3.

Solving the system of differential equations (corresponding to the diagram of Figure 5.5.2) for the probabilities $p_m(t)$ that at time t the system is in state H_m, subject to the obvious initial conditions

$$p_0(0) = 1, \qquad p_m(0) = 0, 1 \leq m \leq N$$

in the case of distinct ω_i, we obtain

$$p_m(t) = \prod_{j=0}^{m-1} \omega_j \sum_{i=0}^{m} \frac{e^{-\omega_i t}}{\prod\limits_{\substack{l=0 \\ l \neq i}}^{m} (\omega_l - \omega_i)}, \qquad 0 \leq m \leq N - 1, \qquad (5.5.1)$$

$$p_N(t) = 1 - \prod_{j=0}^{N-1} \omega_j \sum_{i=0}^{N-1} \frac{e^{-\omega_i t}}{\omega_i \prod\limits_{\substack{l=0 \\ l \neq i}}^{N-1} (\omega_l - \omega_i)}. \qquad (5.5.2)$$

$$H_0 \xrightarrow{\omega_0} H_1 \xrightarrow{\omega_1} H_2 \longrightarrow \cdots \xrightarrow{\omega_{2j-1}} H_{2j} \xrightarrow{\omega_{2j}} H_{2j+1} \xrightarrow{\omega_{2j+1}} \cdots \xrightarrow{\omega_{N-2}} H_{N-1} \xrightarrow{\omega_{N-1}} H_N$$

Figure 5.5.2 Transformed Diagram of Passages—"Death Process."

If all replacements of a failed unit by a stand-by are made with the same intensity μ independent of the number of failures of the system, and the intensities of ensuing failures are all equal to λ (thus do not depend on the number of previous failures)—that is,

$$\omega_{2j+1} = \mu, \qquad 0 \le j \le \frac{N-3}{2},$$

$$\omega_{2j} = \lambda, \qquad 0 \le j \le \frac{N-1}{2},$$

then in this case (apparently the most typical in practice) we obtain

$$
p_m(t) = \frac{(\lambda\mu)^{m/2}}{(\mu - \lambda)^m} \left\{ e^{-\lambda t} \sum_{i=0}^{m/2} (-1)^i C_{(m/2)-1+i}^{(m/2)-1} \frac{[(\mu - \lambda)t]^{(m/2)-1}}{\left(\dfrac{m}{2} - i\right)!} \right.
$$
$$
\left. + (-1)^{(m/2)-1} e^{-\mu t} \times \sum_{i=0}^{(m/2)-1} C_{(m/2)+i}^{m/2} \frac{[(\mu - \lambda)t]^{(m/2)-i-1}}{\left(\dfrac{m}{2} - i - 1\right)!} \right\}, \quad (5.5.3)
$$

if m is even, $0 \le m \le N - 1$;

$$
p_m(t) = \frac{\lambda(\lambda\mu)^{(m-1)/2}}{(\mu - \lambda)^m} \left\{ e^{-\lambda t} \sum_{i=0}^{(m-1)/2} (-1)^i C_{[(m-1)/2]+i}^{(m-1)/2} \frac{[(\mu - \lambda)t]^{(m-1)/2 - i}}{\left(\dfrac{m-1}{2} - i\right)!} \right.
$$
$$
\left. + e^{-\mu t} \cdot (-1)^{(m+1)/2} \times \sum_{i=0}^{(m-1)/2} C_{[(m-1)/2]+i}^{(m-1)/2} \frac{[(\mu - \lambda)t]^{[(m-1)/2]-i}}{\left(\dfrac{m-1}{2} - i\right)!} \right\}, \quad (5.5.4)
$$

if m is odd, $1 \le m \le N$;

$$
P_N(t) = 1 - \frac{1}{(\mu - \lambda)^{n-1}} \left[\mu^{n-1} e^{-\lambda t} \sum_{i=0}^{n-1} \frac{(\lambda t)^{n-i-1}}{(n-i-1)!} \right.
$$
$$
\times \sum_{j=0}^{l} (-1)^j C_{n-2+j}^{n-2} \left(\frac{\lambda}{\mu - \lambda}\right)^j + (-1)^n \frac{\lambda^n}{\mu - \lambda} e^{-\mu t}
$$
$$
\left. \times \sum_{i=0}^{n-2} \frac{(\mu t)^{n-i-2}}{(n-i-2)!} \sum_{j=0}^{i} C_{n-1+j}^{n-1} \left(\frac{\mu}{\mu - \lambda}\right)^j \right]. \quad (5.5.5)
$$

Making use of Formulas (5.5.1)–(5.5.5) for the appropriate cases, we can easily determine the probability $K(t)$ that the system is in one of the operational states at the moment of time t. It is

$$K(t) = \sum_{\substack{m=0 \\ \{\text{over } m \text{ even}\}}}^{N-1} p_m(t). \qquad (5.5.6)$$

Obviously, given the conditions of the model under consideration, the readiness coefficient is

$$K = \lim_{t \to \infty} K(t) = 0. \qquad (5.5.7)$$

The mean operating time T_N of the system up to its final failure (entering state H_N) and the mean time T of occupancy of the system in the operational states are determined by the obvious considerations to be

$$T_N = \sum_{i=0}^{N-1} \frac{1}{\omega_i}, \qquad (5.5.8)$$

$$T = \sum_{\substack{i=0 \\ \{\text{over } i \text{ even}\}}}^{N-1} \frac{1}{\omega_i}. \qquad (5.5.9)$$

The probability that the system is still functioning in the time interval $[0, t]$ (is either operating or is having a failed unit replaced)—that is, the probability that it does not enter the last state H_N in this time interval— is determined by the formula

$$S_N(t) = 1 - p_N(t), \qquad (5.5.10)$$

where the probabilities $p_N(t)$ are determined by Formulas (5.5.2) and (5.5.5) for the corresponding cases.

Example 5.5.1 Suppose we have a system with one basic device and one stand-by device in a nonoperating state. Suppose the failure intensity of the operating device is $\lambda = 0.02 \, \dfrac{1}{\text{hr}}$, and the intensity of replacement of the failed operating device by the stand-by device (repair intensity) is $\mu = 2 \, \dfrac{1}{\text{hr}}$.

We wish to determine the probability that the system is operational at time $t = 24$ hours.

Solution.   The system has four states:

H_0, H_2 are the operational states;
H_1, H_3 are the failed states, where
    H_0 is the state in which the basic device operates failure-free;
    H_1 is the state in which the basic device has failed and substitution of the stand-by device has been initiated (repair of the system);
    H_2 is the state in which substitution of the stand-by device has been made and failure-free operation of the system continues;
    H_3 is the state of final failure of the system, wherein the stand-by device has failed in service.

Since, in our case, $n = 2$, $N = 2n - 1 = 3$, Formulas (5.5.3)–(5.5.5) for the probabilities $p_m(t)$ that the system is in state H_m ($m = 0, 1, 2, 3$) at time t yield

$$p_0(t) = e^{-\lambda t},$$

$$p_1(t) = \frac{\lambda}{\mu - \lambda} (e^{-\lambda t} - e^{-\mu t}),$$

$$p_2(t) = \frac{\lambda \mu}{(\mu - \lambda)^2} \{e^{-\lambda t}[(\mu - \lambda)t - 1] + e^{-\mu t}\},$$

$$p_3(t) = 1 - \frac{1}{\mu - \lambda} \left[\mu e^{-\lambda t} \left(1 + \lambda t - \frac{\lambda}{\mu - \lambda}\right) + \frac{\lambda^2}{\mu - \lambda} e^{-\mu t} \right].$$

The probability that at time t the system is operational is, according to Formula (5.5.6),

$$K(t) = p_0(t) + p_2(t)$$
$$= e^{-\lambda t} + \frac{\lambda \mu}{(\mu - \lambda)^2} \{e^{-\lambda t}[(\mu - \lambda)t - 1] + e^{-\mu t}\}.$$

At time $t = 24$ hours this probability is

$$K(t = 24 \text{ hr}) = e^{-0.02 \cdot 24}$$
$$+ \frac{0.02 \cdot 2}{(2 - 0.02)^2} \{e^{-0.02 \cdot 24}[(2 - 0.02) \cdot 24 - 1] + e^{-2 \cdot 24}\} \approx 0.91.$$

The probability that the system is already in the final state of failure (state H_3) at time $t = 24$ hours is

$$P_N(t) = P_3(t = 24 \text{ hr})$$

$$= 1 - \frac{1}{\mu - \lambda} \left[\mu e^{-\lambda t} \left(1 + \lambda t - \frac{\lambda}{\mu - \lambda} \right) + \frac{\lambda^2}{\mu - \lambda} e^{-\mu t} \right]$$

$$= 1 - \frac{1}{2 - 0.02} \left[2e^{-0.02 \cdot 24} \left(1 + 0.02 \cdot 24 - \frac{0.02}{2 - 0.02} \right) \right.$$

$$\left. + \frac{(0.02)^2}{2 - 0.02} e^{-2 \cdot 24} \right] \approx 0.08.$$

5.6 REPAIRABLE REDUNDANT SYSTEMS OF TWO UNITS WITH SWITCHING

In investigating a repairable redundant system consisting of two units with switching, one should first determine the influence of switch failure on the reliability of the system. Depending on the failure of the switch, switching structures can be divided into two types: compulsory and optional (system failure) structures.

The first type contains switches whose failure at any moment of time leads to failure of the system. This is observed in those cases where the switching structure consists basically of a power (current-carrying) unit.

The second type of structure contains switches that lead to the failure of the system only if switching from a failed operating unit to a stand-by unit is required at a moment of switch failure. This is the case if the basic part of the switching structure consists of an apparatus for checking the operationality of the operating unit and an apparatus for controlling the current-carrying part.

Computations for a redundant system consisting of two units with compulsory switching present no difficulties if one considers that such a system is a series connection of a repairable redundant system of two units without a switch characterized by its own reliability measures (see Chapter 4) with a switching structure whose reliability characteristics are known. In this case we can use the methods presented in Chapter 6 for the computations. (Incidentally, computations can be carried out analogously for more complex systems of m operating and $n - 1$ stand-by units with switching.)

However, precise computations for redundant systems consisting of two units with optional switching involves rather cumbersome expressions for the reliability measures. At the same time, in the majority of cases we deal with systems for which the distribution of failure-free operating time and repair time of individual units is exponential and the condition

$$\gamma_i = \frac{\lambda_i}{\mu_i} \ll \frac{1}{N}$$

is satisfied, where $N = m + n - 1$.

In this case the total failure intensity of the system can be determined approximately as

$$\Lambda \approx \Lambda_{\text{stand-by}} + k\lambda\gamma_S$$

where $\Lambda_{\text{stand-by}}$ is the failure intensity of the stand-by system without consideration of the switching structure, $\gamma_S = \lambda_S/\mu_S$ is the value of γ for the switching structure, and λ is the failure intensity of an operating unit.

The mean repair time of such a system can be approximately determined as

$$\tau = \tau_{\text{stand-by}} \frac{\Lambda_{\text{stand-by}}}{\Lambda} + \frac{k\gamma\gamma_S}{\Lambda} \cdot \frac{1}{1 + \frac{\mu_S}{\mu}}.$$

In Table 5.6.1 we give formulas for the reliability measures of a redundant repairable system consisting of two units with optional switching for operating and nonoperating stand-by.

A switching structure is usually more complex than the two basic types considered above. The composition of a switching structure involves such components as a device for checking the operationality of the operating unit, devices for properly controlling the switchboard (switch action selector), and the current-carrying line of the switchboard. If part of the switch can be considered as the compulsory part (current-carrying line) and part as optional (checking and controlling structures), then we can consider that we have a series connection of a redundant system consisting of two units with an optional switch and a compulsory switch.

5.7   BRIDGE CIRCUIT

Often a so-called bridge circuit (Figure 5.7.1) is used to raise reliability. In particular, such a connection of units is encountered in information networks when one investigates the probability of transmission of information from one point to another, taking account of the reliability of the communication channel.

In computing the reliability measures of these circuits, such as the probability of failure-free operation and the readiness coefficient, one essen-

Table 5.6.1 Redundant System of Two Units with Optional Switching: Operating Stand-by with Repair

MEASURE	EXACT VALUE	APPROXIMATE VALUE	ERROR
T_1	—	$\dfrac{1}{\lambda(\gamma_s + 2\gamma)}$	$\delta_+ \sim \dfrac{1}{\lambda}$
T_2	$\dfrac{\gamma_s(1+\gamma)\left(1+\frac{\mu_s}{\mu}\right) + (1 + 2\gamma + \gamma\gamma_s)\left(1 + \frac{\mu_s}{\mu} + \gamma\right)}{\lambda\left[\frac{\mu_s}{\mu}\gamma\gamma_s + (2\gamma + \gamma_s + \gamma\gamma_s)\left(1 + \frac{\mu_s}{\mu} + \gamma\right)\right]}$	$\dfrac{1}{\lambda(\gamma_s + 2\gamma)}$	$\delta_+ \sim \dfrac{1}{\lambda}$
τ	$\dfrac{\gamma_s(1+\gamma) + \gamma(1+\gamma_s)\left(1+\frac{\mu_s}{\mu}+\gamma\right)}{\mu\left[\frac{\mu_s}{\mu}\gamma\gamma_s + (2\gamma + \gamma_s + \gamma\gamma_s)\left(1 + \frac{\mu_s}{\mu} + \gamma\right)\right]}$	$\dfrac{\gamma_s + \gamma\left(1 + \frac{\mu_s}{\mu}\right)}{\mu\left(1 + \frac{\mu_s}{\mu}\right)(\gamma_s + 2\gamma)}$	$\delta_+ \sim \dfrac{\max\{\gamma, \gamma_s\}}{\mu}$
$P(t_0)$	—	$e^{-\lambda(\gamma_s + 2\gamma)t_0}$	$\delta \sim \max\{\gamma, \gamma_s\}$
K	$\dfrac{\gamma_s(1+\gamma)\left(1+\frac{\mu_s}{\mu}\right) + (1 + 2\gamma + \gamma\gamma_s)\left(1 + \gamma + \frac{\mu_s}{\mu}\right)}{(1+\gamma)^2(1+\gamma_s)\left(1 + \frac{\mu_s}{\mu} + \gamma_s\right)}$	$1 - \gamma\left(\gamma + \dfrac{\gamma_s}{1 + \frac{\mu_s}{\mu}}\right)$	$\delta_+ \sim \max\{\gamma^3, \gamma_s^3\}$
$R(t_0)$	—	$Ke^{-\lambda(\gamma_s + 2\gamma)t_0}$	$\delta \sim \max\{\gamma, \gamma_s\}$

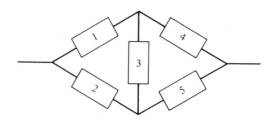

Figure 5.7.1   Bridge Circuit.

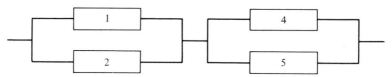

Figure 5.7.2   Series-Parallel Connection.

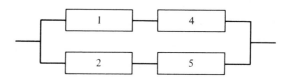

Figure 5.7.3   Parallel-Series Connection.

tially has to resort to a direct enumeration of states (for more details see Chapter 7). However, we note that in a given bridge circuit we can simplify the problem significantly if we consider only two classes of states. In the first class the bridging unit is operational (the probability of this event is r), and in the second class it fails (the probability of this event is q). Then, for the first class of states it suffices to consider a series-parallel connection (Figure 5.7.2), and for the second class, a parallel-series connection (Figure 5.7.3).

Table 5.7.1   Nonrepairable Bridge Circuit

MEASURE	EXACT VALUE	APPROXIMATE VALUE	ERROR
T	$2.88\dfrac{1}{\lambda}$	—	—
$P(t_0)$	$r(1 - q^2)^2 + q[1 - (1 - r^2)^2]$	$1 - 2q^2$	$\delta_- \sim 2q^3$
$Q(t_0)$	$rq^2(2 - q^2) + q(1 - r^2)^2$	$2q^2$	$\delta_+ \sim 2q^3$

Table 5.7.2 Bridge Circuit: Unrestricted Repair

MEASURE	EXACT VALUE	APPROXIMATE VALUE	ERROR
T_1	—	$\dfrac{1}{4\lambda\gamma}$	$\delta_+ \sim \dfrac{1}{\lambda}$
T_2	$\dfrac{1 + 5\gamma + 8\gamma^2 + 2\gamma^3}{2\lambda\gamma(2 + 9\gamma + 2\gamma^2)}$	$\dfrac{2 + \gamma}{8\lambda\gamma}$	$\delta_+ \sim \dfrac{19}{16}\dfrac{1}{\mu}$
τ	$\dfrac{2 + 8\gamma + 5\gamma^2 + \gamma^3}{2\mu(2 + 9\gamma + 2\gamma^2)}$	$\dfrac{2 - \gamma}{4\mu}$	$\delta_+ \sim \dfrac{15}{8}\dfrac{\gamma^2}{\mu}$
$P(t_0)$	—	$e^{-4\lambda\gamma t_0}$	$\delta \sim 4\gamma$
K	$\dfrac{1 + 5\gamma + 8\gamma^2 + 2\gamma^3}{(1 + \gamma)^5}$	$1 - 2\gamma^2$	$\delta_+ \sim 2\gamma^3$
$R(t_0)$	—	$K \cdot e^{-4\lambda\gamma t_0}$	$\delta \sim 4\gamma$

Formulas for computing the reliability measures are given in Tables 5.7.1 and 5.7.2.

REFERENCES

1. VASIL'EV, B. V., KOZLOV, B. A., AND TKACHENKO, L. G., *Reliability and Efficiency of Radioelectronic Structures*, chap. 2. Moscow: Soviet Radio, 1964.
2. EPISHIN, Y. G., "Dependence of Reliability on Completeness of Monitoring of Stand-by," in the collection *Cybernetics in the Service of Communism*. Moscow: Energiya, 1964.
3. SEDYAKIN, N. M., *Elements of the Theory of Random Impulse Currents*, chap. 9. Moscow: Soviet Radio, 1965.

6

SYSTEMS OF SERIES-CONNECTED UNITS

A series connection of units is one in which failure of even one unit leads to failure of the entire connection as a whole. A block diagram of a series connection is shown in Figure 6.0.1.

In this sense a series connection does not always coincide with a physical series connection of units. Thus, if we consider a group of capacitors connected in parallel (Figure 6.0.2) such that failure of a capacitor is equivalent to a "short circuit," then, for purposes of reliability computation, such a connection has a block diagram equivalent to that shown in Figure 6.0.1, since shorting of any one of them leads to shorting of the entire circuit. However, in the reliability sense, a series connection coincides with a physical series connection for an "open-circuit" failure.

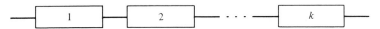

Figure 6.0.1 Block Diagram of the Model Used for Reliability Computations for Series-Connected Units.

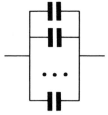

Figure 6.0.2 Group of Capacitors Connected in Parallel (Electrical Connection).

238

Below we always assume that failures of units are independent: failure of any set of units has no influence on the probability characteristics of the remaining units.

Recall that we are using the word unit in a broad sense, namely, as an independent part of the series connection. More generally, a unit itself may comprise stand-by circuits for which computation methods have been given in Chapters 3, 4, and 5.

6.1 SYSTEMS OF SERIES-CONNECTED NONREPAIRABLE UNITS

The duration of failure-free operation of a system of series-connected nonrepairable units is defined as the minimum duration of failure-free operation of any of its units (Figure 6.1.1)—that is,

$$\theta^{(1)} = \min_{i} \theta_i^{(1)}.$$

We assume that we know expressions for the probability $P_i(t)$ of failure-free operation of the individual units.

Table 6.1.1 shows the basic reliability measures for a system of series-connected nonrepairable units for arbitrary functions $P_i(t)$. The approximate expressions for the reliability measures are given subject to the condition that $\max_{i} Q_i(t_0) \ll 1/m$.

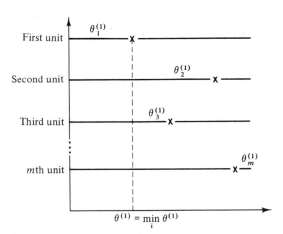

Figure 6.1.1 Time Diagram Illustrating the Failure-Free Operation Time of a System of m Series-Connected Nonrepairable Units.

Table 6.1.1 System of Series-Connected Nonrepairable Units: Arbitary Distribution Law of Operating Time of Individual Units up to Failure

MEASURE	EXACT VALUE	APPROXIMATE VALUE	ERROR
$P(t_0)$	$\displaystyle\prod_{i=1}^{m} P_i(t_0)$	$\displaystyle 1 - \sum_{i=1}^{m} Q_i(t_0)$	$\delta_+ < \dfrac{m^2}{2}\left[\max_i Q_i(t_0)\right]^2$
$Q(t_0)$	$\displaystyle 1 - \prod_{i=1}^{m} P_i(t_0)$	$\displaystyle \sum_{i=1}^{m} Q_i(t_0)$	$\delta_- < \dfrac{m^2}{2}\left[\max_i Q_i(t_0)\right]^2$
$P(t; t+t_0)$	$\displaystyle\prod_{i=1}^{m} P_i(t; t+t_0) = \frac{P(t+t_0)}{P(t)}$	$1 - Q(t+t_0) + Q(t) = P(t+t_0) + Q(t)$	$\delta_+ = Q(t)Q(t+t_0) < [Q(t+t_0)]^2$
$Q(t; t+t_0)$	$\displaystyle 1 - \prod_{i=1}^{m} P_i(t; t+t_0) = 1 - \frac{P(t+t_0)}{P(t)}$	$Q(t+t_0) - Q(t)$	$\delta_- = Q(t)Q(t+t_0) < Q[(t+t_0)]^2$
T	$\displaystyle\int_0^\infty P(t)\, dt$	—	—

Table 6.1.2 System of Series-Connected Nonrepairable Units: Exponential Distribution Law of Operating Time of Individual Units up to Failure

$$\Lambda = \sum_{i=1}^{m} \lambda_i$$

MEASURE	EXACT VALUE	APPROXIMATE VALUE	ERROR
$P(t_0)$	$\displaystyle\prod_{i=1}^{m} e^{-\lambda_i t_0} = e^{-t_0 \sum_{i=1}^{m} \lambda_i} = e^{-\Lambda t_0}$	$1 - \Lambda t_0$	$\delta_+ < \dfrac{(\Lambda t_0)^2}{2}$
$Q(t_0)$	$1 - e^{-\Lambda t_0}$	Λt_0	$\delta_- < \dfrac{(\Lambda t_0)^2}{2}$
$P(t; t + t_0)$	$e^{-\Lambda t_0}$	$1 - \Lambda t_0$	$\delta_+ < \dfrac{(\Lambda t_0)^2}{2}$
$Q(t; t + t_0)$	$1 - e^{-\Lambda t_0}$	Λt_0	$\delta_- < \dfrac{(\Lambda t_0)^2}{2}$
T	$\dfrac{1}{\Lambda}$	—	—

In Table 6.1.2 we give the basic reliability measures for a system of series-connected nonrepairable units where each of the units has an exponential distribution law for the operating time up to failure—that is,

$$P_i(t) = e^{-\lambda_i t}.$$

The approximate expressions for the reliability measures are given subject to the condition that

$$\Lambda t_0 = t_0 \sum_{i=1}^{m} \lambda_i \ll 1.$$

Example 6.1.1 Suppose we have a system of two units whose block diagram, for the purposes of reliability computation, is shown in Figure 6.1.2. The first unit, in turn, consists of two series-connected components and one universal stand-by component in an operating state, while the

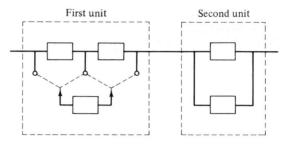

Figure 6.1.2   Block Diagram of the Model Used for Reliability Computations for the System of Example 6.1.1.

second unit is a redundant connection consisting of two units with the stand-by component in a nonoperating state.

All the components (parts of units) are assumed to be identical and have an exponential distribution of time of failure-free operation with parameter $\lambda = 0.005 \dfrac{1}{\text{hr}}$.

We wish to determine the basic reliability measures of the system.

Solution.   1.   The probability of initial failure-free operation of the system is

$$P(t_0) = P_1(t_0)P_2(t_0).$$

The quantities P_1 and P_2 are found from the formulas in Table 3.3.2 to be

$$P_1(t_0) = e^{-3\lambda t_0} + 3e^{-2\lambda t_0}(1 - e^{-\lambda t_0})$$
$$= 3e^{-2\lambda t_0} - 2e^{-3\lambda t_0} = 3e^{-2 \cdot 0.005 \cdot 50} - 2e^{-3 \cdot 0.005 \cdot 50} = 0.8743,$$
$$P_2(t_0) = e^{-\lambda t_0}(1 + \lambda t_0) = e^{-0.005 \cdot 50}(1 + 0.005 \cdot 50) = 0.9735.$$

Hence we obtain

$$P(t_0) = 0.8743 \cdot 0.9735 = 0.8511.$$

An approximate value is given by

$$\tilde{P}(t_0) = 1 - [Q_1(t_0) + Q_2(t_0)].$$

The quantities $Q_1(t_0)$ and $Q_2(t_0)$ are determined to be

$$Q_1(t_0) = 1 - P_1(t_0) = 1 - 0.8743 = 0.1257,$$
$$Q_2(t_0) = 1 - P_2(t_0) = 1 - 0.9735 = 0.0265.$$

Hence we find

$$\tilde{P}(t_0) = 1 - (0.1257 + 0.0265) = 0.8478.$$

2. By using the preceding results, we find the probability of failure in the initial $t_0 = 50$ hours to be

$$Q(t_0) = 1 - P(t_0) = 1 - 0.8511 = 0.1489.$$

An approximate value is

$$\tilde{Q}(t_0) = Q_1(t_0) + Q_2(t_0) = 0.1257 + 0.0265 = 0.1522.$$

3. The probability of failure-free operation in the interval from $t = 25$ hours to $t + t_0 = 50$ hours is obtained as follows:
First we compute the value $P(t)$. We find

$$P_1(t) = 3e^{-2\lambda t} - 2e^{-3\lambda t} = 3e^{-2 \cdot 0.005 \cdot 25} - 2e^{-3 \cdot 0.005 \cdot 25} = 0.9618,$$
$$P_2(t) = e^{-\lambda t}(1 + \lambda t) = e^{-0.005 \cdot 25}(1 + 0.005 \cdot 25) = 0.9934.$$

Hence we obtain

$$P(t) = P_1(t) \cdot P_2(t) = 0.9618 \cdot 0.9934 = 0.9552.$$

The desired value is found by the formula

$$P(t, t + t_0) = \frac{P(t + t_0)}{P(t)} = \frac{0.8511}{0.9552} = 0.8893.$$

An approximate value is

$$\tilde{P}(t, t + t_0) = P(t + t_0) + [1 - P(t)] = 0.8511 + (1 - 0.9552) = 0.8959.$$

4. The probability of failure in the time interval from $t = 25$ hours to $t + t_0 = 50$ hours is computed next. We make use of the preceding computations to obtain

$$Q(t, t + t_0) = 1 - P(t, t + t_0) = 1 - 0.8893 = 0.1107.$$

An approximate value is

$$\tilde{Q}(t, t + t_0) = Q(t + t_0) - Q(t) = 0.1489 - 0.0448 = 0.1041.$$

5. The mean time of failure-free operation T is obtained as follows. We first express $P(t)$ in the form

$$P(t) = (3e^{-2\lambda t} - 2e^{-3\lambda t})e^{-\lambda t}(1 + \lambda t)$$
$$= 3e^{-3\lambda t} - 2e^{-4\lambda t} + 3\lambda t e^{-3\lambda t} - 2\lambda t e^{-4\lambda t}.$$

Further, using the formula

$$T = \int_0^\infty P(t)\, dt,$$

we obtain

$$T = 3 \int_0^\infty e^{-3\lambda t}\, dt - 2 \int_0^\infty e^{-4\lambda t}\, dt + 3\lambda \int_0^\infty t e^{-3\lambda t}\, dt - 2\lambda \int_0^\infty t e^{-4\lambda t}\, dt,$$

whence, using an integral table (see Appendix 2.4), we find:

$$T = \frac{1}{\lambda} - \frac{1}{2\lambda} + \frac{1}{3\lambda} - \frac{1}{8\lambda} = \frac{17}{24\lambda}.$$

Substituting the numerical value $\lambda = 0.005 \dfrac{1}{\text{hr}}$, we have

$$T = \frac{17}{24 \cdot 0.005} \approx 144 \quad \text{hr}.$$

Example 6.1.2   Suppose we are given a system of five independent series-connected units (Figure 6.1.3), where each of the units has an exponential distribution of the failure-free operating time with parameters $\lambda_1 = 0.05 \dfrac{1}{\text{hr}}$, $\lambda_2 = 0.02 \dfrac{1}{\text{hr}}$, $\lambda_3 = 0.01 \dfrac{1}{\text{hr}}$, $\lambda_4 = 0.01 \dfrac{1}{\text{hr}}$, $\lambda_5 = 0.08 \dfrac{1}{\text{hr}}$.
We wish to determine the basic reliability measures of the system.
Solution.   First we find the value Λ:

$$\Lambda = \lambda_1 + \lambda_2 + \lambda_3 + \lambda_4 + \lambda_5 = 0.05 + 0.02 + 0.01 + 0.01 + 0.08$$
$$= 0.17 \frac{1}{\text{hr}}.$$

1.   The probability of initial failure-free operation for $t_0 = 0.5$ hour is

$$P(t_0) = e^{-\Lambda t_0} = e^{-0.17 \cdot 0.5} = e^{-0.085} = 0.9187.$$

An approximate value is

$$\tilde{P}(t_0) = 1 - \Lambda t_0 = 1 - 0.085 = 0.915.$$

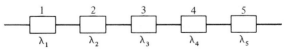

Figure 6.1.3   Block Diagram of the Model Used for Reliability Computations for the System in Example 6.1.2.

2. The probability of failure in the initial $t_0 = 0.5$ hours is

$$Q(t_0) = 1 - e^{-\Lambda t_0} = 1 - 0.9187 = 0.0813.$$

An approximate value is

$$\tilde{Q}(t_0) = \Lambda t_0 = 0.17 \cdot 0.5 = 0.085.$$

3. The mean time of failure-free operation is

$$T = \frac{1}{\Lambda} = \frac{1}{0.17} = 5.9 \quad \text{hr.}$$

6.2 SYSTEMS OF SERIES-CONNECTED REPAIRABLE UNITS

Failure-free operation of a system of series-connected repairable units is determined by various factors.

1. If the individual units of a system have arbitrary distribution of operating time up to failure, then the operating time of the system up to the first failure will be different from the operating time between the first and second failures, and so on. Thus, not only the mean values of the failure-free operation intervals but the distribution laws of the corresponding random variables will be different.

We illustrate this for the case where repair of failed units in the system is carried out instantaneously. Suppose that for each i we know the quantities $\theta_i^{(1)}$, $\theta_i^{(2)}$, ..., $\theta_i^{(j)}$, ..., where $\theta_i^{(j)}$ is the random variable representing the length of the jth interval of failure-free operation of the ith unit (Figure 6.2.1). For simplicity, consider a system consisting

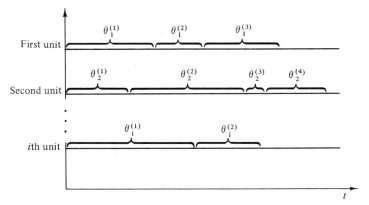

Figure 6.2.1 Time Diagram of Random Realizations of Failure-Free Operation Time of a System of Series-Connected Units with Instantaneous Repair.

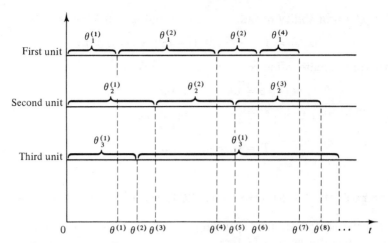

Figure 6.2.2   Illustration of the Mechanism of Formation of Failures in a System of Three Series-Connected Repairable Units (with Instantaneous Repair).

entirely of three units (Figure 6.2.2). Then if $\theta^{(j)}$ denotes the length of the jth failure-free operating interval of the system, we can write

$$\theta^{(1)} = \min \{\theta_1^{(1)}, \theta_2^{(1)}, \theta_3^{(1)}\}.$$

Suppose (in accordance with Figure 6.2.2) that $\theta_1^{(1)}$ turned out to be the smallest of all the $\theta_i^{(1)}$; then

$$\theta^{(2)} = \min \{\theta_1^{(2)}, \theta_2^{(1)} - \theta_1^{(1)}, \theta_3^{(1)} - \theta_1^{(1)}\},$$

—that is, failure is determined either by the condition that again the first unit fails earliest of all, or that the second or third unit fails first.

Consideration of such operating conditions for arbitrary distribution laws of the random intervals of failure-free operation leads to results that are essentially inapplicable in practice. Therefore, below we shall consider the case where all units have exponential distribution laws:

$$P_i(t) = \mathcal{P}\{\theta_i \geq t\} = e^{-\lambda_i t}.$$

2.   The reliability measures depend on the operating conditions of the remaining units of the system when one of the failed units is being repaired.

We can consider the following two basic operating conditions of a system:

    (a)   When any one of the units fails, the remaining units of the system are switched—that is, the failure intensity of the operational units is assumed to be zero at this time (Figure

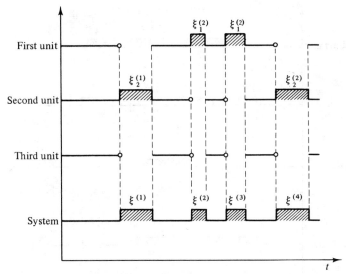

Figure 6.2.3 Time Diagram for the Operating Condition of a System Where the Remaining Units Are Switched When One Unit Fails. Notation: ——× is the moment of failure of a unit; ——○ is the moment of starting repair on a failed unit; $\xi_i^{(j)}$ is the jth interval of repair of the ith unit; $\xi^{(j)}$ is the jth repair interval of the system.

6.2.3). This means that when one of the units is being repaired, none of the remaining units can fail. A diagram of transitions for this case is given in Figure 6.2.4.

(b) When any one of the units fails, the remaining units of the system remain in an operating condition—that is, the failure intensity of each of the units remains unchanged except for dependence on the states of the other units (Figure 6.2.5).

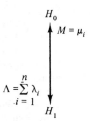

Figure 6.2.4 Diagram of Passages of a System of n Series-Connected Repairable Units for the Case Where the Remaining Units Are Switched When One Unit Fails.

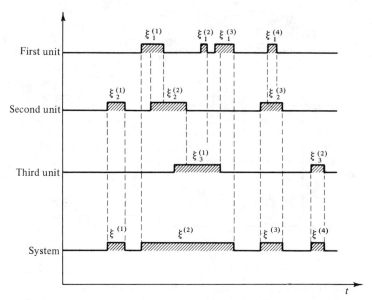

Figure 6.2.5   Time Diagram of the Operating Condition of a System Where the Remaining Units Stay in an Operating State When One Unit Fails. (See Figure 6.2.3 for the notation.)

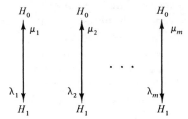

Figure 6.2.6   Diagram of Passages of a Collection of Units, Determining the States of a System of m Series-Connected Repairable Units for the Case Where, When One Unit Fails, the Remaining Units Stay in an Operating Condition.

We assume that repair is unrestricted in this case, so that repair of any failed unit in the system is initiated immediately. This allows us to assume the units independent in the repair process also; we need consider only the collection of states of individual units (Figure 6.2.6).

Table 6.2.1 shows the reliability measures of a system consisting of m series-connected repairable units with failure intensities λ_i and repair intensities μ_i, $i = 1, 2, \ldots, m$.

Analogous measures are presented in Table 6.2.2 for the case where all the units of a system are identical, characterized by the identical value $\lambda_i = \lambda$ and $\mu_i = \mu$.

Remark. In Tables 6.2.1 and 6.2.2 the measures corresponding to the second operating condition of the system have an asterisk as upper index; the formula for the precise value for $K^*(t)$ is not given because it is too unwieldy.

Example 6.2.1 Suppose we have a system of three independent, series-connected units, where each of the units has an exponential distribution for the failure-free operation time as well as the repair time. Suppose the intensities of failures and repairs of units are, respectively,

$$\lambda_1 = 0.05 \frac{1}{\text{hr}}; \quad \mu_1 = 1 \frac{1}{\text{hr}}; \quad \lambda_2 = 0.02; \quad \mu_2 = 0.5; \quad \lambda_3 = 0.01; \quad \mu_3 = 1.$$

We want to find the basic reliability measures of the system for operation for an initial period of $t_0 = 0.5$ hours; the readiness coefficient for operating conditions where, when one of the units fails, the system is switched; and the mean value of failure-free operating time and repair time (under the same operating conditions).

Solution.   First we determine the values

$$\Lambda = \lambda_1 + \lambda_2 + \lambda_3 = 0.05 + 0.02 + 0.01 = 0.08 \ \frac{1}{\text{hr}},$$

$$\gamma_1 = \frac{\lambda_1}{\mu_1} = 0.05, \quad \gamma_2 = \frac{\lambda_2}{\mu_2} = 0.04, \quad \gamma_3 = \frac{\lambda_3}{\mu_3} = 0.01.$$

1.   The mean time of failure-free operation is

$$T = \frac{1}{\Lambda} = \frac{1}{0.08} = 12.5 \ \text{hr}.$$

2.   The mean repair time is

$$\tau = \frac{1}{\Lambda} (\gamma_1 + \gamma_2 + \gamma_3)$$

$$= \frac{1}{0.08} (0.05 + 0.04 + 0.01) = 1.25 \ \text{hr}.$$

Table 6.2.1   System of m Series-Connected Repairable Units: Exponential Distribution Laws for Operating Time of Individual Units up to Failure, $F_i(t) = 1 - e^{-\lambda_i t}$ and Repair Time of Individual Units $F_{R_i} = 1 - e^{-\mu_i t}$

$$\gamma_i = \frac{\lambda_i}{\mu_i}$$
$$\Lambda = \sum_{i=1}^{m} \lambda_i$$

MEASURE	EXACT VALUE	APPROXIMATE VALUE	CONDITION	ERROR
T	$\dfrac{1}{\Lambda} = \left(\displaystyle\sum_{i=1}^{m} \lambda_i\right)^{-1}$	—	—	—
τ	$\dfrac{1}{\Lambda}\displaystyle\sum_{i=1}^{m}\gamma_i$	—	—	—
τ^*	$\dfrac{1}{\Lambda}\left(\displaystyle\prod_{i=1}^{m}(1+\gamma_i) - 1\right)$	$\dfrac{1}{\Lambda}\displaystyle\sum_{i=1}^{m}\gamma_i$	$\displaystyle\max_i \gamma_i \ll \frac{1}{m}$	$\delta_- < \dfrac{1}{\Lambda}\displaystyle\sum_{j=1}^{m}\sum_{i=1}^{j}\gamma_i\gamma_j \leq \dfrac{m^2}{2\Lambda}\left(\max_i \gamma_i\right)^2$
$P(t_0)$	$e^{-\Lambda t_0}$	$1 - \Lambda t_0$	$\Lambda t_0 \ll 1$	$\delta_+ < \dfrac{(\Lambda t_0)^2}{2}$

K	$\left(1 + \displaystyle\sum_{i=1}^{m} \gamma_i\right)^{-1}$	$1 - \displaystyle\sum_{i=1}^{m} \gamma_i$	$\max_i \gamma_i \ll \dfrac{1}{m}$	$\delta_+ < \left(\displaystyle\sum_{i=1}^{m} \gamma_i\right)^2 < m^2 (\max_i \gamma_i)^2$
K^*	$\displaystyle\prod_{i=1}^{m} (1 + \gamma_i)^{-1}$	$1 - \displaystyle\sum_{i=1}^{m} \gamma_i$	$\max_i \gamma_i \ll \dfrac{1}{m}$	$\delta_+ < \displaystyle\sum_{j=1}^{m}\sum_{i=1}^{j} \gamma_i\gamma_j \leq \dfrac{m^2}{2}(\max_i \gamma_i)^2$
$K(t)$	$K + (1 - K)e^{-[\Lambda+(1/\tau)]t}$	$1 - \Lambda\tau(1 - e^{-t/\tau})$	$\Lambda\tau \ll 1$	$\delta_+ \sim (\Lambda\tau)^2(1 - e^{-t/\tau})$
$K^*(t)$	Exact value not given	$1 - \Lambda\tau^*(1 - e^{-t/\tau^*})$	$\Lambda\tau^* \ll 1$	$\delta_+ \sim (\Lambda\tau^*)^2(1 - e^{-t/\tau^*})$
$R(t_0)$	$Ke^{-\Lambda t_0}$	$1 - \Lambda(\tau + t_0)$	$\Lambda t_0 \ll 1,\ \Lambda\tau \ll 1$	$\delta_+ < 2.5\,[\Lambda \max(t_0, \tau)]^2$
$R^*(t_0)$	$K^* e^{-\Lambda t_0}$	$1 - \Lambda(\tau^* + t_0)$	$\Lambda t_0 \ll 1,\ \Lambda\tau^* \ll 1$	$\delta_+ < 2.5\,[\Lambda \max(t_0, \tau^*)]^2$

Table 6.2.2 System of Series-Connected Repairable Units: Exponential Distribution Law for Operating Time up to Failure for Each Unit, $F_i(t) = 1 - e^{-\lambda t}$, and Exponential Distribution Law for Repair Time of Each Unit, $F_{R_i}(t) = 1 - e^{-\mu t}$

MEASURE	EXACT VALUE	APPROXIMATE VALUE	CONDITION	ERROR
T	$\dfrac{1}{m\lambda}$	—	—	—
τ	$\dfrac{1}{\mu}$	—	—	—
τ^*	$\dfrac{1}{m\lambda}[(1+\gamma)^m - 1]$	$\dfrac{1}{\mu}$	$\gamma \ll \dfrac{1}{m}$	$\delta_+ < \dfrac{m-1}{2}\cdot\lambda$
$P(t_0)$	$e^{-m\lambda t_0}$	$1 - m\lambda t_0$	$m\lambda t_0 \ll 1$	$\delta_+ < \dfrac{(m\lambda t_0)^2}{2}$
K	$(1+m\gamma)^{-1}$	$1 - m\gamma$	$\gamma \ll \dfrac{1}{m}$	$\delta_+ < (m\gamma)^2$
K^*	$(1+\gamma)^{-m}$	$1 - m\gamma$	$\gamma \ll \dfrac{1}{m}$	$\delta_+ < \dfrac{m(m-1)}{2}\gamma^2$
$K(t)$	$K + (1-K)e^{-t[m\lambda+(1/\tau)]}$	$1 - m\lambda\tau(1 - e^{-t/\tau})$	$m\lambda\tau \ll 1$	$\delta_+ \sim (m\lambda\tau)^2(1 - e^{-t/\tau})$
$K^*(t)$	Exact value not given	$1 - m\lambda\tau^*(1 - e^{-t/\tau^*})$	$m\lambda\tau^* \ll 1$	$\delta_+ \sim (m\lambda\tau^*)^2(1 - e^{-t/\tau^*})$
$R(t_0)$	$Ke^{-m\lambda t_0}$	$1 - m\lambda(\tau + t_0)$	$m\lambda t_0 \ll 1,\ m\lambda\tau \ll 1$	$\delta_+ < 2.5\,[m\lambda \max(t_0, \tau)]^2$
$R^*(t_0)$	$K^*e^{-m\lambda t_0}$	$1 - m\lambda(\tau^* + t_0)$	$m\lambda t_0 \ll 1,\ m\lambda\tau^* \ll 1$	$\delta_+ < 2.5\,[m\lambda \max(t_0, \tau^*)]^2$

3. The probability of failure-free operation of the system for initial time of $t_0 = 0.5$ hour is given by

$$P(t_0) = e^{-\Lambda t_0} = e^{-0.08 \cdot 0.5} = 0.962.$$

An approximate value for this quantity is

$$\tilde{P}(t_0) = 1 - \Lambda t_0 = 1 - 0.08 \cdot 0.5 = 0.96.$$

4. The readiness coefficient of the system is

$$K = \frac{1}{1 + \gamma_1 + \gamma_2 + \gamma_3} = \frac{1}{1 + 0.05 + 0.04 + 0.01} = 0.909.$$

An approximate value of the readiness coefficient is

$$\tilde{K} = 1 - (\gamma_1 + \gamma_2 + \gamma_3) = 1 - (0.05 + 0.04 + 0.01) = 0.9.$$

Remark. If we first compute the values of Λ and τ, then it is often more convenient to find the readiness coefficient by the formula

$$K = \frac{1}{1 + \Lambda\tau} \approx 1 - \Lambda\tau$$

(approximately true for $\Lambda\tau \ll 1$).

5. The reliability coefficient of the system for a time interval of $t_0 = 0.5$ hour is

$$R(t_0) = Ke^{-\Lambda t_0} = KP(t_0) = 0.909 \cdot 0.962 = 0.874.$$

An approximate value of the reliability coefficient is

$$\tilde{R}(t_0) = 1 - \Lambda(\tau + t_0) = 1 - 0.08(1.25 + 0.5) = 0.86.$$

Example 6.2.2 Suppose we have a system of 100 independent, series-connected units, in which all the units are identical and each has an exponential distribution for failure-free operating time and repair time with parameters $\lambda = 1 \cdot 10^{-5} \dfrac{1}{\text{hr}}$ and $\mu = 0.2 \dfrac{1}{\text{hr}}$.

We have to find the basic reliability measures for the operation of the system for $t_0 = 10$ hours; the readiness coefficient under the condition that when one of the units fails the remaining units of the system continue to be in the operating condition; and the mean value of failure-free operation time and repair time (under the same condition).

Solution.   First we determine the value $\gamma = \dfrac{1 \cdot 10^{-5}}{0.2} = 5 \cdot 10^{-5}$:

1.   The mean time of failure-free operation is

$$T = \frac{1}{m\lambda} = \frac{1}{100 \cdot 10^{-5}} = 1000 \quad \text{hr.}$$

2.   The mean repair time is

$$\tau^* = \frac{1}{m\lambda} [(1 + \gamma)m - 1] = \frac{1}{100 \cdot 10^{-5}} [(1 + 5 \cdot 10^{-5})^{100} - 1] = 5.01 \quad \text{hr.}$$

An approximate value for this quantity is

$$\tau^* = \frac{1}{\mu} = \frac{1}{0.2} = 5 \quad \text{hr.}$$

3.   The probability of failure-free operation of the system for an initial time of $t_0 = 10$ hours is found to be

$$P(t_0) = e^{-m\lambda t_0} = e^{-100 \cdot 10^{-5} \cdot 10} = e^{-0.01} = 0.9901.$$

An approximate value is

$$\tilde{P}(t_0) = 1 - m\lambda t_0 = 1 - 100 \cdot 10^{-5} \cdot 10 = 0.99.$$

4.   The readiness coefficient of the system is

$$K^* = \frac{1}{(1 + \gamma)^m} = \frac{1}{(1 + 5 \cdot 10^{-5})^{100}} = 0.99503.$$

An approximate value of the readiness coefficient is

$$\tilde{K}^* = 1 - m\gamma = 1 - 100 \cdot 5 \cdot 10^{-5} = 0.995.$$

Remark.   If we first compute the value τ^*, then it is often more convenient to find the readiness coefficient from the formula

$$K = \frac{1}{1 + m\lambda\tau^*} \approx 1 - m\lambda\tau^*$$

(which holds approximately for $m\lambda\tau^* \ll 1$).

5.   The reliability coefficient of the system for a time interval of $t_0 = 10$ hours is

$$R^*(t_0) = K^* e^{-m\lambda t} = K^* P(t_0) = 0.9950 \cdot 0.9901 = 0.9851.$$

An approximate value is

$$\tilde{R}^*(t_0) = 1 - m\lambda(\tau + t_0) = 1 - 100 \cdot 10^{-5}(5 + 10) = 0.985.$$

REFERENCES

1. BAZOVSKY, I., *Reliability: Theory and Practice*. Englewood Cliffs, N.J.: Prentice-Hall, Inc., 1961.
2. GNEDENKO, B. V., BELYAEV, Y. K., AND SOVOL'EV, A. D., *Mathematical Methods in Reliability Theory*. New York: Academic Press, Inc., 1968.
3. LLOYD, D. K., AND LIPOW, M., *Reliability: Management, Methods, and Mathematics*. Englewood Cliffs, N.J.: Prentice-Hall, Inc., 1962.

7

DETERMINATION OF THE OPERATING EFFICIENCY OF SYSTEMS

When redundancy is present in the structure of a complex system, the occurrence of failures of individual units or of significant changes in their operating parameters may not lead to the total loss of operation of the system, but only to a partial deterioration of the operating quality of the system as a whole. In order to estimate the quality of operation of such systems it is appropriate to introduce a qualitative measure of operating efficiency that takes into account the effect of such partial failures. We claim that measures of efficiency characterize the operating quality of systems in concrete problems more precisely and concretely than do the corresponding reliability measures. Thus, if we are confronted with the problem of traveling a long distance, most people prefer a less reliable airplane to a highly reliable railroad train, which is comparatively inefficient in the use of travel time.

It should be kept in mind that the choice of an appropriate measure of operating efficiency is, in each concrete case, determined by the type of system, its purpose, the type of problem, and the nature of the various external conditions.

To determine operating efficiency, we consider two types of systems: (1) sustained-action systems, (2) short-time action systems.

By a sustained-action system we mean one that can accomplish a required task over a protracted time interval of length t_0 starting from some t. The operating efficiency of such a system depends essentially on the collection of states and on the process that governs the transitions from state to state during the time of fulfilling the task.

By a short-time action system we mean one designed to perform tasks whose duration t_0 is such that the system almost certainly remains in

256

one and the same state for the duration of any given task. In the ideal case we can assume that the duration t_0 is equal to zero. A system of short-time action is a very special case of a sustained-action system.

At some arbitrary moment of time each of the units of the system may be in a certain state—for example, an operational state or a failed state. The collection of states of the units of a system uniquely determines the state of the system as a whole.

Each state of a short-time action system can be characterized by a completely determined conditional measure of operating efficiency. This conditional measure quantitatively characterizes the quality with which the system fulfills its functions, under the condition that it is then in the given state.

In the course of time the system undergoes a change of state, owing to state changes of its constituent units (failure of units, repair, and so on). Each realization of the process of passing from state to state by a sustained-action system can be characterized by a completely determined conditional efficiency measure. This conditional measure quantitatively characterizes the quality with which the system fulfills its functions, given a particular realization of the stochastic process of transitions of the system from state to state.

7.1 A GENERAL METHOD FOR ESTIMATING OPERATING EFFICIENCY OF SHORT-TIME ACTION SYSTEMS

If we let $h_s(t)$ denote the probability that a short-time action system is in the sth state at the moment of time t and Φ_s be the conditional measure of operating efficiency of the system in the sth state, then the measure of operating efficiency of a short-time action system is determined by the formula

$$E(t) = \sum_s h_s(t)\Phi_s, \qquad (7.1.1)$$

where summation is taken over all states of the system.

For a system consisting of m mutually independent units, each of which can be in one of only two states (operational and failed), the probabilities $h_s(t)$ are easily expressed in terms of the reliability measures of the units of the system.

The probability that all units of the system are operational is

$$h_0(t) = r_1(t)r_2(t)\cdots r_m(t) = \prod_{i=1}^{m} r_i(t). \qquad (7.1.2)$$

The probability that only the ith unit of the system is in a failed state is

$$h_i(t) = r_1(t) \cdots r_{i-1}(t) q_i(t) r_{i+1}(t) \cdots r_m(t)$$

$$= \frac{q_i(t)}{r_i(t)} \prod_{l=1}^{m} r_l(t). \qquad (7.1.3)$$

The probability that only the ith and jth units of the system are in a failed state is

$$h_{ij}(t) = \frac{q_i(t)}{r_i(t)} \frac{q_j(t)}{r_j(t)} \prod_{l=1}^{m} r_l(t), \qquad (7.1.4)$$

and so on.

Here $r_i(t)$ denotes the probability that the ith unit of the system is in an operational state at time t and $q_i(t) = 1 - r_i(t)$.

If the condition

$$\max q_i(t) \ll \frac{1}{m} \qquad (7.1.5)$$

holds, an estimate of the operating efficiency of the system can be constructed by the approximate formula

$$\tilde{E}(t) \approx 1 - \sum_{i=1}^{m} q_i(t)(\Phi_0 - \Phi_i). \qquad (7.1.6)$$

Formula (7.1.6) yields a lower bound for the true value.

The error of the efficiency estimate according to Formula (7.1.6) is of approximate order

$$\delta \sim \frac{m(m-1)}{2} [\max q_i(t)]^2. \qquad (7.1.7)$$

Example 7.1.1   A radar system in a commercial airport, designed to scan a sector of 180°, is assumed to be served by two identical stations a and b. Station a covers the sector 0–110°, and station b, the sector 70°–180°. The mean time of failure-free operation for each station is taken to be 95 hours, and the mean time of idleness to be 5 hours. Then the readiness coefficient of an individual station is $K = 0.95$.

The probability of detecting an object in the scan zone of a single station is $p = 0.9$, and in the simultaneous scan zone of both stations (overlap zone) it is

$$\mathcal{P} = 1 - (1 - p)^2 = 0.99.$$

We have to find the probability of detecting an approaching plane at a given distance; the plane is assumed to appear with equal probability at any azimuthal direction in the sector 0–180° at an arbitrary moment of time.

Solution. Consider the state S_0. The probability that the system is in this state at a certain moment of time is

$$h_0 = K^2 = 0.9025.$$

Thus, both stations operate in the sector from 70° to 110°—that is, in a 40° range—and one station at a time operates in the remaining sectors 0–70° and 110–180°—that is, in a 140° range. Then the conditional measure of efficiency of the state S_0, determined as a weighted-mean value, is

$$\Phi_0 = \frac{40°}{180°} \cdot 0.99 + \frac{140°}{180°} \cdot 0.9 = 0.92.$$

Multiplying the values h_0 and Φ_0, we obtain

$$h_0 \Phi_0 = 0.828.$$

The probability of the state S_a is

$$h_a = K(1 - K) = 0.0495.$$

In this state one station operates in the sector 0–110° while none is operating in the sector 110–180°—that is,

$$\Phi_a = \frac{110°}{180°} \cdot 0.9 + \frac{70°}{180°} \cdot 0 = 0.55.$$

Consequently the value of $h_a \Phi_a$ is

$$h_a \Phi_a = 0.028.$$

Since the states S_a and S_b are absolutely identical, we can write

$$h_b \Phi_b = 0.028.$$

As a result we have

$$E = h_0 \Phi_0 + 2 h_a \Phi_a = 0.828 + 0.056 = 0.883.$$

7.2 A GENERAL METHOD FOR ESTIMATING THE EFFICIENCY OF SUSTAINED-ACTION SYSTEMS

If we let $dh_\pi(t,\, t + t_0)$ denote the probability element of the stochastic process of transitions from state to state for a sustained-action system in the interval $[t,\, t + t_0]$ evaluated at the πth realization of the process, and

Φ_π denote the conditional measure of the operating efficiency of the system for this realization of the process, then the measure of operating efficiency of the system is determined by the formula

$$E(t,\, t + t_0) = \int_{G_\pi} \Phi_\pi \, dh_\pi(t,\, t + t_0), \tag{7.2.1}$$

where integration is performed over the space of all possible realizations of the transition process of the system from state to state in the time interval $[t,\, t + t_0]$.

For a system consisting of m independent nonrepairable units, each of which can be in only two states (operational and failed), Formula (7.2.1) can be written in the form

$$
\begin{aligned}
E(t,\, t + t_0) = \Phi_0 h_0 &+ \sum_{i=1}^{m} h_i^* \int_t^{t+t_0} \Phi_i(t_i) f_i(t_i) \, dt_i \\
&+ \sum_{1 \le i < j \le m} h_{ij}^* \int_t^{t+t_0} f_i(t_i) \, dt_i \int_t^{t+t_0} \Phi_{ij}(t_i,\, t_j) f_j(t_j) \, dt_j \\
&+ \sum_{1 \le i < j < l \le m} h_{ijl}^* \int_t^{t+\tau} f_i(t_i) \, dt_i \int_t^{t+t_0} f_j(t_j) \, dt_j \\
&\quad \times \int_t^{t+t_0} \Phi_{ijl}(t_1,\, t_j,\, t_l) f_l(t_l) \, dt_l + \cdots, \tag{7.2.2}
\end{aligned}
$$

where $f_i(t_i)$ is the probability density of failure of the ith unit at the moment of time t_i; h_0 is the probability that no unit of the system fails during the interval $[t,\, t + t_0]$:

$$h_0 = \prod_{i=1}^{m} r_i(t,\, t + t_0); \tag{7.2.3}$$

h_i^* is the probability that none of the units, except possibly the ith, fails during the interval $[t,\, t + t_0]$:

$$h_i^* = \prod_{\substack{l=1 \\ l \ne i}}^{m} r_l(t,\, t + t_0) = \frac{1}{r_i(t,\, t + t_0)} h_0; \tag{7.2.4}$$

h_{ij}^* is the probability that none of the units, except possibly the ith and jth, fails during the interval $[t,\, t + t_0]$:

$$h_{ij}^* = \frac{1}{r_i(t,\, t + t_0) \cdot r_j(t,\, t + t_0)} h_0; \tag{7.2.5}$$

Φ_0 is the conditional measure of operating efficiency of the system under the condition that none of the units has failed in the interval $[t, t + t_0]$; $\Phi_i(t_i)$ is the conditional measure of operating efficiency under the condition that only the ith unit has failed, and its failure occurred at the moment of time t_i $(t < t_i < t + t_0)$; $\Phi_{ij}(t_i, t_j)$ is the conditional measure of operating efficiency of the system given that only the ith and jth units have failed and their failures occurred at the moments of time t_i and t_j, respectively $(t < t_i < t + t_0, t < t_j < t + t_0)$.

If the condition

$$\max_{1 \le i \le m} q_i(t, t + t_0) = \max_{1 \le i \le m} \int_t^{t+t_0} f_i(t_i)\, dt_i \ll \frac{1}{m}, \qquad (7.2.6)$$

is satisfied, then it is possible to obtain the approximation

$$\tilde{E} = \Phi_0 - \sum_{i=1}^m \left[\Phi_0 q_i(t, t + t_0) - \int_t^{t+t_0} \Phi_i(t_i) f_i(t_i)\, dt_i \right]. \qquad (7.2.7)$$

The error of this approximation does not exceed

$$\delta \sim \frac{m(m-1)}{2} \max_{1 \le i \le m} [q_i(t, t + t_0)]^2. \qquad (7.2.8)$$

Example 7.2.1   We consider an information storage system consisting of two identical receivers a and b. When both receivers are operational, the receiving capacity of the system is determined as some quantity A. Suppose that with failure of one of the receivers the receiving capacity of the system falls to, say, $B = 0.3A$. When two receivers fail, the reception of information ceases. It is assumed that the probability of failure-free operation of each receiver has an exponential distribution and is characterized by the failure intensity λ; that is, $r = e^{-\lambda t}$.

The operating time (duration) of the system is taken to be $t = 0.1\dfrac{1}{\lambda}$.

Receiver failures are assumed independent. For the various realizations of the process of passing from state to state by the system we use the following conditional measures of efficiency:

$$\Phi_0 = At,$$
$$\Phi_i(t_i) = Bt + (A - B)t_i \qquad (i = a, b),$$
$$\Phi_{ij}(t_i, t_j) = Bt_j + (A - B)t_i \qquad (t_j > t_i;\ i, j = a, b).$$

In other words, the efficiency measure of the system for each path is defined as the amount of stored information (product of receiving capacity and operating time). We wish to determine the efficiency measure of the system.

Solution. A determination of the efficiency measure is made by Formula (7.2.2). For the given case this formula takes the form

$$E = r^2 At + 2r \int_0^t [Bt + (A - B)x)]\lambda e^{-\lambda x}\, dx$$
$$+ 2 \int_0^t \lambda e^{-\lambda x}\, dx \int_0^x [Bx + (A - B)y]\lambda e^{-\lambda y}\, dy.$$

After computing the integrals, we obtain

$$E = r^2 At + 2r \left\{ Bt_q + (A - B)\frac{1}{\lambda}[1 - r(1 - \lambda t)] \right\}$$
$$+ 2 \left\{ Bt[\tfrac{3}{4} - r(1 + \lambda t) + \tfrac{1}{4}r^2(1 + 2\lambda t)] \right.$$
$$\left. + (A - B)\frac{1}{\lambda}[\tfrac{1}{4} - r + (\tfrac{3}{4} + \tfrac{1}{2}\lambda t)r^2] \right\}.$$

Substituting numerical values, for example, $r = 0.905$ and $t = 0.1\,\dfrac{1}{\lambda}$, we obtain

$$E = (0.812 + 0.111 + 0.004)At = 0.927At.$$

7.3 DETERMINATION OF EFFICIENCY OF SYSTEMS WITH ADDITIVE EFFICIENCY MEASURES

Many systems are characterized by a very simple form of the conditional measure of operating efficiency. The property of the system that leads to this characterization is: each unit of the system carries its own definite and independent share of the total output. This type of conditional efficiency measure is characteristic of systems that are, for example, a collection of transporting units (such as trucks in a freighting firm).

If the ith unit of such a system has a certain share φ_i of the total output, then we can write

$$\Phi_i = \Phi_0 - \varphi_i,$$
$$\Phi_{ij} = \Phi_0 - (\varphi_i + \varphi_j), \tag{7.3.1}$$
$$\cdots\cdots\cdots\cdots\cdots\cdots$$

For short-time action systems of this type we may write

$$E = \sum_{i=1}^n \varphi_i r_i. \tag{7.3.2}$$

Example 7.3.1 Consider a system consisting of three units designed to store information. Suppose the receiving capacity of each unit is $\varphi_1 =$

100 bit/sec, $\varphi_2 = 200$ bit/sec, and $\varphi_3 = 250$ bit/sec (a bit is a binary unit of information), and the failure intensity of these units is $\lambda_1 = 0.01 \dfrac{1}{hr}$, $\lambda_2 = 0.03 \dfrac{1}{hr}$, and $\lambda_3 = 0.04 \dfrac{1}{hr}$.

The duration of failure-free operation of each unit is assumed to have an exponential distribution.

We want to determine, taking failures into account, the mean receiving capacity of the system at the moment of time $t = 10$ hours and the amount of information collected by the system as a result of continuous operation under full load for 50 hours.

Solution. The mean receiving capacity of the system at an arbitrary moment of time has the form

$$E(t) = \sum_{i=1}^{3} \varphi_i e^{-\lambda_i t}$$

or, after substitution of numerical values for $t = 10$ hours,

$$E(t) = 100e^{-0.1} + 200e^{-0.3} + 250e^{-0.4} = 405.5 \quad \text{bit/sec.}$$

The amount of information collected by the moment of time $t = 50$ hours can be defined as

$$E_t = \int_0^t E(x)\,dx = \int_0^t \sum_{i=1}^{3} \varphi_i e^{-\lambda_i x}\,dx$$

$$= \sum_{i=1}^{3} \frac{\varphi_i}{\lambda_i}(1 - e^{-\lambda_i t})$$

or, after converting φ and λ to identical time units and substituting numerical values,

$$E_t = \left(\frac{100}{0.01} \cdot 0.39 + \frac{200}{0.03} \cdot 0.78 + \frac{250}{0.04} \cdot 0.86 \right) 3600 \approx 5.2 \cdot 10^7 \quad \text{bit.}$$

7.4 EFFICIENCY ESTIMATION FOR SYSTEMS WITH FUNCTIONAL STAND-BY UNITS

There is a whole class of systems in which fulfillment of the same task may be accomplished, say, by several independent control units. Fulfillment of the task by at least one of the units is sufficient. Suppose the probability of fulfilling the task by a given unit is φ_s when the system is in the sth state. We want to determine the probability of fulfilling the task by at least one of the m units.

This type of problem can be solved for two basic cases.

1.   The executive units of the system fulfill the task simultaneously— that is, for all units the probability of fulfilling the task is equal to the same quantity φ_s.

In this case we can write down the following formula for determining the operating efficiency:

$$E = \sum_s h_s[1 - (1 - \varphi_s)^n], \qquad (7.4.1)$$

where h_s is the probability that the system is in the sth state; $1 - (1 - \varphi_s)^m$ is the conditional probability that a system with m functional stand-by units fulfills the task given that the system is in the sth state.

Example 7.4.1   Suppose a gunner can fire a salvo of two shots at some target. If the fire control system is operational at the moment of the salvo, then the probability of hitting with a given shot is $\varphi_1 = 0.85$, while if the system has failed, the probability of hitting with a given shot is $\varphi_2 = 0.58$. Suppose the probability of failure-free operation of the fire control system is $r = 0.7$.

We want to determine the over-all hit probability.

Solution.   The over-all hit probability is

$$E = r[1 - (1 - \varphi_1)^2] + q[1 - (1 - \varphi_2)^2]$$

or, substituting numerical values,

$$E = 0.7(1 - 0.15^2) + 0.3(1 - 0.42^2) = 0.931.$$

2.   The executive units satisfy the task at different moments of time —that is, for each unit the probability of fulfilling the task is equal to φ_s, corresponding to the state the system is in when the given unit operates during fulfillment of the task.

In this case the formula for determining efficiency is:

$$E = 1 - \left(1 - \sum_s h_s\varphi_s\right)^n. \qquad (7.4.2)$$

Example 7.4.2   We change the conditions of the preceding example somewhat. Suppose two independent salvos are fired at the target (this might be the case if two independent artillery systems fire at the same time).

We need to determine the over-all hit probability.

Solution. The over-all hit probability for this case is

$$E = 1 - [1 - (r\varphi_1 + q\varphi_2)]^2$$

or, substituting the numerical values of the preceding example,

$$E = 1 - [1 - (0.7 \cdot 0.85 + 0.3 \cdot 0.58)]^2 = 0.947.$$

7.5 EFFICIENCY DETERMINATION FOR MULTIFUNCTIONAL SYSTEMS

A multifunctional system is one that can fulfill the same task by various methods characterized by different efficiency measures, where for each state of the system the method most efficient for that state is always chosen.

Consider a system of m units. Suppose a certain task can be fulfilled by various methods, and when the task is fulfilled by the jth method the conditional measure of technical efficiency of the system is Φ_j. For definiteness let $\Phi_1 > \Phi_2 > \cdots > \Phi_l$.

We divide the system into l subsystems $G_1, G_2, \ldots, G_l$, so that in each of the subsystems we have only those units which guarantee fulfillment of the task by the jth method (in general, individual units may be included in several subsystems—that is, they may participate in the fulfillment of the same task by various methods).

Comparatively simple expressions can be written for the operating efficiency for two particular cases:

1. Each unit may be included in only one subsystem. The probability that the task is fulfilled by the jth method in this case is

$$H_j = P_j \prod_{i=1}^{j} Q_i, \tag{7.5.1}$$

where P_j is the probability of failure-free operation for the jth subsystem and Q_j is the probability of failure of the jth subsystem.

The formula for determining the operating efficiency of the system is of the form

$$E = \sum_{j=1}^{l} \Phi_j P_j \prod_{i=1}^{j} Q_i. \tag{7.5.2}$$

2. The subsystems G_j are "imbedded in each other"—that is, in order to fulfill the task by the first method all units must be operational; to fulfill the task by the second method, not all units need be operational; by the third method, still fewer units need be operational, and so on.

We let G_{j-1}^* denote the set of units of a system that belong to the subsystem G_{j-1} and at the same time do not belong to the subsystem G_j.

According to the principle of operation of the given system, the task is fulfilled by the jth method if all units of the subsystem G_j are operational, while in the subsystem G_{j-1} there is at least one failed unit (here it is not important whether there are failed units among the sets G_{j-2}^*, G_{j-3}^*, and so on).

In this case the probability that the system fulfills the task by the jth method is

$$H_j = P_j - P_{j-1}.$$

The formula for estimating the operating efficiency is written as

$$E = \sum_{j=1}^{l} \Phi_j(P_j - P_{j-1}). \tag{7.5.3}$$

Example 7.5.1   Consider a system that is required to perform a task in three successive stages: search, detection, and tracking of some object. At each stage, fulfillment of the functions can be accomplished by various methods. It is possible to combine methods for performing each individual stage, guaranteeing that the system solves its problem; the measures of technical efficiency corresponding to these combinations are listed in Table 7.5.1.

Table 7.5.1   Measures of Technical Efficiency

INDEX OF METHOD OF FULFILLMENT	STAGE			MEASURE OF TECHNICAL EFFICIENCY
	SEARCH	DETECTION	TRACKING	
1	search tracer	indicator	tracking tracer	1.00
2	tracking tracer	indicator	tracking tracer	0.60
3	search tracer	indicator	optical structure and calculator	0.30
4	optical structure	optical structure	optical structure and calculator	0.15
5	optical structure	optical structure	optical structure	0.10

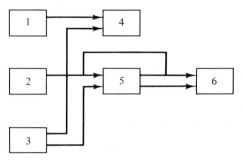

Figure 7.5.1 Block Diagram of a Multi-functional System: 1—search tracer, 2—optical structure, 3—tracking tracer, 4—indicator, 5—switch, 6—control panel.

Furthermore, it is necessary that the control panel be operational in order to perform the task at all.

The values of the probabilities of failure-free operation for the individual units of the system at a certain moment of time (indexed as in Figure 7.5.1) are

$$r_1 = 0.80, \quad r_2 = 0.99, \quad r_3 = 0.80, \quad r_4 = 0.95, \quad r_5 = 0.90, \quad r_6 = 0.95.$$

We wish to compute the over-all probability that the system fulfills its task.

Solution.   The performance of the task by the first method is accomplished with probability

$$h_1 = r_1 r_2 r_3 r_4 r_5 r_6 \approx 0.52$$

(whether the optical structure does or does not work is taken to be irrelevant in this case).

Performance of the task by the second method is accomplished only if the search tracer has failed—that is, with probability

$$h_2 = q_1 r_3 r_4 r_5 r_6 \approx 0.13.$$

Performance of the task by the third method is accomplished only if the tracking tracer has failed—that is, with probability

$$h_3 = q_3 r_1 r_4 r_5 r_6 \approx 0.13.$$

Performance of the task by the fourth method is accomplished only if one of the following has failed:

the indicator;
the indicator and search tracer;
the indicator and tracking tracer;
the indicator, search tracer, and tracking tracer;
the search tracer and tracking tracer.

In all of these cases the optical structure and the calculator are assumed to be operational.

The probability of this event is

$$h_4 = r_2 r_5 r_6 (q_4 r_1 r_3 + q_4 q_1 r_3 + q_4 q_1 q_3 + r_4 q_1 q_3) \approx 0.08.$$

Finally, performance of the task by the fifth method is accomplished only if in addition the calculating structure has failed—that is, with probability

$$h_5 = r_2 q_5 r_6 (q_4 + q_1 q_3 r_4) \approx 0.01.$$

The final value of the over-all probability of performing the task is

$$E = 0.52 \cdot 1 + 0.13 \cdot 0.6 + 0.13 \cdot 0.3 + 0.08 \cdot 0.15 + 0.01 \cdot 0.1 = 0.66.$$

The computation of the probability that no failure occurs in this system yields a value of only 0.51.

7.6 DETERMINATION OF EFFICIENCY FOR BRANCHING SYSTEMS

In practice one often encounters systems that have a branching structure (Figure 7.6.1). They are characterized by the fact that some basic unit controls the operation of a certain number of branches, each of which controls the operation of a certain number of sub-branches, and so on.

Individual units of such systems may cease normal functioning and give rise to the termination of the operation of the succeeding subordinate units.

The basic ("starting") unit of a symmetric branching system is called the unit of rank zero; the units immediately subordinate to it are units of first rank; units directly subordinate to first-rank units are units of second rank, and so on. We say that a system has structure of mth order if the output units of the system are units of rank m.

The number of units of rank m that are controlled by each unit of rank $m - 1$ is called the branch coefficient and denoted by a_m. The total num-

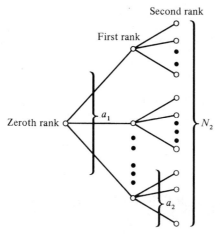

Figure 7.6.1   System with Second-Order
Branching Structure.

ber of units of rank i is denoted by N_i. The probability of an operational
state for each unit of rank i at some moment of time t is denoted by $r_i(t)$,
and the probability of a failed state for a unit of ith rank is $q_i(t)$.

We say that an output unit of the system functions normally if it is
operational and all the units connecting the given output unit with the
basic unit of the system are operational. We assume the units of the
system are mutually independent.

The conditional efficiency measure of such a system depends only on z,
the number of normally functioning output units, and is given in the
form of a function $\Phi(z)$.

The operating efficiency of such a system can be determined by the
formula

$$
E = \sum_{j=0}^{\infty} \frac{M_n^{(j)}}{j!} \frac{d^j \Phi(z)}{dz^j} \bigg|_{z=0} , \tag{7.6.1}
$$

where $M_m^{(j)}$ is the jth moment about the origin of the distribution of the
number of normally functioning output units and

$$
\frac{d^j \Phi(z)}{dz^j} \bigg|_{z=0}
$$

is the jth derivative of $\Phi(z)$ evaluated at $z = 0$. Since z is integer-valued
in general, we can differentiate $\Phi(z)$ only under some further assumptions.

But sometimes the function $\Phi(z)$ is a polynomial of z—that is,

$$\Phi(z) = \sum_{i=1}^{l} c_i z^i,$$

so that the value desired is apparent.

In general the function $\Phi(z)$ can be sufficiently well approximated by a second-order polynomial

$$\Phi(z) = c_1 z + c_2 z^2. \tag{7.6.2}$$

In this case

$$E = c_1 M_n^{(1)} + c_2 M_n^{(2)}, \tag{7.6.3}$$

where

$$M_n^{(1)} = r_0 \prod_{i=1}^{n} a_i r_i, \tag{7.6.4}$$

$$M_n^{(2)} = r_0 \prod_{i=1}^{n} a_i r_i \left\{ \prod_{i=1}^{n} a_i r_i + \sum_{i=1}^{n} q_i \prod_{j=i+1}^{n} a_j r_j \right\}. \tag{7.6.5}$$

Example 7.6.1 We consider various cases of second-order branching systems having six output units (Figure 7.6.2). First we assume that the function $\Phi(z)$ has the form

$$\Phi(z) = cz$$

—that is, the operating efficiency of the system is proportional to the number of normally functioning output units. This kind of system might be a transport system, communications system, or the like.

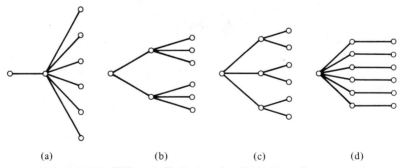

(a)   (b)   (c)   (d)

Figure 7.6.2 Different Variants of a Branching Structure of Second Order with Six Output Units.

Then the efficiency of the system is

$$E = cr_0 \prod_{i=1}^{m} a_i r_i.$$

Noting that $\prod_{i=1}^{m} a_i = N_m$, we have

$$E = cN_m \prod_{i=1}^{m} r_i.$$

From this formula we can draw the interesting conclusion that all the systems pictured in Figure 7.6.2 are equivalent in efficiency for the given measure $\Phi(z)$.

Now suppose that the function $\Phi(z)$ has the form

$$\Phi(z) = cz^2,$$

—that is, the operating efficiency is proportional to the square of the number of normally operating output units. This occurs, for example, in certain game situations. In this case the operating efficiency is

$$E = cM_m^{(2)}.$$

For high values of the probability of an operational state of the units (that is, $q_i \ll 1/N$) we can, using the formula for the second-order moment about the origin, write approximately

$$E \approx 2cN_m^2 \left[1 - \sum_{i=1}^{m} q_i \left(1 - \frac{1}{N_i} \right) \right].$$

For simplicity, letting $q_0 = q_1 = q_2 = q$, we obtain

$$E_a \approx 2cN^2(1 - 0.83q),$$
$$E_b \approx 2cN^2(1 - 1.33q),$$
$$E_c \approx 2cN^2(1 - 1.5q),$$
$$E_d \approx 2cN^2(1 - 1.67q).$$

Consequently, in this case the system having the greatest value of a_2/a_1 is preferable.

REFERENCES

1. GNEDENKO, B. V., BELYAEV, Y. K., AND SOLOV'EV, A. D., *Mathematical Methods in Reliability Theory*, chap. 2. New York: Academic Press, Inc., 1968.

2.   KOVALENKO, I. N., "Some Problems in Reliability Theory of Complex Systems," in the collection *Cybernetics in the Service of Communism*. Moscow: Energiya, 1964.
3.   MOSKATOV, G. K., "On the Problem of Efficiency and Invariance of Multifunctional Controlling Systems," in the collection *Cybernetics in the Service of Communism*. Moscow: Energiya, 1964.
4.   USHAKOV, I. A., "Estimating Efficiency of Complex Systems," in the collection *Reliability of Radioelectronic Equipment*. Moscow: Soviet Radio, 1964.
5.   SHISHONOK, N. A., REPKIN, V. F., AND BARCINSKII, L. L., *Foundations of Reliability Theory and Use of Radioelectronic Equipment*, chap. 14. Moscow: Soviet Radio, 1964.
6.   USHAKOV, I. A., AND KONYONKOV, Y. K., "Estimating Operating Efficiency of Complex Branching Systems with Reliability Considerations," in the collection *Cybernetics in the Service of Communism*, vol. 2. Moscow: Energiya, 1964.

8

PROBLEMS OF OPTIMAL USE

8.1 METHODS OF OPTIMAL REDUNDANCY

In studying the use of redundancy to increase the reliability of various systems, we encounter the problem of not only guaranteeing certain minimum reliability standards, but also of designing the systems so as to effect this reliability criterion as economically as possible, with the least total expenditure on stand-by units for the system as a whole.

In practice such expenditures may be numerical measures of various characteristics of the system such as its cost, weight, or size. The selection of the characteristics is determined by the concrete form of the system and its intent. It is usually possible to distinguish one most important characteristic, which, for convenience, we shall call "weight," no matter what the actual physical nature of the characteristic is.

We consider a system consisting of several series-connected, mutually independent units.

As the basic system reliability measures that are to be improved by the use of redundancy, we take the probability of initial failure-free operation of the system for a length of time t_0, the readiness coefficient, or the reliability coefficient. (For repairable systems we here assume unrestricted repair—that is, the number of repair teams for the system is equal to the number of series-connected units in the initial variant of the system before redundancy is imposed, which is needed to guarantee the mutual independence of the units of the system that are in the repair process.)

We can state the following two problems of optimal redundancy for unrepairable systems that are subject to a constraint:

1.   By imposing redundancy on each part of a system consisting of n parts, achieve the minimal "weight" of the entire system under the constraint that the probability of failure-free operation of the system during a given time t_0 is not less than $P_{RE}(t_0)$ [or that the probability of failure of the system is not greater than $Q_{RE}(t_0)$].

2.   By imposing redundancy on each part of a system consisting of n parts, achieve the maximal possible probability of failure-free operation of the system (or minimal possible probability of failure of the system) during a given time t_0 under the constraint that the "weight" of the entire system does not exceed W_{RE}.

These two problems of optimal redundancy can also be formulated for repairable systems.

1.   By imposing redundancy on each part of a system consisting of n parts, achieve the minimal "weight" of the entire system under the constraint that the readiness coefficient (or the reliability coefficient for a given time t_0) is not less than K_{RE} [or $R_{RE}(t_0)$].

2.   By imposing redundancy on each part of a system consisting of n parts, achieve the maximal possible readiness coefficient (or reliability coefficient for a given time t_0) under the constraint that the "weight" of the entire system does not exceed W_{RE}.

The entire presentation below will be made in terms of the probability of failure-free operation for unrepairable systems. However, one can also solve the optimization problem for repairable systems simply by replacing the given quantities by their corresponding coefficients of readiness (or reliability coefficients).

8.1.1   A method for determining the optimal number of stand-by units by a modified method of dynamic programming

In many practical cases the statistical characteristics of the reliability of individual units of the system and the nature of the dependence of the reliability measures on an increase in the number of stand-by units are comparatively well known. In such situations, when extreme computational difficulties do not arise because of the resulting number of computations, it is convenient to determine exactly the optimal number of stand-by units by a modified method of dynamic programming. The method is as follows:

1.   For the ith part of the redundant system for some fixed time interval of duration t_0 with various numbers x_i of stand-by units, we compute the values of the probabilities of failure-free operation and denote them by $P_i(x_i)$, where $x_i = 0, 1, 2, \ldots$.

2.   For convenience we construct a table of values of $P_i(x_i)$ for various x_i (Table 8.1.1).

Table 8.1.1   Combined Table of Values
$P_i(x_i)$

x_i	$P_1(x_1)$	$P_2(x_2)$	$\cdots$	$P_i(x_i)$	$\cdots$	$P_m(x_m)$
0	$P_1(0)$	$P_2(0)$	$\cdots$	$P_i(0)$	$\cdots$	$P_m(0)$
1	$P_1(1)$	$P_2(1)$	$\cdots$	$P_i(1)$	$\cdots$	$P_m(1)$
2	$P_1(2)$	$P_2(2)$	$\cdots$	$P_i(2)$	$\cdots$	$P_m(2)$
.						
.						
x	$P_1(x)$	$P_2(x)$	$\cdots$	$P_i(x)$	$\cdots$	$P_m(x)$
.						
.						

3.   For two arbitrary parts of the system, say the mth and $(m-1)$st, we set up a table in the form of Table 8.1.2.

In the square at the intersection of the x_mth column and the x_{m-1}st row we write the values

$$P^*_{m-1}(x_{m-1},\, x_m) = P_{m-1}(x_{m-1}) P_m(x_m) \tag{8.1.1}$$

and

$$w^*_{m-1}(x_{m-1},\, x_m) = w_{m-1}x_{m-1} + w_m x_m + W^0_{m-1} + W^0_m, \tag{8.1.2}$$

where W^0_i is the initial weight of the ith part of the system.

4.   The values $P^*_{m-1}(x_{m-1}, x_m)$ and the corresponding values $w^*_{m-1}(x_{m-1}, x_m)$ are placed in Table 8.1.3 in increasing order of the quantities $w^*_{m-1}(x_{m-1}, x_m)$.

Table 8.1.2   Composition of $(m-1)$st and mth Parts

x_{m-1} ＼ x_m	0	1	2	$\cdots$
0	$P^*_{m-1}(0,\,0)$ $w^*_{m-1}(0,\,0)$	$P^*_{m-1}(0,\,1)$ $w^*_{m-1}(0,\,1)$	$P^*_{m-1}(0,\,2)$ $w^*_{m-1}(0,\,2)$	$\cdots$
1	$P^*_{m-1}(1,\,0)$ $w^*_{m-1}(1,\,0)$	$P^*_{m-1}(1,\,1)$ $w^*_{m-1}(1,\,1)$	$P^*_{m-1}(1,\,2)$ $w^*_{m-1}(1,\,2)$	$\cdots$
2	$P^*_{m-1}(2,\,0)$ $w^*_{m-1}(2,\,0)$	$P^*_{m-1}(2,\,1)$ $w^*_{m-1}(2,\,1)$	$P^*_{m-1}(2,\,2)$ $w^*_{m-1}(2,\,2)$	$\cdots$
$\cdots$	$\cdots$	$\cdots$	$\cdots$	$\cdots$

Table 8.1.3   Ordered Pairs $\{w^*_{m-1}, P^*_{m-1}\}$

w^*_{m-1}	$w^*_{m-1}(0, 0)$	$w^*_{m-1}(x^{\mathrm{I}}_{m-1}, x^{\mathrm{I}}_m)$	$w^*_{m-1}(x^{\mathrm{II}}_{m-1}, x^{\mathrm{II}}_m)$	$\cdots$
P^*_{m-1}	$P^*_{m-1}(0, 0)$	$P^*_{m-1}(x^{\mathrm{I}}_{m-1}, x^{\mathrm{I}}_m)$	$P^*_{m-1}(x^{\mathrm{II}}_{m-1}, x^{\mathrm{II}}_m)$	$\cdots$

5. Eliminate all columns from Table 8.1.3 containing values P^*_{m-1} not exceeding their left neighbors.

6. Enter the sequence P^*_{m-1} remaining after this procedure (called domination) in the new Table 8.1.4.

Here we conditionally let $x^*_{m-1} = 1$ denote the fact that the system has x^1_{m-1} stand-by units for the $(m-1)$st part and x^1_m for the mth, and $x^*_{m-1} = 2$ the fact that the system has x^2_{m-1} stand-by units for the $(m-1)$st part and x^2_m for the mth, and so on. In other words, x^*_{m-1} is a vector with components x_{m-1} and x_m; that is, $x^*_{m-1} = (x_{m-1}, x_m)$.

7. We place the derived values $P^*_{m-1}(x^*_{m-1})$ in Table 8.1.1 in place of the mth and $(m-1)$st columns. Thus as the result of combining the two parts into one, $m-1$ parts remain in the given system rather than the original m.

8. Continue this procedure until finally, after $m-1$ steps, we have constructed the final dominating sequence $P^*_1(x^*_1)$ and the corresponding quantities $w^*_1(x^*_1)$.

9. In the final dominating sequence we seek a solution x^*_1 such that for the first problem of optimal stand-by

$$P^*_1(x^*_1 - 1) < P_{RE} \leq P^*_1(x^*_1) \tag{8.1.3}$$

or for the second problem

$$w^*_1(x^*_1) \leq W_{RE} < w^*_1(x^*_1 + 1). \tag{8.1.4}$$

10. The x^*_1 thus obtained contains all unknown optimal values x_i, which can be found as follows:

$$x^*_1 = (x_1, x^*_2)$$

—that is, we can find x_1 and x^*_2 on the basis of x^*_1. Further,

$$x^*_2 = (x_2, x^*_3), \quad \text{and so on.}$$

Table 8.1.4 $(m-1)$*st Dominating Sequence

x^*_{m-1}	0	1	2	$\cdots$
w^*_{m-1}	$w^*_{m-1}(0, 0)$	$w^*_{m-1}(x^1_{m-1}, x^1_m)$	$w^*_{m-1}(x^2_{m-1}, x^2_m)$	$\cdots$
P^*_{m-1}	$P^*_{m-1}(0, 0)$	$P^*_{m-1}(x^1_{m-1}, x^1_m)$	$P^*_{m-1}(x^2_{m-1}, x^2_m)$	$\cdots$

Remarks. 1. From the computational point of view it is usually more convenient to make a pairwise composition of all m parts of the system first, and then a pairwise composition of all the obtained parts, and so forth—that is, the method of composition by scheme (b) may turn out to be preferable to the method of composition by scheme (a) (Figure 8.1.1).

2. In making computations in practice it is often helpful, for the purpose of decreasing the amount of subsequent computation, to exclude from consideration several of the terms of the dominating sequences. If, for example, two neighboring terms differ insignificantly in weight, then we can neglect those among them which are characterized by a smaller reliability measure. Analogously, if we observe an insignificant difference in the reliability measures of two neighboring terms of a dominating sequence, then we can neglect those among them with the greater weight.

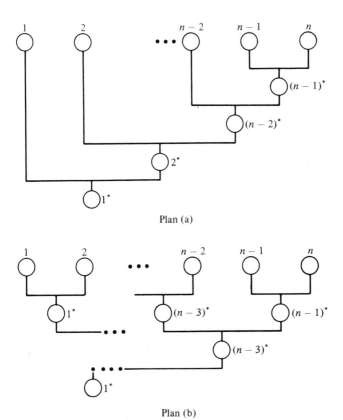

Plan (a)

Plan (b)

Figure 8.1.1.

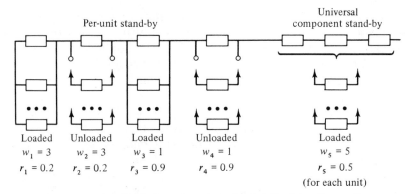

Figure 8.1.2   Block Diagram of the Model Used for Computing Reliability for the System of Example 8.1.1.

Example 8.1.1   Suppose a system consists of units with different reliability measures, and for the purpose of increasing reliability we use different methods of redundancy for different parts of the system (see Figure 8.1.2).

We want to determine the optimal stand-by assignment for each part for two cases (types of constraints): (a) the total cost of the system is not to exceed 50 cost units; (b) the probability of failure-free operation is to be not less than 0.45.

Solution.   In Table 8.1.5 values $P_i(x_i)$ are given for various x_i.

For the first and third parts, the computation of $P_i(x_i)$ is made according to the formula

$$P_i(x_i) = 1 - q_i x_i + 1;$$

Table 8.1.5   Values $P_i(x_i)$ for Different Parts

NUMBER OF STAND-BY UNITS FOR A PART	$P_i(x_i)$				
	$i = 1$	$i = 2$	$i = 3$	$i = 4$	$i = 5$
0	0.2000	0.2000	0.9000	0.9000	0.1250
1	0.3600	0.5249	0.9900	0.9953	0.3125
2	0.4880	0.7834	0.9990	0.9998	0.5000
3	0.5904	0.9212	0.9999	1	0.6563
4	0.6724	0.9763	1	—	0.7734
5	0.7379	0.9940	—	—	0.8553
6	0.7902	0.9987	—	—	0.9103
7	0.8319	0.9997	—	—	0.9453
8	0.8658	0.9999	—	—	0.9673
9	0.8924	1	—	—	0.9807
10	0.9139	—	—	—	0.9888

for the second and fourth parts, by the formula

$$P_i(x_i) = e^{-\lambda_i t} \sum_{j=0}^{x_i} \frac{(\lambda_i t)^j}{j!};$$

and for the fifth part, by the formula (for $n = 3$)

$$P_i(x_i) = \sum_{j=n}^{n+x_i} c_{n+x_i}^j \tau^j q^{n+x_i-j}.$$

In Tables 8.1.6 and 8.1.7 we give the compositions of the fifth with the fourth and the second with the third parts, respectively. The numbers in the lower right corners show the order index of the given number in the dominating sequence; and here, as is easily seen, we have used the procedure of discarding terms that are close.

Table 8.1.8 gives the composition of two newly formed parts, and Table 8.1.9 shows the final dominating sequence.

Table 8.1.6 Table of Optimal Redundancy Algorithm

x_4 \ x_5	0	1	2	3	4	5	...
0	0.1125 16 **1**	0.2812 21 **3**	0.4500 26 **5**	0.5907 31 **7**	0.6961 36 **9**	0.7698 41	...
1	0.1244 17 **2**	0.3109 22 **4**	0.4976 27 **6**	0.6531 32 **8**	0.7695 37 **10**	0.8510 42 **11**	...
2	0.1250 18	0.3125 23	0.4999 28	0.6562 33	0.7733 38	0.8551 43	...
3	0.1250 19	0.3125 24	0.5000 29	0.6563 34	0.7734 39	0.8553 44	...
4	0.1250 20	0.3125 25	0.5000 30	0.6563 35	0.7734 40	0.8553 45	...
5	0.1250 21	0.3125 26	0.5000 31	0.6563 36	0.7734 41	0.8553 46	...
...	...	...	...	...	...	...	...

Table 8.1.7   Table of Optimal Redundancy Algorithm

x_2 \ x_3	0	1	2	3	$\cdots$
0	0.1800 3 **1**	0.1980 4 **2**	0.1998 5	0.2000 6	$\cdots$
1	0.4724 5 **3**	0.5197 6 **4**	0.5244 7	0.5249 8	$\cdots$
2	0.7051 7 **5**	0.7756 8 **6**	0.7826 9	0.7833 10	$\cdots$
3	0.8291 9 **7**	0.9120 10 **8**	0.9203 11	0.9211 12	$\cdots$
4	0.8887 11	0.9666 12 **9**	0.9753 13	0.9762 14	$\cdots$
5	0.8946 13	0.9841 14 **10**	0.9930 15 **11**	0.9939 16	$\cdots$
6	0.8988 15	0.9888 16	0.9977 17 **12**	0.9986 18	$\cdots$
7	0.8997 17	0.9898 18	0.9987 19 **13**	0.9996 20	$\cdots$

We denote for our example $y_1 = (x_4, x_5)$, $y_2 = (x_2, x_3)$, $y_3 = (y_1, y_2)$, and $y_4 = (x_1, y_3)$. Thus the index in the right corners of Table 8.1.6 is y_1, in Table 8.1.7, y_2, in Table 8.1.8, y_3, and in Table 8.1.9, y_4.

From Table 8.1.9 we find that the solution for the first problem (cost of the system must not exceed 50 units) is $y_4 = 8$; that is, $x_1 = 3$ and $y_3 = 19$. Then from Table 8.1.8 we find the value $y_3 = 19$ and immediately find $y_1 = 6$ and $y_2 = 8$. Taking $y_1 = 6$, in Table 8.1.6 we determine $x_4 = 1$ and $x_5 = 2$, and taking $y_2 = 8$ from Table 8.1.7 we find $x_2 = 3$

Table 8.1.8 Optimal Redundancy Algorithm

y_1 \ y_2	1	2	3	4	5	6	7	8	9	10	11	12
1	0.0203 / 19 **1**	0.0223 / 20	0.0532 / 21 **3**	0.0585 / 22	0.0793 / 23	0.0873 / 24	0.0933 / 25	0.1024 / 26	0.1087 / 28	0.1107 / 30	0.1117 / 31	0.1122 / 33
2	0.0224 / 20 **2**	0.0246 / 21	0.0588 / 22 **4**	0.0647 / 23	0.0877 / 24 **6**	0.0965 / 25 **7**	0.1031 / 26	0.1135 / 27	0.1202 / 29	0.1224 / 31	0.1235 / 32	0.1241 / 34
3	0.0506 / 24	0.0558 / 25	0.1328 / 26 **8**	0.1461 / 27	0.1983 / 28 **10**	0.2181 / 29	0.2331 / 30	0.2565 / 31	0.2716 / 33	0.2767 / 35	0.2792 / 36	0.2806 / 38
4	0.0560 / 25	0.0616 / 26	0.1469 / 27 **9**	0.1616 / 28	0.2192 / 29 **11**	0.2411 / 30 **12**	0.2578 / 31 **13**	0.2835 / 32 **14**	0.3005 / 34	0.3060 / 36	0.3087 / 37	0.3102 / 39
5	0.0810 / 29	0.0891 / 30	0.2126 / 31	0.2339 / 32	0.3173 / 33 **15**	0.3490 / 34	0.3731 / 35	0.4104 / 36	0.4350 / 38	0.4428 / 40	0.4469 / 41	0.4490 / 43
6	0.0896 / 30	0.0985 / 31	0.2351 / 32	0.2586 / 33	0.3509 / 34 **16**	0.3859 / 35 **17**	0.4126 / 36 **18**	0.4538 / 37 **19**	0.4810 / 39 **20**	0.4897 / 41	0.4941 / 42	0.4965 / 44
7	0.1063 / 34	0.1170 / 35	0.2790 / 36	0.3070 / 37	0.4165 / 38	0.4581 / 39	0.4897 / 40	0.5387 / 41	0.5710 / 43	0.5813 / 45	0.5866 / 46	0.5893 / 48
8	0.1176 / 35	0.1293 / 36	0.3085 / 37	0.3394 / 38	0.4605 / 39	0.5065 / 40 **21**	0.5415 / 41 **22**	0.5956 / 42 **23**	0.6313 / 44 **24**	0.6427 / 46 **25**	0.6485 / 47	0.6516 / 49
9	0.1253 / 39	0.1378 / 40	0.3288 / 41	0.3618 / 42	0.4908 / 43	0.5399 / 44	0.5771 / 45	0.6348 / 46	0.6729 / 48	0.6850 / 50	0.6912 / 51	0.6945 / 53
10	0.1385 / 40	0.1524 / 41	0.3635 / 42	0.3999 / 43	0.5426 / 44	0.5968 / 45	0.6380 / 46	0.7018 / 47 **26**	0.7438 / 49 **27**	0.7573 / 51 **28**	0.7641 / 52	0.7677 / 54
11	0.1532 / 45	0.1685 / 46	0.4020 / 47	0.4423 / 48	0.6000 / 49	0.6600 / 50	0.7056 / 51	0.7761 / 52 **29**	0.6226 / 54 **30**	0.8375 / 56	0.8450 / 57	0.8490 / 59

Table 8.1.9 Optimal Redundancy Algorithm

x_1 \ y_3	18	19	20	21	22	23	24	25	26
0	0.0825 / 39 **1**	0.0908 / 40 **2**	0.0962 / 42	0.1013 / 43	0.1083 / 44	0.1191 / 45	0.1263 / 47	0.1285 / 49	0.1404 / 50
1	0.1485 / 42 **3**	0.1634 / 43 **4**	0.1732 / 45	0.1823 / 46	0.1949 / 47	0.2144 / 48	0.2273 / 50	0.2314 / 52	0.2526 / 53
2	0.2013 / 45 **5**	0.2215 / 46 **6**	0.2347 / 48	0.2472 / 49	0.2643 / 50	0.2907 / 51	0.3081 / 53	0.3136 / 55	0.3425 / 56
3	0.2436 / 48 **7**	0.2679 / 49 **8**	0.2840 / 51 **9**	0.2900 / 52	0.3197 / 53 **11**	0.3516 / 54 **12**	0.3727 / 56 **13**	0.3795 / 58	0.4143 / 59
4	0.2774 / 51	0.3051 / 52 **10**	0.3234 / 54	0.3406 / 55	0.3641 / 56	0.4000 / 57 **14**	0.4245 / 59 **15**	0.4322 / 61	0.4719 / 62
5	0.3044 / 54	0.3349 / 55	0.3549 / 57	0.3737 / 58	0.3996 / 59	0.4395 / 60 **16**	0.4658 / 62	0.4742 / 64	0.5179 / 65
6	0.3260 / 57	0.3586 / 58	0.3801 / 60	0.4000 / 61	0.4279 / 62	0.4706 / 63	0.4989 / 65	0.5079 / 67	0.5546 / 68
7	0.3432 / 60	0.3775 / 61	0.4000 / 63	0.4214 / 64	0.4505 / 65	0.4955 / 66	0.5252 / 68	0.5347 / 70	0.5888 / 71

and $x_3 = 1$. Here the probability of failure-free operation of the system is 0.2679.

In an analogous fashion we find from Table 8.1.9 that the solution for the second problem (the probability of failure-free operation must not be lower than 0.45) is $y_4 = 17$. Also, successively, we find that to $y_4 = 17$ correspond the values $x_1 = 4$, $x_2 = 3$, $x_3 = 1$, $x_4 = 1$, and $x_5 = 4$. Here the cost of the system is 62 units.

8.1.2 A method for determining the optimal number of stand-by units using the method of steepest descent

If we do not know the precise statistical characteristics of the different units, or if the mathematical model is a very crude approximation to the real system, then the use of exact computational methods to determine the number of stand-by units does not make sense, since the imprecision of a method is only a secondary source of error in this case.

Generally speaking, the method of steepest descent does not lead to all possible variants of the optimal distribution of stand-by units; however, all the solutions obtained by this method are optimal.

The method of steepest descent is convenient in that it requires significantly less computation than the method of dynamic programming.

The method consists of the following steps:

1. For the ith part of the redundant system for some fixed time interval of duration t_0 and for various numbers of stand-by units x_i, we compute the values of the probability of failure-free operation $P_i(x_i)$, where $x_i = 0, 1, 2, \ldots$.

2. We set up a table of the values[1] of $\log P_i(x_i)$ for the various x_i (Table 8.1.10).

---

[1] The logarithm may be taken to any base.

Table 8.1.10   Table of Values $\log P_i(x_i)$

x_i	$\log P_1(x_1)$	$\log P_2(x_2)$	$\cdots$	$\log P_i(x_i)$	$\cdots$	$\log P_m(x_m)$
0	$\log P_1(0)$	$\log P_2(0)$	$\cdots$	$\log P_i(0)$	$\cdots$	$\log P_m(0)$
1	$\log P_1(1)$	$\log P_2(1)$	$\cdots$	$\log P_i(1)$	$\cdots$	$\log P_m(1)$
2	$\log P_1(2)$	$\log P_2(2)$	$\cdots$	$\log P_i(2)$	$\cdots$	$\log P_m(2)$
.						
.						
.						
x	$\log P_1(x)$	$\log P_2(x)$	$\cdots$	$\log P_i(x)$	$\cdots$	$\log P_m(x)$
.						
.						
.						

3.   On the basis of Table 8.1.10 and the known values of the weights of the units w_i, we set up a table of values $g_i(x_i)$ computed from the formula

$$g_i(x_i) = \frac{\log P_i(x_i) - \log P_i(x_i - 1)}{w_i} \qquad (8.1.5)$$

for all i and the various values x_i (Table 8.1.10).

Remark.   For values $P_i(x_i)$ close to unity, computation of the quantities $g_i(x_i)$ can be accomplished by the approximate formula

$$g_i(x_i) \approx \frac{Q_i(x_i - 1) - Q(x_i)}{w_i}. \qquad (8.1.6)$$

4.   All the values $g_i(x_i)$ in Table 8.1.12 are renumbered in decreasing order.

5.   We now consider a multistep process.

First step:

Choose $g^{(1)} = g_j^{(1)}$—the maximum of the quantities $g_i(1)$.
From Table 8.1.11 find the corresponding quantity $P_j(1)$.
Compute the value

$$\log P^{(1)} = \log P^{(0)} - \log P_j(0) + \log P_j(1), \qquad (8.1.7)$$

where $P^{(0)} = \sum_{l=1}^{n} \log P_l(0)$ is the initial value of the logarithm of the probability of failure-free operation of the system.
Compute the value

$$W^{(1)} = W_0 + w_j, \qquad (8.1.8)$$

where W_0 is the initial weight of the system.

Second step:

Choose $g^{(2)}$—the maximum of the remaining $g_i(1)$ for $i \neq j$ or $g_j(2)$.
From Table 8.1.11 find the corresponding quantity $\log P_i(1)$ [or $\log P_j(2)$ if $g_j(2)$ has index 2].
Compute the value

$$\begin{aligned} \log P^{(2)} &= \log P^{(1)} - \log P_i(0) + \log P_i(1) \\ [\text{or } \log P^{(2)} &= \log P^{(1)} - \log P_j(1) + \log P_j(2)]. \end{aligned} \qquad (8.1.9)$$

Compute the value

$$W^{(2)} = W^{(1)} + w_i \qquad (8.1.10)$$

[or $W^{(2)} = W^{(1)} + w_j$ if $g_j(2)$ has index 2].

Table 8.1.11   Values $g_i(x_i)$

x_i	$g_1(x_1)$	$g_2(x_2)$	$\cdots$	$g_i(x_i)$	$\cdots$	$g_m(x_m)$
0	—	—	$\cdots$	—	$\cdots$	—
1	$g_1(1)$	$g_2(1)$	$\cdots$	$g_i(1)$	$\cdots$	$g_m(1)$
2	$g_2(2)$	$g_2(2)$	$\cdots$	$g_i(2)$	$\cdots$	$g_m(2)$
.						
.						
.						
x	$g_1(x)$	$g_2(x)$	$\cdots$	$g_i(x)$	$\cdots$	$g_m(x)$
.						
.						
.						

Table 8.1.12   Values $P_i(x_i)$

x_i	$P_1(x_1)$	$P_2(x_2)$	$\cdots$	$P_i(x_i)$	$\cdots$	$P_n(x_n)$
0	$P_1(0)$	$P_2(0)$	$\cdots$	$P_i(0)$	$\cdots$	$P_n(0)$
1	$P_1(1)$	$P_2(1)$	$\cdots$	$P_i(1)$	$\cdots$	$P_n(1)$
2	$P_1(2)$	$P_2(2)$	$\cdots$	$P_i(2)$	$\cdots$	$P_n(2)$
.						
.						
.						
x	$P_1(x)$	$P_2(x)$	$\cdots$	$P_i(x)$	$\cdots$	$P_n(x)$
.						
.						
.						

The process is terminated at that step $N \left(N = \sum_{i=1}^{m} x_i \right)$ where for the first problem of optimal stand-by the following condition is satisfied:

$$\log P^{(N-1)} < \log P_{RE} < \log P^{(N)} \qquad (8.1.11)$$

or for the second problem the following condition is satisfied:

$$W^{(N)} \leq W_{RE} < W^{(N+1)}. \qquad (8.1.12)$$

8.1.3   A method for determining the optimal number of stand-by units using the method of steepest descent (nonstrict algorithm)

The strict algorithm based on the method of steepest descent uses as the criterion function

$$\sum_{i=1}^{m} P_i(x_i).$$

If we use operating or nonoperating redundancy of units, then for the computations we can use tabulated gamma and beta functions. In this case the additional operation of taking logarithms is undesirable.

The nonstrict algorithm of steepest descent uses as the criterion function

$$\prod_{i=1}^{m} P_i(x_i)$$

—that is, this method does not require the additional computation of logarithms.

The method consists of the following steps:

1.   For the ith part of the redundant system, for some fixed time interval of duration t_0 and various numbers of stand-by units x_i, compute the values of the probability of failure-free operation $P_i(x_i)$, where $x_i = 0, 1, 2, \ldots$.

2.   Set up a table of values of $P_i(x_i)$ for the various x_i (Table 8.1.12).

3.   On the basis of Table 8.1.12 and the known values of the "weights" of the units w_i set up a table of values $g_i(x_i)$ computed by the formula

$$g_i(x_i) = \frac{P_i(x_i + 1) - P_i(x_i - 1)}{w_i P_i(x_i - 1)} \tag{8.1.13}$$

for all i and the various values of x_i (Table 8.1.14).

Remark.   For values $P_i(x_i)$ close to unity, computations of the quantities $g_i(x_i)$ can be made by the approximate formula

$$g_i(x_i) \approx \frac{Q_i(x_i - 1) - Q_i(x_i)}{w_i}. \tag{8.1.14}$$

In this case the strict and nonstrict algorithms using the method of steepest descent coincide exactly.

4.   All the values $g_i(x_i)$ in Table 8.1.13 are renumbered in decreasing order $g^{(1)}, g^{(2)}, \ldots, g^{(N)}, \ldots$.

5.   We now consider a multistep process.

First step:

Choose $g^{(1)} = g_j^{(1)}$—the maximum of the quantities $g_i(1)$.
From Table 8.1.12 find the corresponding quantity $P_j(1)$.
Compute the value

$$P^{(1)} = \frac{P_j(1)}{P_j(0)} P^{(0)}, \tag{8.1.15}$$

where $P^{(0)} = \prod_{k=1}^{n} P_k(0)$ is the initial value of the probability of failure-free operation of the system.

Table 8.1.13 Values $g_i(x_i)$

x_i	$g_1(x_1)$	$g_2(x_2)$	$\cdots$	$g_i(x_i)$	$\cdots$	$g_n(x_n)$
0	—	—	$\cdots$	—	$\cdots$	—
1	$g_1(1)$	$g_2(1)$	$\cdots$	$g_i(1)$	$\cdots$	$g_n(1)$
2	$g_1(2)$	$g_2(2)$	$\cdots$	$g_i(2)$	$\cdots$	$g_n(2)$
.						
.						
.						
x	$g_1(x)$	$g_2(x)$	$\cdots$	$g_i(x)$	$\cdots$	$g_n(x)$
.						
.						
.						

Table 8.1.14 Values $g_i(x_i)$ for Various Parts

NUMBER OF STAND-BY UNITS FOR PART (x_i)	$g_i(x_i)$				
	$i = 1$	$i = 2$	$i = 3$	$i = 4$	$i = 5$
0					
1	0.300[No.2]	0.812[No.1]	0.100[No.8]	0.100[No.7]	0.300[No.3]
2	0.118[No.6]	0.246[No.4]	0.009	0.004	0.120[No.5]
3	0.060[No.11]	0.087[No.9]	0.001	—	0.062[No.10]
4	0.046[No.12]	0.030[No.15]	—	—	0.035[No.13]
5	0.033[No.14]	0.009	—	—	0.021[No.17]
6	0.024[No.16]	0.002	—	—	0.013[No.20]
7	0.017[No.18]	—	—	—	0.008
8	0.014[No.19]	—	—	—	0.005
9	0.010[No.21]	—	—	—	0.003
10	0.008	—	—	—	0.001

Compute the value

$$W^{(1)} = W_0 + w_j, \qquad (8.1.16)$$

where W_0 is the initial weight of the system.

Second step:

Choose $g^{(2)}$—the maximum of the remaining $g_k(1)$ for $k \neq j$ or $g_j(2)$.

From Table 8.1.12 find the corresponding quantity $P_k(1)$ [or $P_j(2)$ if $g_j(2)$ has index 2].

Compute the value

$$P^{(2)} = \frac{P_k(1)}{P_k(0)} P^{(1)} \qquad (8.1.17)$$

(or $P^{(2)} = [P_j(2)/P_j(1)]P^{(1)}$ if $g_j(2)$ has index 2).

Compute the value

$$W^{(2)} = W^{(1)} + w_k \qquad (8.1.18)$$

[or $W^{(2)} = W^{(1)} + w_j$ if $g_j(2)$ has index 2].

The process is terminated at that step $N \left(N = \sum_{i=1}^{n} x_i \right)$ where for the first problem of optimal stand-by the following condition is satisfied:

$$P^{(N-1)} < P_{RE} \le P^{(N)} \qquad (8.1.19)$$

or for the second problem the following condition is satisfied:

$$W^{(N)} \le W_{RE} < W^{(N+1)}. \qquad (8.1.20)$$

Example 8.1.2   Consider Example 8.1.1, and this time to obtain the solution we use the non-strict algorithm based on the method of steepest descent.

Solution.   For this solution we make use of the values for $P_i(x_i)$ in Table 8.1.5.

On the basis of Table 8.1.5 we set up Table 8.1.15, where we put the values $g_i(x_i)$ computed by Formula (8.1.13). The final computed results are given in Table 8.1.16, which allows us to solve both the direct and the inverse problems of optimal redundancy. From this table it can be seen that:

(a)   The process must be terminated at the ninth step. Here the probability of failure-free operation of the system is 0.221, and the corresponding values of x_i are $x_1 = 2$, $x_2 = 3$, $x_3 = 1$, $x_4 = 1$, $x_5 = 2$;

(b)   The process must be terminated at the thirteenth step, at which point it is seen that 62 weight units are expended on the system. The values x_i are $x_1 = 4$, $x_2 = 3$, $x_3 = 1$, $x_4 = 1$, $x_5 = 4$.

8.1.4   The distribution of the remaining resources

The following situation may arise in practice in many cases: the $(N - 1)$st step does not yet yield the desired solution (for example, the entire "weight" has not yet been used up or the desired value of the probability of failure-free operation has not yet been reached), while the Nth step of the optimal procedure requires introducing a unit with too great a "weight."

Table 8.1.15   Resulting Table for Example 8.1.2

STEP NUMBER	NUMBER OF STAND-BY UNITS FOR PARTS AT THE NTH STEP					PROBABILITY OF FAILURE-FREE OPERATION OF THE SYSTEM $P^{(N)}$	"WEIGHT" OF THE SYSTEM $W^{(N)}$
	x_1	x_2	x_3	x_4	x_5		
0	0	0	0	0	0	0.004	22
1	0	1	0	0	0	0.011	24
2	1	1	0	0	0	0.019	27
3	1	1	0	0	1	0.048	32
4	1	2	0	0	1	0.071	34
5	1	2	0	0	2	0.114	39
6	2	2	0	0	2	0.155	42
7	2	2	0	1	2	0.171	43
8	2	2	1	1	2	0.188	44
9	2	3	1	1	2	0.221	46
10	2	3	1	1	3	0.290	51
11	3	3	1	1	3	0.351	54
12	4	3	1	1	3	0.400	57
13	4	3	1	1	4	0.471	62
14	5	3	1	1	4	0.517	65
15	5	4	1	1	4	0.548	67
16	6	4	1	1	4	0.587	70
17	6	4	1	1	5	0.649	75
18	7	4	1	1	5	0.683	78
19	8	4	1	1	5	0.711	81

Thus in solving the first problem the introduction of an optimal unit yields a probability of failure-free operation of the system a fortiori greater than is required. At the same time the redundancy of other parts of the system (not optimal at the given step under the criterion being used) may guarantee the required value of the probability of failure-free operation with less "weight" expenditure.

When such situations arise, for solving the first problem we recommend using the following two methods for distributing the remaining resources.

1.   If the Nth step of the optimal procedure would lead to an essential overexpenditure of "weight" (or to attaining an unnecessarily high value of P), then at the Nth step we exclude from consideration a part having maximal value $g^{(N)}$. If, after doing this, we still have an essential over-expenditure of "weight" or achieve an unnecessarily high value P, then the procedure should be continued in the same direction—that is, at the Nth step we exclude from consideration the two parts having the largest values $g^{(N)}$.

Table 8.1.16 Poisson Distribution

k	0.1	0.2	0.3	0.4	0.5	1.0	1.5	2.0	2.5	3.0	4.0
0	0.905	0.819	0.741	0.670	0.606	0.368	0.223	0.135	0.082	0.050	0.018
1	0.090	0.164	0.222	0.268	0.303	0.368	0.335	0.271	0.205	0.149	0.073
2	0.0^2452	0.016	0.033	0.058	0.076	0.184	0.251	0.271	0.257	0.224	0.146
3	0.0^3151	0.0^2109	0.0^2333	0.0^2715	0.013	0.061	0.126	0.180	0.214	0.224	0.195
4	0.0^5400	0.0^4550	0.0^3250	0.0^3715	0.0^2158	0.015	0.047	0.090	0.137	0.168	0.195
5	—	0.0^5200	0.0^4150	0.0^4570	0.0^3158	0.0^2307	0.014	0.036	0.067	0.102	0.156
6	—	—	0.0^5100	0.0^5400	0.0^4130	0.0^3511	0.0^2353	0.012	0.028	0.050	0.104
7	—	—	—	—	0.0^5100	0.0^4730	0.0^3756	0.0^2344	0.0^2994	0.022	0.060
8	—	—	—	—	—	0.0^5900	0.0^3142	0.0^3859	0.0^2311	0.0^2810	0.030
9	—	—	—	—	—	0.0^5100	0.0^4240	0.0^3191	0.0^3863	0.0^2270	0.013
10	—	—	—	—	—	—	0.0^5400	0.0^4380	0.0^3216	0.0^3810	0.0^2529
11	—	—	—	—	—	—	—	0.0^5700	0.0^4490	0.0^3221	0.0^2192
12	—	—	—	—	—	—	—	0.0^5100	0.0^4100	0.0^4550	0.0^3642
13	—	—	—	—	—	—	—	—	0.0^5200	0.0^4130	0.0^3197
14	—	—	—	—	—	—	—	—	—	0.0^5300	0.0^4560
15	—	—	—	—	—	—	—	—	—	0.0^5100	0.0^4150
16	—	—	—	—	—	—	—	—	—	—	0.0^5400
17	—	—	—	—	—	—	—	—	—	—	0.0^5100

Table 8.1.16 Poisson Distribution (*Continued*)

k	a = 5.0	6.0	7.0	8.0	9.0	10.0
0	0.0^2674	0.0^2248	0.0^3912	0.0^3335	0.0^3123	0.0^4450
1	0.034	0.0149	0.0^2638	0.0^2268	0.0^3111	0.0^3454
2	0.084	0.045	0.022	0.0107	0.0^2500	0.0^2227
3	0.140	0.089	0.052	0.029	0.015	0.0^2757
4	0.175	0.134	0.091	0.057	0.034	0.019
5	0.175	0.161	0.128	0.092	0.061	0.038
6	0.146	0.161	0.149	0.122	0.091	0.063
7	0.104	0.138	0.149	0.140	0.117	0.090
8	0.065	0.103	0.130	0.140	0.132	0.112
9	0.036	0.069	0.101	0.124	0.132	0.125
10	0.018	0.041	0.071	0.099	0.118	0.125
11	0.0^2824	0.022	0.045	0.072	0.097	0.114
12	0.0^2343	0.011	0.026	0.048	0.073	0.095
13	0.0^2132	0.0^2520	0.014	0.030	0.050	0.073
14	0.0^3472	0.0^2223	0.0^2709	0.017	0.032	0.052

k	a = 5.0	6.0	7.0	8.0	9.0	10.0
15	0.0^3157	0.0^3891	0.0^3331	0.0^3903	0.019	0.035
16	0.0^4490	0.0^3334	0.0^2145	0.0^2451	0.011	0.022
17	0.0^4140	0.0^3118	0.0^3596	0.0^2212	0.0^2579	0.013
18	0.0^5400	0.0^4390	0.0^3232	0.0^3944	0.0^2289	0.0^2709
19	0.0^5100	0.0^4120	0.0^4850	0.0^3397	0.0^2137	0.0^2373
20	—	0.0^5400	0.0^4300	0.0^3159	0.0^3617	0.0^2187
21	—	0.0^5100	0.0^4100	0.0^4610	0.0^3264	0.0^3889
22	—	—	0.0^5300	0.0^4220	0.0^3108	0.0^3404
23	—	—	0.0^5100	0.0^5800	0.0^4420	0.0^3176
24	—	—	—	0.0^5300	0.0^4160	0.0^4730
25	—	—	—	0.0^5100	0.0^5600	0.0^4290
26	—	—	—	—	0.0^5200	0.0^4110
27	—	—	—	—	0.0^5100	0.0^5400
28	—	—	—	—	—	0.0^5100
29	—	—	—	—	—	0.0^5100

Table 8.1.16   Poisson Distribution (*Continued*)

k	a 15.0	a 20.0	k	a 15.0	a 20.0
0	—	—	23	0.013	0.067
1	0.0^5500	—	24	0.0^2830	0.056
2	0.0^4340	—	25	0.0^2498	0.044
3	0.0^3172	0.0^5300	26	0.0^2287	0.034
4	0.0^3645	0.0^4140	27	0.0^2160	0.025
5	0.0^2194	0.0^4550	28	0.0^3855	0.018
6	0.0^2484	0.0^3183	29	0.0^3442	0.012
7	0.010	0.0^3523	30	0.0^3221	0.0^2834
8	0.019	0.0^2131	31	0.0^3107	0.0^2538
9	0.032	0.0^2291	32	0.0^4500	0.0^2336
10	0.047	0.0^2582	33	0.0^4230	0.0^2204
11	0.066	0.011	34	0.0^4100	0.0^2120
12	0.083	0.018	35	0.0^5400	0.0^3685
13	0.096	0.027	36	0.0^5200	0.0^3381
14	0.102	0.039	37	0.0^5100	0.0^3206
15	0.102	0.052	38	—	0.0^3108
16	0.096	0.064	39	—	0.0^4560
17	0.085	0.076	40	—	0.0^4280
18	0.071	0.084	41	—	0.0^4140
19	0.056	0.089	42	—	0.0^5600
20	0.042	0.089	43	—	0.0^5300
21	0.030	0.085	44	—	0.0^5100
22	0.020	0.077	45	—	0.0^5100

2.   If the Nth step of the optimal procedure leads to an essential over-expenditure of "weight" (or to an unnecessarily high value P), then we should "remove" the stand-by unit added at the $(N - 1)$st step of the optimal procedure.

Then after the $(N - 2)$nd step continue the procedure, excluding from consideration the part having maximal value $g^{(N-1)}$. If in this case we also have an essential overexpenditure of "weight" or achieve an unnecessarily high value P, then the procedure should be continued in the same direction—that is, after the $(N - 2)$nd step exclude from consideration the two parts having the greatest values $g^{(N-1)}$, and so on.

8.1.5   A simplified method of achieving optimal redundancy

If we wish to achieve a high value of the probability of failure-free operation of the system as a result of imposing redundancy—that is, achieve the condition

$$1 - P_{RE} = Q_{RE} \ll 1, \tag{8.1.21}$$

then we can use a method of optimal stand-by allocation based on an approximate solution of the problem.

1.   The determination of the optimal stand-by allocation for attaining the required value of the probability of failure-free operation of the system P_{RE} (or an admissible value of the probability of failure of the system Q_{RE}) is carried out as follows:

(a)   Determine the quantity g from

$$g = \frac{Q_{RE}}{\sum\limits_{i=1}^{n} w_i}. \tag{8.1.22}$$

(b)   Determine the quantity c_i from

$$c_i = w_i g = \frac{w_i}{\sum\limits_{=1}^{n} w_i} Q_{RE}. \tag{8.1.23}$$

(c)   From the equation

$$Q_i(x_i) = c_i \tag{8.1.24}$$

find the value x_i.

Remark.   Usually the computed values of x_i are fractions. In this case for all practical purposes we may round these values to the nearest integers.

In cases where the fractional parts of the quantities x_i are close to 0.5, we recommend rounding the smaller quantities x_i upwards, and the larger ones downwards.

In many cases where computational speed is important, it is more convenient to round all computed x_i values to the next larger integer.

Example 8.1.3   Suppose a system consists of five independent parts, each of whose units are characterized by the following readiness coefficients and measures of "weight," respectively:

$$K_1 = 0.97, \quad w_1 = 1; \qquad K_2 = 0.97, \quad w_2 = 5;$$
$$K_3 = 0.90, \quad w_3 = 1; \qquad K_4 = 0.90, \quad w_4 = 5;$$
$$K_5 = 0.995, \quad w_5 = 0, 1.$$

In the system we use only operating stand-by units for each of the parts, and each unit of the system can be repaired independently.

We wish to raise the readiness coefficient of the system to a value not less than $K_{RE} = 0.999$ in an optimal way.

Solution.   It is evident from the problem that condition (8.1.24) is satisfied, since $0.001 \ll 1$. From Formula (8.1.22) we determine the value

g to be

$$g = \frac{1 - K_{RE}}{\sum\limits_{i=1}^{n} w_i} = \frac{0.001}{1 + 5 + 1 + 5 + 0.1} = 8.264 \cdot 10^{-5}.$$

Further, we find the values c_i from Formula (8.1.23) for all parts of the system to be

$$c_1 = 8.264 \cdot 10^{-5},$$
$$c_2 = 4.132 \cdot 10^{-4},$$
$$c_3 = 8.264 \cdot 10^{-5},$$
$$c_4 = 4.132 \cdot 10^{-4},$$
$$c_5 = 8.264 \cdot 10^{-6}.$$

In subsequent computations we use the formula

$$c_i = (1 - r_i)^{x_i + 1},$$

whence[1]

$$x_i = \frac{\log c_i}{\log q_i} - 1.$$

As a result we find (using, say, the logarithm to base ten):

$$x_1 = \frac{-4.0828}{-1.5229} - 1 = 1.68,$$

$$x_2 = \frac{-3.3840}{-1.5229} - 1 = 1.22,$$

$$x_3 = \frac{-4.0828}{-1} - 1 = 3.08,$$

$$x_4 = \frac{-3.3840}{-1} - 1 = 2.38,$$

$$x_5 = \frac{-5.0828}{-2.301} - 1 = 1.21.$$

From the computed values it follows that

$$1 \leq x_1 \leq 2,$$
$$1 \leq x_2 \leq 2,$$
$$3 \leq x_3 \leq 4,$$
$$2 \leq x_4 \leq 3,$$
$$1 \leq x_5 \leq 2.$$

[1] The logarithm may be taken to any base.

We take $x_1 = 2$, $x_2 = 2$, $x_3 = 4$, $x_4 = 3$, and $x_5 = 2$. We compute the probability of failure for each of these quantities x_i as follows:

$$1 - K \approx \sum_{i=1}^{5} (1 - K_i)^{x_i+1}$$
$$= 0.03^3 + 0.03^3 + 0.1^5 + 0.1^4 + 0.005^3$$
$$= 5.04 \cdot 10^{-4}$$

or

$$K = 1 - 5.04 \cdot 10^4 \approx 0.9995.$$

Here the "weight" of the system is 43.3 units in the scale of measure for "weight" we assume for this problem.

2.   The determination of optimal stand-by allocation under the condition of not exceeding a given system "weight" is carried out as follows:

(a)   We choose an arbitrary value for the quantity $g^{(1)}$. For this value we compute the quantities $x_1^{(1)}$, $x_2^{(1)}$, ..., $x_n^{(1)}$, using Formulas (8.1.22)–(8.1.24).

(b)   We control the resulting "weight" of the system by the formula

$$W^{(1)} = \Sigma \, x_i^{(1)} w_i + w_0. \qquad (8.1.25)$$

If $W^{(1)} > W_{RE}$, then we choose a new value $g^{(2)} > g^{(1)}$ and for it, as before, determine the quantities $x_1^{(2)}$, $x_2^{(2)}$, ..., $x_n^{(2)}$. If $W^{(1)} < W_{RE}$, we choose a new value $g^{(2)}$ such that $g^{(2)} < g^{(1)}$ and for it analogously determine the quantities $x_1^{(2)}$, $x_2^{(2)}$, ..., $x_n^{(2)}$.

The process is continued until at the Nth step of the process we have found a quantity $W^{(N)}$ that practically coincides with the quantity W_{RE}.

Remark.   In practice the process of searching for a solution can be shortened to from three to five steps, if, in computing the subsequent value $W^{(k+1)}$, we use the method of linear interpolation when

$$W^{(k-1)} < W_{RE} < W^{(k)} \qquad (8.1.26)$$

or

$$W^{(k)} < W_{RE} < W^{(k-1)}, \qquad (8.1.27)$$

or linear extrapolation when

$$W^{(k-1)} < W_{RE}, \qquad W^{(k)} < W_{RE} \qquad (8.1.28)$$

or

$$W^{(k-1)} > W_{RE}, \qquad W^{(k)} > W_{RE}. \qquad (8.1.29)$$

In the case when conditions (8.1.26) or (8.1.28) are satisfied, the subsequent value $g^{(k+1)}$ should be chosen on the basis of the expression

$$g^{(k+1)} = g^{(k)} + (g^{(k-1)} - g^{(k)}) \frac{W^{(k)} - W_{RE}}{W^{(k)} - W^{(k-1)}}, \qquad (8.1.30)$$

and when conditions (8.1.27) or (8.1.29) are satisfied, on the basis of the expression

$$g^{(k+1)} = g^{(k-1)} + (g^{(k)} - g^{(k-1)}) \frac{W^{(k-1)} - W_{RE}}{W^{(k-1)} - W^{(k)}}. \qquad (8.1.31)$$

Example 8.1.4 Consider the system of Example 8.1.3. For this system we want to raise the readiness coefficient in an optimal way so that the "weight" of the system does not exceed 60 units in the scale of "weight" measure assumed for the problem.

Solution. We use the results of the preceding example as the solution for the first step. Recall that $W^{(1)} = 37.2$ units and

$$g^{(1)} = 8.264 \cdot 10^{-5} \; \frac{1}{\text{unit of weight}}.$$

Since $W^{(1)} < W_{RE}$, we should choose a value $g^{(2)}$ such that $g^{(2)} < g^{(1)}$. We take, for example,

$$g^{(2)} = 1 \cdot 10^{-8} \; \frac{1}{\text{unit of weight}}.$$

Then

$$c_1 = 1 \cdot 10^{-8},$$
$$c_2 = 1 \cdot 10^{-8},$$
$$c_3 = 1 \cdot 10^{-8},$$
$$c_4 = 1 \cdot 10^{-8},$$
$$c_5 = 1 \cdot 10^{-9}.$$

Hence, by the method described earlier in this section, we find

$$x_1 = \frac{-8}{-1.5229} - 1 = 4.25,$$

$$x_2 = \frac{-7.3010}{-1.5229} - 1 = 3.80,$$

$$x_3 = \frac{-8}{-1} - 1 = 7.0,$$

$$x_4 = \frac{-7.3010}{-1} - 1 = 6.3,$$

$$x_5 = \frac{-9}{-2.301} - 1 = 2.9.$$

As solutions we take the values $x_1 = 5$, $x_2 = 4$, $x_3 = 7$, $x_4 = 7$, $x_5 = 3$. Then the "weight" of the system is

$$W^{(2)} = 6\cdot 1 + 5\cdot 5 + 8\cdot 1 + 8\cdot 5 + 4\cdot 0.1 = 79.4 \quad \text{units of weight.}$$

Next, since condition (8.1.26) is satisfied, we use Formula (8.1.30) and find

$$g^{(3)} = 1\cdot 10^{-8} + (8.26\cdot 10^{-5} - 1\cdot 10^{-8})\cdot \frac{79.4 - 60}{79.4 - 43.2}$$

$$\approx 8.26\cdot 10^{-5}\cdot \frac{19.4}{36.2} = 4.4\cdot 10^{-5} \quad \frac{1}{\text{unit of weight}}\cdot$$

By analogy with the above, we find the values $x_1 = 2$, $x_2 = 3$, $x_3 = 4$, $x_4 = 3$, and $x_5 = 2$. Then the "weight" of the system is

$$W^{(3)} = 3\cdot 1 + 4\cdot 5 + 5\cdot 1 + 4\cdot 5 + 3\cdot 0.1 = 48.3 \quad \text{units of weight.}$$

The result obtained can be considered fully acceptable.

8.1.6   A method for determining the optimal number of required spare unrepairable units

We assume that the spare units are initially in a nonoperating state, and when they are put into service the operating time up to failure has an exponential distribution. A failed unit can be replaced only by a good one, but cannot itself be repaired.

For a certain use period t we want to guarantee the optimal number of spare units for the system. The order of computation reduces to the following.

1.   Determine the total number of units of ith type in the system.

2.   Determine the total failure intensities of units of ith type by the formula

$$\Lambda_i = m_i\lambda_i.$$

3.   Determine the mathematical expectation of the number of failures of units of ith type for a time of duration t by the formula

$$a_i = \Lambda_i t = m_i\lambda_i t.$$

4.   For each unit of type i set up tables of values $g_i^{(k)}$ for various $k = 1$, $2,\ldots$ by the formula

$$g_i^{(k)} = \frac{p_k(a_i)}{w_i[1 - P_k(a_i)]},$$

where $p_k(a_i)$ is found from Table 8.1.17 and $P_k(a_i)$ from Table 8.1.18.

We now repeat the procedure described in Section 8.1.3.

Remark.   In Tables 8.1.17 and 8.1.18 we give values only for values of the parameter a from 0.1 to 20. However, in practice one is sometimes

Table 8.1.17 Summed Values of the Poisson Function

k	\multicolumn{11}{c}{a}
	0.1	0.2	0.3	0.4	0.5	1.0	1.5	2.0	2.5	3.0	4.0
0	1.0	1.0	1.0	1.0	1.0	1.0	1.0	1.0	1.0	1.0	1.0
1	0.0951	0.181	0.259	0.330	0.393	0.632	0.777	0.865	0.918	0.950	0.982
2	0.0^2467	0.0175	0.0369	0.0615	0.0902	0.264	0.442	0.594	0.713	0.801	0.908
3	0.0^3155	0.0^2115	0.0^236	0.0^2793	0.0143	0.0803	0.191	0.323	0.456	0.577	0.762
4	0.0^5400	0.0^4570	0.0^3266	0.0^3776	0.0^2175	0.0190	0.0656	0.143	0.242	0.353	0.566
5	—	0.0^52	0.0^416	0.0^461	0.0^3172	0.0^2366	0.0186	0.0526	0.109	0.185	0.371
6	—	—	0.0^51	0.0^54	0.0^414	0.0^3594	0.0^2446	0.0166	0.042	0.0839	0.215
7	—	—	—	—	0.0^51	0.0^483	0.0^3926	0.0^2453	0.0141	0.0335	0.111
8	—	—	—	—	—	0.0^41	0.0^317	0.0^2110	0.0^2425	0.0119	0.0511
9	—	—	—	—	—	0.0^51	0.0^428	0.0^3237	0.0^2114	0.0^2380	0.0214
10	—	—	—	—	—	—	0.0^54	0.0^445	0.0^3277	0.0^2110	0.0^2813
11	—	—	—	—	—	—	0.0^51	0.0^58	0.0^462	0.0^3292	0.0^2284
12	—	—	—	—	—	—	—	0.0^51	0.0^413	0.0^471	0.0^3915
13	—	—	—	—	—	—	—	—	0.0^52	0.0^416	0.0^3274
14	—	—	—	—	—	—	—	—	—	0.0^53	0.0^476
15	—	—	—	—	—	—	—	—	—	0.0^51	0.0^420
16	—	—	—	—	—	—	—	—	—	—	0.0^55
17	—	—	—	—	—	—	—	—	—	—	0.0^51

Table 8.1.17 Summed Values of the Poisson Function
(Continued)

k			a			
	5.0	6.0	7.0	8.0	9.0	10.0
0	1.0	1.0	1.0	1.0	1.0	1.0
1	0.993	0.997	0.9^3088	0.9^3665	0.9^3877	0.9^455
2	0.960	0.983	0.9^2270	0.9^2698	0.9^2877	0.9^3501
3	0.875	0.938	0.9704	0.9862	0.9^2377	0.9^2723
4	0.733	0.849	0.9182	0.9576	0.9788	0.9897
5	0.560	0.713	0.827	0.9001	0.9450	0.9707
6	0.384	0.554	0.699	0.809	0.884	0.9329
7	0.237	0.394	0.550	0.687	0.793	0.869
8	0.133	0.256	0.401	0.547	0.676	0.779
9	0.0680	0.153	0.271	0.407	0.544	0.667
10	0.0318	0.0839	0.169	0.283	0.412	0.542
11	0.0137	0.0426	0.0985	0.184	0.294	0.417
12	0.0^2545	0.0201	0.0533	0.112	0.197	0.303
13	0.0^2202	0.0^2883	0.0270	0.0638	0.124	0.208
14	0.0^3698	0.0^2363	0.0128	0.0342	0.0738	0.136

k			a			
	5.0	6.0	7.0	8.0	9.0	10.0
15	0.0^3226	0.0^2140	0.0^2572	0.0172	0.0415	0.0834
16	0.0^469	0.0^3509	0.0^2241	0.0^2823	0.0220	0.0487
17	0.0^420	0.0^3175	0.0^3958	0.0^2372	0.0111	0.0270
18	0.0^55	0.0^457	0.0^3362	0.0^3159	0.0^2532	0.0143
19	0.0^51	0.0^418	0.0^3130	0.0^3650	0.0^2243	0.0^2719
20	—	0.0^55	0.0^444	0.0^3253	0.0^2105	0.0^2345
21	—	0.0^51	0.0^414	0.0^494	0.0^3439	0.0^2159
22	—	—	0.0^55	0.0^433	0.0^3175	0.0^3700
23	—	—	0.0^51	0.0^411	0.0^467	0.0^3296
24	—	—	—	0.0^54	0.0^425	0.0^3120
25	—	—	—	0.0^51	0.0^59	0.0^447
26	—	—	—	—	0.0^53	0.0^418
27	—	—	—	—	0.0^51	0.0^56
28	—	—	—	—	—	0.0^52
29	—	—	—	—	—	0.0^51

Table 8.1.17   Summed Values of the Poisson Function (*Continued*)

k	a 15.0	a 20.0	k	a 15.0	a 20.0
0	—	—	23	0.0327	0.279
1	1.0	—	24	0.0195	0.212
2	0.9^55	—	25	0.0112	0.157
3	0.9^461	1.0	26	0.0^2618	0.112
4	0.9^3789	0.9^57	27	0.0^2331	0.0779
5	0.9^3143	0.9^483	28	0.0^2172	0.0525
6	0.9^2721	0.9^428	29	0.0^3861	0.0343
7	0.9^2237	0.9^3745	30	0.0^3418	0.0218
8	0.9820	0.9^3221	31	0.0^3197	0.0135
9	0.9625	0.9^2791	32	0.0^490	0.0^2809
10	0.9301	0.9^2500	33	0.0^440	0.0^2472
11	0.881	0.9892	34	0.0^417	0.0^2269
12	0.815	0.9786	35	0.0^57	0.0^2149
13	0.732	0.9610	36	0.0^53	0.0^3804
14	0.637	0.9339	37	0.0^51	0.0^3423
15	0.534	0.895	38	—	0.0^3217
16	0.432	0.843	39	—	0.0^3109
17	0.336	0.779	40	—	0.0^453
18	0.251	0.703	41	—	0.0^425
19	0.180	0.618	42	—	0.0^412
20	0.120	0.530	43	—	0.0^55
21	0.0830	0.441	44	—	0.0^52
22	0.0531	0.356	45	—	0.0^51

faced with the problem of determining the quantities

$$p_k(a) = \frac{a^k}{k!} \cdot e^{-a} \quad \text{and} \quad P_{(k)}(a) = \sum_{s=k}^{\infty} \frac{a^s}{s!} \cdot e^{-a}$$

for values of the parameter $a < 0.1$ and $a > 30$. In both of these cases the computations can be carried out by the following approximate formulas.

Case $a < 0.1$.   (1) $p_k(a) \approx a^k/k!$.

Values of $k!$ for $k \leq 10$:

k	1	2	3	4	5	6	7	8	9	10
k!	1	2	6	24	$1.2 \cdot 10^2$	$7.2 \cdot 10^2$	$5.0 \cdot 10^3$	$4.0 \cdot 10^4$	$3.6 \cdot 10^5$	$3.6 \cdot 10^6$

Table 8.1.18 Values g_i

step N	$g_1^{(N)}$	$g_2^{(N)}$	$g_3^{(N)}$	$g_4^{(N)}$	$g_5^{(N)}$
1	$3.0 \cdot 10$	1	2.1	2.5	2
2	$1.1 \cdot 10$	$8.7 \cdot 10^{-1}$	1.6	2.5	$9.1 \cdot 10^{-1}$
3	5.2	$8.1 \cdot 10^{-1}$	1.2	2.5	$5.5 \cdot 10^{-1}$
4	2.6	$7.5 \cdot 10^{-1}$	$9.1 \cdot 10^{-1}$	2.5	$3.7 \cdot 10^{-1}$
5	1.2	$6.9 \cdot 10^{-1}$	$6.2 \cdot 10^{-1}$	2.5	$2.5 \cdot 10^{-1}$
6	$5.4 \cdot 10^{-1}$	$6.3 \cdot 10^{-1}$	$3.9 \cdot 10^{-1}$	2.5	$1.9 \cdot 10^{-1}$
7	$2.2 \cdot 10^{-1}$	$5.5 \cdot 10^{-1}$	$2.4 \cdot 10^{-1}$	2.5	$1.4 \cdot 10^{-1}$
8	$8.1 \cdot 10^{-2}$	$5.1 \cdot 10^{-1}$	$1.3 \cdot 10^{-1}$	2.5	$1.0 \cdot 10^{-1}$
9	$2.7 \cdot 10^{-2}$	$4.6 \cdot 10^{-1}$	$6.0 \cdot 10^{-2}$	2.5	$7.6 \cdot 10^{-2}$
10	$8.1 \cdot 10^{-3}$	$4.0 \cdot 10^{-1}$	$2.6 \cdot 10^{-2}$	2.1	$5.4 \cdot 10^{-2}$
11	$2.2 \cdot 10^{-3}$	$3.5 \cdot 10^{-1}$	$1.0 \cdot 10^{-2}$	1.9	$3.9 \cdot 10^{-2}$
12	$5.5 \cdot 10^{-4}$	$3.2 \cdot 10^{-1}$	$3.8 \cdot 10^{-3}$	1.6	$2.7 \cdot 10^{-2}$
13	$1.3 \cdot 10^{-4}$	$2.6 \cdot 10^{-1}$	$1.2 \cdot 10^{-3}$	1.4	$1.8 \cdot 10^{-2}$
14	$3.0 \cdot 10^{-5}$	$2.1 \cdot 10^{-1}$	$4.0 \cdot 10^{-4}$	1.3	$1.2 \cdot 10^{-2}$
15	$1.0 \cdot 10^{-5}$	$1.7 \cdot 10^{-1}$	$1.2 \cdot 10^{-4}$	1.1	$7.6 \cdot 10^{-3}$
16		$1.4 \cdot 10^{-1}$	$3.0 \cdot 10^{-5}$	$9.9 \cdot 10^{-1}$	$4.6 \cdot 10^{-3}$
17		$1.3 \cdot 10^{-1}$	$8.0 \cdot 10^{-6}$	$8.8 \cdot 10^{-1}$	$2.6 \cdot 10^{-3}$
18		$1.2 \cdot 10^{-1}$	$2.0 \cdot 10^{-6}$	$7.8 \cdot 10^{-1}$	$1.4 \cdot 10^{-3}$
19		$6.6 \cdot 10^{-2}$		$6.9 \cdot 10^{-1}$	$7.4 \cdot 10^{-4}$
20		$4.8 \cdot 10^{-2}$		$5.9 \cdot 10^{-1}$	$3.7 \cdot 10^{-4}$
21		$3.3 \cdot 10^{-2}$		$5.4 \cdot 10^{-1}$	$1.8 \cdot 10^{-4}$
22		$2.1 \cdot 10^{-2}$		$4.8 \cdot 10^{-1}$	$8.0 \cdot 10^{-5}$
23		$1.3 \cdot 10^{-2}$		$4.2 \cdot 10^{-1}$	$3.6 \cdot 10^{-5}$
24		$8.3 \cdot 10^{-3}$		$3.8 \cdot 10^{-1}$	$1.5 \cdot 10^{-5}$
25		$5.0 \cdot 10^{-3}$		$3.2 \cdot 10^{-1}$	$5.8 \cdot 10^{-6}$
26		$2.9 \cdot 10^{-3}$		$2.8 \cdot 10^{-1}$	$2.2 \cdot 10^{-6}$
27		$1.6 \cdot 10^{-3}$		$2.4 \cdot 10^{-1}$	$8.0 \cdot 10^{-7}$
28		$8.6 \cdot 10^{-4}$		$2.1 \cdot 10^{-1}$	$2.0 \cdot 10^{-7}$
29		$4.4 \cdot 10^{-4}$		$1.8 \cdot 10^{-1}$	
30		$2.2 \cdot 10^{-4}$		$1.5 \cdot 10^{-1}$	
31		$1.1 \cdot 10^{-4}$		$1.3 \cdot 10^{-1}$	
32		$5.0 \cdot 10^{-5}$		$1.1 \cdot 10^{-1}$	
33		$2.3 \cdot 10^{-5}$		$8.7 \cdot 10^{-2}$	

Table 8.1.18   Values g_i (*Continued*)

STEP N	$g_1^{(N)}$	$g_2^{(N)}$	$g_3^{(N)}$	$g_4^{(N)}$	$g_5^{(N)}$
34		$4 \cdot 10^{-6}$		$7.1 \cdot 10^{-2}$	
35		$2 \cdot 10^{-6}$		$5.6 \cdot 10^{-2}$	
36		$1 \cdot 10^{-6}$		$4.5 \cdot 10^{-2}$	
37				$3.5 \cdot 10^{-2}$	
38				$2.6 \cdot 10^{-2}$	
39				$2.0 \cdot 10^{-2}$	
40				$1.5 \cdot 10^{-2}$	
41				$1.0 \cdot 10^{-2}$	
42				$7.3 \cdot 10^{-3}$	
43				$5.1 \cdot 10^{-3}$	
44				$3.5 \cdot 10^{-3}$	
45				$2.3 \cdot 10^{-3}$	
46				$1.6 \cdot 10^{-3}$	
47				$9.6 \cdot 10^{-4}$	
48				$6.0 \cdot 10^{-4}$	
49				$3.7 \cdot 10^{-4}$	
50				$2.2 \cdot 10^{-4}$	
51				$1.3 \cdot 10^{-4}$	
52				$7.5 \cdot 10^{-5}$	
53				$4.2 \cdot 10^{-5}$	
54				$2.4 \cdot 10^{-5}$	
55				$1.3 \cdot 10^{-5}$	
56				$7.0 \cdot 10^{-6}$	
57				$4.0 \cdot 10^{-6}$	
58				$2.0 \cdot 10^{-6}$	
59				$1.0 \cdot 10^{-6}$	

For $k > 10$ the values $k!$ can be computed approximately by Stirling's formula (see Appendix A2.2).

(2) $P_k(a) \approx a^k/k!$. Here, if $k > 1$ for practically any $a < 0.1$, in computing the optimal number of necessary spare units we can assume that $1 - P_k(a) \approx 1$.

Case $a > 30$. In this case the Poisson distribution can be successfully approximated by the normal distribution with mean a and standard deviation $\sigma = \sqrt{a}$.

(1) $p_k(a) \approx \varphi(|k - a|/\sqrt{a})$, where $\varphi(u)$ is the value of the standard normal density given in Table A3.2.

(2) $P_k(a) \approx 1 - \Phi[(k - a)/\sqrt{a}]$, where $\Phi(u)$ is the normal distribution function, whose values are given in Table A3.1.

Here, for $(k - a)/\sqrt{a} > 2 \div 2.5$ we can assume that $1 - P_k(a) \approx 1$.

Example 8.1.5 Suppose a system is put together from five types of units. The units of the first type in the system comprise 2000 elements, the second, 250 elements, the third 400, the fourth 300, and the fifth 100. The failure intensities of the various types of units are

$$\lambda_1 = 1.5 \cdot 10^{-6} \frac{1}{hr}, \qquad \lambda_2 = 6 \cdot 10^{-5} \frac{1}{hr}, \qquad \lambda_3 = 1 \cdot 10^{-5} \frac{1}{hr},$$

$$\lambda_4 = 1 \cdot 10^{-4} \frac{1}{hr}, \qquad \lambda_5 = 1 \cdot 10^{-4} \frac{1}{hr}.$$

The price of one unit of each type is:

$$w_1 = \$0.10, \quad w_2 = \$1, \quad w_3 = \$0.50, \quad w_4 = \$1, \quad w_5 = \$5.$$

We want to determine the optimal number of spare units for the system in a use period of 1000 hours for the following two cases: (a) to achieve the maximal possible probability that there will be a sufficient number of spare units given that their total cost will not exceed 100 dollars; (b) with minimal expense to have the probability that the spare units guarantee the desired reliability equal to 0.995.

Solution. For each type of unit we determine the quantities a_i:

$$a_1 = 2000 \cdot 1.5 \cdot 10^{-6} \cdot 1000 = 3,$$
$$a_2 = 250 \cdot 6 \cdot 10^{-5} \cdot 1000 = 15,$$
$$a_3 = 400 \cdot 1 \cdot 10^{-5} \cdot 1000 = 4,$$
$$a_4 = 300 \cdot 1 \cdot 10^{-4} \cdot 1000 = 30,$$
$$a_5 = 100 \cdot 1 \cdot 10^{-4} \cdot 1000 = 10.$$

Using Tables 8.1.17 and 8.1.18, we compute the values $g_i^{(k)}$, which we put in Table 8.1.19.

Table 8.1.19 Values x_i

$g^{(N)}$	x_1	x_2	x_3	x_4	x_5	$P^{(N)}$	$W^{(N)}$
$1 \cdot 10^{-1}$	7	18	8	32	8	0.093	94.7
$1 \cdot 10^{-2}$	9	23	11	41	14	0.80	140.4
$1 \cdot 10^{-3}$	11	27	13	46	18	0.978	170.6
$1 \cdot 10^{-4}$	13	31	15	51	21	0.999	195.8
$1 \cdot 10^{-5}$	15	33	16	55	24	0.99981	217.5

From Table 8.1.19 we make initial approximate determinations of certain levels, noting that at the Nth step the optimal process of solution leads to certain close values $g_i^{(N)}$ for all $i = 1, 2, \ldots, n$. The results of initial approximate determinations are given in Table 8.1.20.

1.   From Table 8.1.20 we find that for $g = 1 \cdot 10^{-1}$ the price of additional units is 94.7 dollars, and the optimal solution is given by the values $x_1 = 7$, $x_2 = 18$, $x_3 = 8$, $x_4 = 32$, and $x_5 = 8$ (that is, this already corresponds to the 73rd step of the optimal procedure). We set up a table, starting with the 74th step of the procedure (Table 8.1.21), from which we can see the final results.

If we require that the price of spare units is strictly less than 100 dollars, then after the 75th step we can use one of the methods of distributing remainders (for example, the first one in Section 8.1.4). In this case the table takes the form of Table 8.1.22. We note that after the 79th step only the first and third parts remain to be considered, and after the 81st step only the first part remains.

Remark.   Obviously in this case the control values $P^{(N)}$ do not have to be computed.

2.   From Table 8.1.20 we find that for $g \approx 10^{-4}$ the desired probability is 0.999, and the optimal solution corresponds to the values $x_1 = 13$, $x_2 = 31$, $x_3 = 15$, $x_4 = 51$, and $x_5 = 21$ (that is, this corresponds to the 131st step of the optimal procedure). Since the probability value obtained at this step exceeds the requirement, it is necessary to take units away

Table 8.1.20   Resulting Table

STEP N	x_1	x_2	x_3	x_4	x_5	$P^{(N)}$	$W^{(N)}$
74	7	18	8	33	8	—	95.7
75	8	18	8	33	8	—	95.8
76	8	18	8	33	9	0.160	105.8

Table 8.1.21   Resulting Table

STEP N	x_1	x_2	x_3	x_4	x_5	$P^{(N)}$	$W^{(N)}$
75	8	18	8	33	8	—	95.8
76	8	18	8	34	8	—	96.8
77	8	19	8	34	8	—	97.8
78	8	19	9	34	8	—	98.3
79	8	19	9	35	8	—	99.3
80	9	19	9	35	8	—	99.4
81	9	19	10	35	8	—	99.9
82	10	19	10	35	8	0.142	100

Table 8.1.22   Resulting Table

STEP N	x_1	x_2	x_3	x_4	x_5	$P^{(N)}$	$W^{(N)}$
131	13	31	15	51	21	0.99788	—
130	13	30	15	51	21	0.99776	—
129	13	30	14	51	21	0.99760	—
128	12	30	14	51	21	0.99755	—
127	12	30	14	50	21	0.99733	—
126	12	30	14	50	20	0.99546	—
125	12	30	14	49	20	0.99509	186.7
124	12	30	13	49	20	0.99489	186.2

from those parts of the system for which the values $g^{(N-1)}$ turn out to be smallest.

Starting with the 131st step of the procedure (see Table 8.1.23), we set up a table from which we can see the final results.

Remark.   Obviously in this case the control values $W^{(N)}$ do not have to be computed.

8.2   METHODS FOR THE OPTIMAL LOCATION AND DETECTION OF FAILURES IN A SYSTEM

8.2.1   The "information" method (exact solution)

When seeking a single failed unit in a structure, we are reduced to the use of control checks such that each check allows us to divide the entire collection of units in the structure into two parts, one having no failed unit and the other containing the failed unit. The use of a particular sequence of such checks leads to the precise determination of the failed unit with the least expenditure (of equipment or time).

The precise solution is applicable in the case where the durations of all the checks are equal.

We assume the failures of the individual units of the structure to be independent.

For optimal search for a failed unit, we propose the following algorithm.

1.   For each unit of the structure we compute the conditional probability of failure, given that precisely one unit has failed, by the formula

$$q_i^* = \frac{q_i}{r_i} \left(\sum_{i=1}^{n} \frac{q_i}{r_i} \right)^{-1}. \tag{8.2.1}$$

2.   The units are arranged in decreasing (nonincreasing) order of the quantities q_i^*.

3.   The two last units in the sequence are grouped into a new compound unit, for which the conditional failure probability is

$$q^* = q_n^* + q_{n-1}^*.$$ (8.2.2)

4.   This compound unit is inserted into the sequence of units according to the order of q^* among the remaining q_i^*'s.

5.   This process is repeated until all the units are grouped into a single compound unit.

The best system for checking a structure with the aim of finding a single failed unit is a collection of checks that allows us to partition the collection of units making up the final compound unit, obtained by the method described above, into subcollections of compound units in the order precisely opposite to that in which this unit was obtained.

Example 8.2.1   Suppose a structure consists of seven units, for each of which we know the values:

$$q_1^* = 0.3, \quad q_2^* = 0.3, \quad q_3^* = 0.1, \quad q_4^* = 0.1, \quad q_5^* = 0.1, \quad q_6^* = 0.05, \quad q_7^* = 0.05.$$

We seek a system of control checks allowing us to find a failed unit in the shortest time.

Solution.   According to the method described above we set up a diagram of the process of optimal checks (Figure 8.2.1). From this diagram we can see that the first check must allow us to separate the entire collection of units into two parts: the first and second on one side, and third to seventh on the other side. If the first or second unit is the failed one, then we need only make check 2 to determine which one it is. If the failed unit is one of 3–7, then check 3 must allow us to separate the third and fourth units from the fifth, sixth, and seventh, and so on.

8.2.2   The "information" method (an approximate solution)

We shall use approximate methods in those cases where different checks have different durations (τ_i) and also in those cases where the system being checked is not sufficiently flexible to allow the use of the procedure described in Section 8.2.1.

We propose the following algorithm for an optimal search for a single failed unit.

1.   Consider the original collection of n units composing the structure. For each possible ith check covering some collection A_i of units of the structure, compute the quantities

$$g_i = \frac{\Delta \mathcal{H}_i}{\tau_i}.$$ (8.2.3)

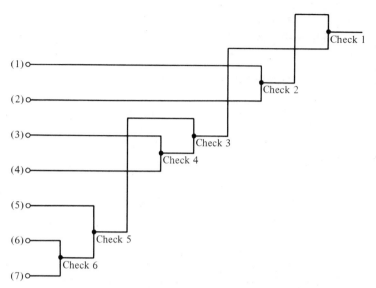

Figure 8.2.1   Diagram of the Process of Optimal Checks for Example 8.2.1.

Here

$$\Delta \mathfrak{K}_i = -[P_i \ln P_i + Q_i \ln Q_i], \tag{8.2.4}$$

where Q_i is the probability that the failed unit is one of the units in the collection A_i, and $P_i = 1 - Q_i$.

2.   Choose as the first check the one with maximum g_i value.

3.   Repeat the same procedure on the remaining checks, using for the φ_i's the conditional probabilities that the failed unit is in the corresponding A_i, given that it is in the collection covered by the first check. Checks are selected and performed by this method until the failed unit is located.

Example 8.2.2   Suppose we have the system of Example 8.2.1, but we can make only the following checks:

> Check 1 separates units 1, 5 from units 2, 3, 4, 6, 7;
> Check 2 separates units 1, 2, 4 from units 3, 5, 6, 7;
> Check 3 separates units 6, 7 from units 1, 2, 3, 4;
> Check 4 separates units 2, 7 from units 1, 3, 4, 5, 6;
> Check 5 separates units 2, 3, 5 from units 1, 4, 6, 7;

where $\tau_1 = 3$ min, $\tau_2 = 1.5$ min, $\tau_3 = 1$ min, $\tau_4 = 2$ min, $\tau_5 = 3$ min. We wish to find the system of optimal checks.

Solution.   From Formula (8.2.3) we determine the quantities for the first step of the search process for the failed unit (using logarithms to the base ten):

$$g_1^{(1)} = \frac{1}{3}(0.4 \cdot 0.398 + 0.6 \cdot 0.222) = 0.098,$$

$$g_2^{(1)} = \frac{1}{1.5}(0.7 \cdot 0.155 + 0.3 \cdot 0.523) = 0.177,$$

$$g_3^{(1)} = 1 \cdot (0.1 \cdot 1 + 0.9 \cdot 0.056) = 0.150,$$

$$g_4^{(1)} = \frac{1}{2}(0.35 \cdot 0.456 + 0.65 \cdot 0.187) = 0.142,$$

$$g_5^{(1)} = \frac{1}{3}(0.5 \cdot 0.301 + 0.5 \cdot 0.301) = 0.100.$$

As the first check we use check 2.

Suppose the failed unit is in the group of units 1, 2, and 4 (A_2).

In this case check 1 allows us to separate unit 1 from units 2, 4; check 3 does not allow us to further localize the failed unit; check 4 and check 5 separate unit 2 from units 1, 4. We must select the next check from among 1, 4, and 5.

Thus we first compute the conditional probabilities q_1^*, q_2^*, and q_4^* under the condition that the failed unit is among these units:

$$q_1^* = \frac{0.3}{0.3 + 0.3 + 0.1} = \tfrac{3}{7} = 0.43,$$

$$q_2^* = \frac{0.3}{0.3 + 0.3 + 0.1} = \tfrac{3}{7} = 0.43,$$

$$q_4^* = \frac{0.1}{0.3 + 0.3 + 0.1} = \tfrac{1}{7} = 0.14.$$

In this case

$$g_1^{(1)} = \tfrac{1}{3}(0.43 \cdot 0.366 + 0.57 \cdot 0.244) = 0.099,$$
$$g_2^{(1)} = 0,$$
$$g_3^{(1)} = 0,$$
$$g_4^{(1)} = \tfrac{1}{2}(0.43 \cdot 0.366 + 0.57 \cdot 0.244) = 0.148,$$
$$g_5^{(1)} = \tfrac{1}{3}(0.43 \cdot 0.366 + 0.57 \cdot 0.244) = 0.099.$$

It can be seen that we should use check 4 at the second step of the process of searching for the failed unit.

If the failed unit turns out to be unit 2, then the procedure is terminated. If the failed unit is either unit 1 or unit 4, the process is continued and, as is easily seen, we can use only check 1 at the third step.

8.2.3   A method of successive checks for finding several failed units

We assume that in some system—for example, in a process that is not under constant observation—several failures may occur. It is assumed that it is possible to make a check of the complete system in a certain fixed length of time and a check of the ith individual unit in the length of time τ_i.

The checking process goes as follows. We do a check of the complete system, and if the system turns out to have failed, then a sequential check of individual units is begun and is continued until we find a failed unit. The discovered unit is restored (repaired or replaced), and we again do a check of the complete system. If the system continues to be failed—that is, there is still at least one failed unit—then the procedure is repeated. This procedure is continued until the system check establishes that the system is operating properly.

We wish to find an order for checking the units of the system such that the process of finding all failed units takes a minimum amount of time on the average.

We propose the following algorithm.

1.   For each unit we compute the values

$$g_i = \frac{\tau_i r_i}{q_i}. \tag{8.2.5}$$

2.   The units are indexed in increasing order of the quantities g_i. The obtained order is also the order of the optimal checking sequence.

Example 8.2.3   Consider a system consisting of five units characterized by the following measures.

$$
\begin{aligned}
r_1 &= 0.2, & \tau_1 &= 10 \text{ min}; \\
r_2 &= 0.3, & \tau_2 &= 20 \text{ min}; \\
r_3 &= 0.4, & \tau_3 &= 10 \text{ min}; \\
r_4 &= 0.5, & \tau_4 &= 5 \text{ min}; \\
r_5 &= 0.6, & \tau_5 &= 10 \text{ min}.
\end{aligned}
$$

We wish to determine the order in which units should be checked to discover all the failed ones.

Solution.   We compute the quantities g_i by Formula (8.4.5):

$$g_1 = \frac{0.2}{0.8} \cdot 10 = 2.5, \qquad g_2 = \frac{0.3}{0.7} \cdot 20 = 8.6, \qquad g_3 = \frac{0.4}{0.6} \cdot 10 = 6.7,$$

$$g_4 = \frac{0.5}{0.5} \cdot 5 = 5.0, \qquad g_5 = \frac{0.6}{0.4} \cdot 10 = 15.0.$$

It follows from the computations that in order to minimize the mean time of search for failed units we should check the units of the given system in the order 1, 4, 3, 2, 5.

8.2.4   A method of successive checks to detect the improper operation of a system

We consider a system analogous to the one described above. It is assumed that we cannot check the complete system. By checking individual units we look for at least one failed unit in the system, in order to detect and thus guard against the use of an improperly operating system.

We wish to find an order for checking the units such that the procedure for checking for proper operation of the system takes a minimum amount of time on the average.

We propose the following algorithm.

1.   For each unit of the system compute the values

$$g_i = \frac{\tau_i}{q_i}. \tag{8.2.6}$$

Index the units in increasing order of the quantities g_i. The obtained order is also the order of optimal checks.

Example 8.2.4   We consider the system described in Example 8.2.3.

We wish to determine the order in which we should check units with the aim of checking proper operation of the system.

Solution.   We compute the quantities g_i by Formula (8.2.6):

$$g_1 = \frac{1}{0.8} \cdot 10 = 12.5, \qquad g_2 = \frac{1}{0.7} \cdot 20 = 28.6, \qquad g_3 = \frac{1}{0.6} \cdot 10 = 16.7,$$

$$g_4 = \frac{1}{0.5} \cdot 5 = 10.0, \qquad g_5 = \frac{1}{0.4} \cdot 10 = 25.0.$$

From the computations it follows that in order to minimize the mean time of checking for proper operation of the system we should check the units of this system in the order 4, 1, 3, 5, 2.

Remark.   If $g_i \ll 1$, both processes are practically the same.

REFERENCES

1.   BELYAEV, Y. K. AND USHAKOV, I. A., "Mathematical Methods for Problems of Detection and Localization of Failures," in the collection *Cybernetics in the Service of Communism*. Moscow: Energiya, 1964.
2.   MALIKOV, I. M., POLOVKO, A. M., ROMANOV, N. A., AND CHUKREEV, P. A., *Foundations of Theory and Computation of Reliability*, sec. 46, chap. 7. Moscow: Sudpromgiz, 1960.
3.   SHISHONOK, N. A., REPKIN, V. F., AND BARVINSKII, L. L., *Foundations of Reliability Theory and Use of Radioelectronic Technology*, sec. 6, chap. 9, Moscow: Soviet Radio, 1964.

9

ELEMENTARY METHODS FOR PROCESSING EXPERIMENTAL DATA ON RELIABILITY

9.1 BASIC FORMS FOR CALCULATING PERFORMANCE AND RECORDING EQUIPMENT FAILURES; ELEMENTARY METHODS FOR PROCESSING TEST RESULTS

9.1.1 Basic forms for calculating performance and recording failures

The basic form for calculating the performance of equipment is an equipment log. An equipment log is a diary in which the entire "life" history of some object is recorded. This object may be an individual functional structure or an independent structural unit (support or block).

In an equipment log one enters all information needed for a complete and comprehensive account of the utilization of the equipment (quality of fulfillment of the required functions, observed disruptions in operation of the equipment and failures of its individual units, measures taken to repair and maintain operationality, and so on). One of the important functions of an equipment log is the collection of statistical data on reliability.

The nature, amount, and detail of the entries in an equipment log are prescribed by the purpose of the equipment, the importance of the operations being performed, and even the qualifications of the service personnel. The following information should be entered in the equipment log:

1. The date.
2. The times of switching the equipment on and off.
3. The nature of the tasks performed by the equipment (verification of operationality, adjustment, regulation, special tests, form of work being performed, and so on).
4. The quality of operation.

5. The moments of appearance of failure (or other interruption of operation).

6. The nature of a failure (total or partial failure, self-eliminating failure, change of basic parameters, and so on).

7. The location of the failure and the failed unit.

8. The reason for occurrence of a failure.

9. The time needed to eliminate a failure (it is desirable to note separately the time needed for the detection, isolation, and proper replacement or readjustment of a failed unit).

10. The measures taken to eliminate the cause of a failure.

11. The signature of the person working with the equipment.

The equipment log should also contain a complete description of the various regulations and precautions pertaining to the operation and maintenance of the equipment.

However, since an equipment log serves other purposes besides the collection of statistical data on the reliability of the given equipment, and inasmuch as it is a working document that must be kept current for a comparatively long time, its direct use as a source of statistical information is not convenient. Moreover, a given type of equipment is often used at different and rather widely separated points. Thus, since it is of interest to obtain generalized statistical information on reliability, it is convenient to make selected excerpts on failures from the equipment logs and then collect them at the location where the processing and analysis of the statistical data is done.

A convenient form for this would be a notebook with tear-off cards for failures.

Each card of this notebook should contain the following information:

1. The date.

2. The output of the equipment from the preceding failure to the moment of the given failure.

3. The nature of the tasks being fulfilled by the equipment at the moment of occurrence of the failure.

4. The nature of the failure.

5. The location of the failure and the failed unit in the system.

6. The alleged cause of the failure.

7. The time needed to eliminate the failure (it is desirable to note separately the time for detection, isolation, and proper replacement or readjustment of the failed unit).

8. The measures taken to eliminate the cause of the failure.

9. The signature of the person operating the equipment.

In many cases the card is also useful for making a detailed analysis of the failed unit. If, for example, an apparatus consists of easily detached structures (blocks or modules), and a failed structural unit is only replaced at the site of the system, while repair is done elsewhere, then the card noting the failure should first be taken to the repair site of the given detached structure. This is convenient because, first of all, the information contained on the card may shorten the duration of the repair, and second, the repair time of the given detached structural unit can be noted on the same card.

9.1.2   Composition of a combined record of failures, and preliminary processing of failure data

The failure record (see Table 9.1.1) can be used as the basic original material for analyzing equipment reliability.

The combined failure record is put together from the results of tests and the normal usage of selected sample units for which the operational records are available on cards. The material in the combined record should be laid out by equipment complexes in the order shown in Table 9.1.1.

COLUMN 1: failure numbers in order
COLUMN 2: date of occurrence of the failure
COLUMN 3: block number
COLUMN 4: failure characteristic
COLUMN 5: total operating time of the block up to the moment of occurrence of the given failure
COLUMN 6: alleged reason for occurrence of the failure
COLUMN 7: measures taken to remedy the failure.

Remark.   For complicated equipment consisting of a large number of blocks of the same type, it is sometimes more convenient to arrange the material by blocks, and put the index number of the complex of the equipment in column 3. This may facilitate analysis and processing of statistical information. A time diagram of the sequence of occurrence of failures as a function of system operating time is composed from the given combined record for each complex (Figure 9.1.1). The graph shows the operating time (in hours) at which various failures occurred and the time intervals between successive failures. (For clarity all expositions below will be based on consideration of data on five concrete equipment complexes.) The results for working out the graph of the sequence of occurrence of failures (or a combined record of failures), namely, the lengths of the time intervals between failures, are shown in Table 9.1.2.

For simplicity in further processing of statistical data on the basis of Table 9.1.2, we set up the order statistics (lengths of time intervals between failures arranged in order of magnitude) (Table 9.1.3).

Table 9.1.1   Example of Combined Record for Complex
Number 1 of Equipment

FAILURE NUMBER	DATE	BLOCK NUMBER	FAILURE CHARACTERISTIC	EQUIPMENT PRODUCTION UP TO THE MOMENT OF FAILURE IN HOURS	CAUSE OF FAILURE	MEASURES TAKEN
1	2/5/70	01	Output voltage not up to specs	21	Breaking of insulation	Insulation repaired
2	1/10/70	02	Relay R-1 not operating	52	Filament break in tube 6D6A	Tube replaced
2	2/11/70	01	Periodic disappearance of input signal	65	Defect in lacquer covering (lacquer fell on a joint)	Lacquer washed away
4	1/17/70	03	AVC lacking	83	MLT resistor burned	Resistor replaced
5	2/22/70	03	Amplification coefficient fell	98	Loss of emission by tube 6J5P	Tube replaced

In processing statistical data on reliability it should be noted that only
random events and variables comparable in nature should be subject to
simultaneous analysis. Thus it is convenient in many cases to divide the
entire period of observation being considered into portions of adjustment,
extra work, tests, use, and so on and to use the appropriate characteristics
for each of these portions. Moreover, it is sometimes convenient to classify

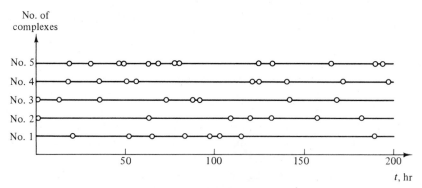

Figure 9.1.1   Time Diagram of the Time Sequence of Failures for
Five Complexes of Equipment.

Table 9.1.2 Production at Failure for Five Different
Equipment Complexes

FAILURE NUMBER	PRODUCTION, HOURS				
	No. 1	No. 2	No. 3	No. 4	No. 5
1	21	1	1	18	19
2	31	63	13	18	12
3	13	45	22	15	16
4	18	11	37	5	1
5	15	12	15	66	15
6	5	25	3	3	5
7	12	25	51	16	10
8	74	64	26	30	1
9	27	3	42	27	46
10	7	24	21	21	7
11	10	55	5	1	33
12	3	48	84	27	25
13	8	72	30	5	4
14	86	18	17	1	9
15	16	3	50	21	1
16	32	45	36	1	10
17	20		18	21	
18	12		11	9	
19	31		2	16	
20	62		6	21	
21	1		7	23	
22	3		16	9	
23	104		27	8	
24			15	11	
25			6	9	
26			31	10	
27			24	1	
28			18		
29			6		
30			6		
31			5		
32			27		
33			5		

failures into, say, "remediable" or "irremediable" ones. Such an approach is well justified in the analysis of experimental data on newly developed equipment, with the aim of predicting reliability characteristics that might later be obtained in actual use. Among "remediable" failures we can consider, for example, failures caused by production defects (defects in assembly and installation); an incompatible set; installation of defective

Table 9.1.3 Order Statistics of Values of Production
for Five Different Equipment
Complexes

NUMBER	PRODUCTION, HOURS				
	No. 1	No. 2	No. 3	No. 4	No. 5
1	1	1	1	1	1
2	3	3	2	1	1
3	3	3	3	1	1
4	5	11	5	1	4
5	7	12	5	3	5
6	8	18	5	5	7
7	10	24	6	5	9
8	12	25	6	8	10
9	12	25	6	9	10
10	13	45	6	9	12
11	15	45	7	9	15
12	16	48	11	10	16
13	18	55	13	11	19
14	20	63	15	15	25
15	21	64	15	16	33
16	27	72	16	16	46
17	31		17	18	
18	31		18	18	
19	32		18	21	
20	62		21	21	
21	74		22	21	
22	86		24	21	
23	104		26	23	
24			27	27	
25			27	27	
26			30	30	
27			31	66	
28			36		
29			37		
30			42		
31			50		
32			51		
33			84		

parts that were not subject to preproduction inspection; violation of
production technology and design (incompleteness of electrical circuits,
construction, and so on). The causes of such failures in assembly-line
production of equipment can be eliminated, thus preventing the occurrence of such defects in later production.

The mixing of statistical information of various kinds without a detailed and complete analysis of the consequences may lead to false and sometimes also absurd quantitative estimates.

9.1.3 Construction of a histogram of the distribution of length of operating time to failure

On the basis of the data given in Table 9.1.3 we construct histograms for each equipment complex (Figure 9.1.2).

Figure 9.1.3 shows a histogram of the distribution of the operating-time intervals between failures for all complexes together. The abscissa is divided into equal intervals. An approximate value of the interval can be determined from the formula

$$\Delta t = \frac{\theta_{\max} - \theta_{\min}}{1 + 3.3 \log N},$$

where $\theta_{\max}$ and $\theta_{\min}$ are the maximal and minimal terms, respectively, of the order statistics, and N is the number of terms of the series of order statistics.

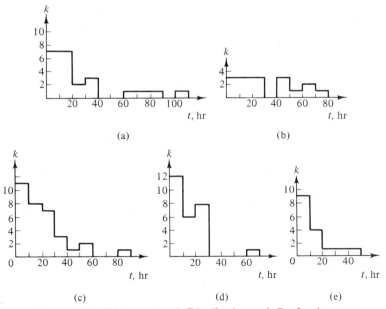

Figure 9.1.2 Histograms of Distributions of Production per Failure for Five Equipment Complexes.

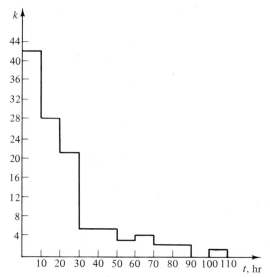

Figure 9.1.3   Summed Histogram of Distribution of Production per Failure for Five Complexes.

In the majority of cases the interval thus defined turns out to be a fractional quantity; however, in constructing a histogram, it is more convenient to use the integer closest to this quantity.

To determine the size Δt of the interval it is sometimes suitable not to take all terms of the series. Thus, in the order statistics series corresponding to complex number 1, it is convenient to restrict consideration to the first 19 terms of the series.

In constructing a histogram for different complexes, one should choose a common unit of time Δt, since this makes it more convenient to compare the histograms, and it also simplifies the construction of the summed histogram. In our example the computed quantity Δt is within the limits 5–15 hours. As the unit of measure we have taken the value $\Delta t = 10$ hours.

From Table 9.1.3 we determine k_i, the number of cases in which the length of time between given failures, which we shall call the operating time to failure, satisfies the condition

$$t_i < \theta \leq t_i + \Delta t.$$

The obtained value of k_i is plotted on the ordinate over the given interval.

For example, for complex number 1 [Figure 9.1.2(a)] we use Table 9.1.3 (column 2) to determine the number of cases in which the operating time to failure is in the interval from 0 to 10 hours inclusive. The obtained quantity (7 units) is put on the ordinate over the given interval. Then we determine the number of cases in which the operating time to failure is in the interval from 10 to 20 hours inclusive. We make the analogous construction, and so forth.

9.1.4 Construction of an empirical distribution function of operating time to failure and of an empirical distribution of the time duration of failure-free operation

The empirical distribution function $\hat{F}$ of the operating time to failure is constructed on the basis of the data in Table 9.1.3. We investigate the construction of the combined distribution function for all five complexes. First we compute the over-all number of failures found in Table 9.1.3:

$$N = N_1 + N_2 + N_3 + N_4 + N_5,$$

where N_i is the number of failures for the ith equipment complex.

For our case

$$N = 23 + 16 + 33 + 27 + 16 = 115.$$

We proceed with the construction as follows. We compute the total number of failures for which the time to failure did not exceed 1 hour. We denote the resulting number by k_1. We plot the quantity k_1/N on the ordinate over the point $t = 1$ (in our case $k_1 = 10$ and $k_1/N = 0.087$). We further compute the number of failures for which the time to failure did not exceed 2 hours. We plot the quantity k_2/N on the ordinate over the point $t = 2$ (in our example $k_2 = 11$ and $k_2/N = 0.096$). This construction is continued until all data are accounted for (see Figure 9.1.4).

The empirical probability of the time duration of failure-free operation $\hat{P}$ is constructed analogously, except that the quantity $1 - k_1/N$ is put over the point $t = 1$, the quantity $1 - k_2/N$ over the point $t = 2$, and so on. In other words, the empirical dependence of the probability of the duration of failure-free operation can be obtained from the empirical distribution function by the equation

$$\hat{P} = 1 - \hat{F}.$$

9.1.5 Computation of the mean and variance of the time of failure-free operation

The mean time of failure-free operation of a structure can be estimated by the formula

$$\hat{T} = \frac{1}{N} \sum_{i=1}^{N} \theta_i.$$

In our example, for complex 5 the value is $\hat{T} = 13.4$ hours.

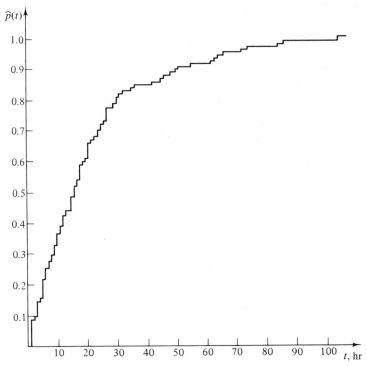

Figure 9.1.4   Empirical Distribution Function of Production at
Failure Constructed from the Order Statistics in
Table 9.1.3.

Remark.   We can compute the mean value in a simpler fashion on the
basis of the series of order statistics. All the quantities are divided into
M sets, for each of which we compute the mean value. Then we compute
$\hat{T}$ by the formula

$$\hat{T} = \sum_{i=1}^{M} \hat{T}_i \frac{k_i}{N},$$

where k_i is the number of terms in the ith set of order statistics and $\hat{T}_i$
is the mean value for the ith group.

If all the k_i are chosen to be identical, then $k_i/N = 1/M$ and

$$\hat{T} = \frac{1}{M} \sum_{i=1}^{M} \hat{T}_i.$$

An approximate value for $\hat{T}$ can be obtained on the basis of the constructed histograms by the formula

$$\tilde{T} = \sum_{k=1}^{L} t^{(i)} \hat{p}_i,$$

where $t^{(i)}$ is the middle of the ith interval of grouped data;

$$t^{(i)} = t_i + \frac{\Delta t}{2};$$

$\hat{p}_i$ is the frequency of the event, $t_i < \theta \le t_i + \Delta t$; that is, $\hat{p}_i = k_i/N$ and L is the total number of intervals of the partition.

For complex 5 in our example we have

$$\tilde{T} = (5 \cdot \tfrac{9}{16} + 15 \cdot \tfrac{4}{16} + 25 \cdot \tfrac{1}{16} + 35 \cdot \tfrac{1}{16} + 45 \cdot \tfrac{1}{16})$$
$$= \tfrac{5}{16}(9 + 12 + 5 + 7 + 9) = \tfrac{5}{16} \cdot 42 = 13.1 \quad \text{hr.}$$

Naturally, the shorter the intervals of the partition, the closer the result will be to the true value of the average.

An unbiased estimate for the variance can be computed by the formula

$$\widehat{\sigma^2} = \frac{1}{N-1} \sum_{i=1}^{N} (\hat{T} - \theta_i)^2.$$

If the quantities θ_i are integers while the obtained value of $\hat{T}$ is not an integer, then, representing $\hat{T}$ in the form $\hat{T} = \hat{T}^* + \epsilon$, we can find $\widehat{\sigma^2}$ by the formula

$$\widehat{\sigma^2} = \frac{1}{N-1} \left[N\epsilon^2 + \sum_{i=1}^{N} (\hat{T}^* - \theta_i)^2 \right].$$

This is convenient in practice, since integer differences $(\hat{T}^* - \theta_i)$ are easier to square.

9.1.6 Determination of the dependence of failure intensity on time

Often we need to determine the dependence of the failure intensity of a structure on time, since on the basis of such investigations we can determine how much longer a structure can be expected to operate, as well as the beginning of the period of aging (wear) characterized by relatively large values of the failure intensity.

The dependence of the failure intensity on time can easily be constructed on the basis of a time diagram of the sequence of occurrences of failures (Figure 9.1.1). For this we take a time interval of length x such that, on the average, at least 5 to 10 failures occur in every interval of time t to $t + x$ [$(t; t + x]$) over the entire equipment complex for a wide range of values of time t. This interval is called the interval of averaging. Then we choose some time interval y [usually such that x is no smaller than $(3 \div 5)y$], called the translation interval.

For the time interval $[jy; jy + x]$ we compute the mean value of the failure intensity by the formula

$$\lambda(jy + x) = \frac{n(jy; x)}{N(jy; x) \cdot x},$$

where $n(jy; x)$ is the total number of failures of all structures in the time interval $[jy; jy + x]$ and $N(jy; x)$ is the number of equipment complexes being tested in the time interval $[jy; jy + x]$.

Usually the value of $N(jy; x)$ is determined as the average of the number of complexes being tested at the moment of time jy and the number of complexes being tested at the moment of time $jy + x$. However, in the majority of cases one may simply take the number of complexes being tested at the moment of time jy.

A more precise estimate of the failure intensity can be computed by the formula

$$\lambda(jy + x) = \frac{n(jy; x)}{t_\Sigma(jy; x)},$$

where $t_\Sigma(jy; x)$ is the summed operating times of all equipment complexes in the time interval $[jy; jy + x]$.

Remark. It is usually convenient to have the quantity x an integral number of times greater than the quantity y. In this case (as will be shown later in an example) the computation of the dependence of the failure intensity on time can be accomplished with a minimum number of computational operations.

We compute Table 9.1.4 on the basis of the data in Table 9.1.2 (or the time diagram in Figure 9.1.1, constructed from this table) for the time interval from 0 to 500 hours.

Values of $n(jy; y)$ and $N(jy; y)$ are given in columns 3 and 4 of Table 9.1.4 for convenience in the computations. Since the value x of the averaging integral was chosen to be an integral number of times greater than

Table 9.1.4   Table for Determining Dependence of
Failure Intensity λ ($y = 20$ hours)

j	$t_j + \Delta t$	n	N	λ ($x = 60$)	λ ($x = 100$)
0	0–20	6	5		
1	20–40	8	5		
2	40–60	5	5	0.0465	—
3	60–80	7	5	0.0500	—
4	80–100	5	5	0.0565	0.0520
5	100–120	3	5	0.0500	0.0460
6	120–140	5	5	0.0435	0.0500
7	140–160	4	5	0.0400	0.0480
8	160–180	3	5	0.0400	0.0400
9	180–200	4	5	0.0365	0.0380
10	200–220	7	5	0.0465	0.0460
11	220–240	6	4	0.0605	0.0510
12	240–260	3	4	0.0605	0.0510
13	260–280	3	4	0.0500	0.0520
14	280–300	2	4	0.0335	0.0490
15	300–320	3	4	0.0335	0.0425
16	320–340	4	4	0.0375	0.0375
17	340–360	3	4	0.0415	0.0375
18	360–380	3	4	0.0415	0.0375
19	380–400	3	4	0.0350	0.0400
20	400–420	6	4	0.0500	0.0475
21	420–440	1	3	0.0445	0.0410
22	440–460	2	3	0.0415	0.0400
23	460–480	5	3	0.0445	0.0390
24	480–500	3	3	0.0555	0.0525

the value of the translation interval, it is easy to compute the desired
means of failure intensity from the data by the formula

$$\lambda(jy + x) = \frac{\displaystyle\sum_{i=j-(x/y)}^{j} n(iy;\, y)}{y \displaystyle\sum_{i=j-(x/y)}^{j} N(iy;\, y)}.$$

Thus for $x = 3y$ and $x + yj = 260$ we have

$$\lambda(260) = \frac{7 + 6 + 3}{20(5 + 4 + 4)} = 0.0615.$$

The results of the computations are shown in columns 5 and 6 of
Table 9.1.4 and are also given in Figure 9.1.5.

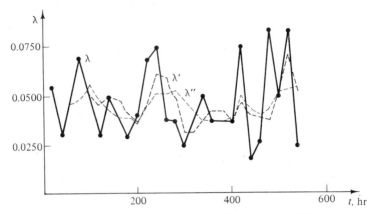

Figure 9.1.5   Dependence of Failure Intensity on Time; λ—
for interval of averaging $x = y$; λ'—for interval
of averaging $x = 3y$; λ''—for interval of averaging
$x = 5y$.

9.2 TESTING A HYPOTHESIS CONCERNING THE NATURE OF THE DISTRIBUTION LAW OF FAILURES

Preliminary remarks

In the preceding section we showed how one should process statistical material on failures of structures and, in particular, how one should construct the empirical distribution of the random variable of operating time of a structure up to failure:

$$\hat{Q}(t) = \hat{\mathscr{P}}\{\theta \leq t\}.$$

Having such a distribution (or the same information in any other form, such as a table), we can test to see whether it is related to any known law —e.g., normal, exponential, Weibull.

It is possible to establish a decision rule such that, on the basis of observed results, we are able to accept or reject a proposed hypothesis about which law corresponds to the empirical distribution we have constructed. To this end we consider a random variable U (goodness-of-fit statistic) characterizing the degree of difference between the theoretical (hypothetical) and empirical distributions. This statistic can be chosen in any number of ways. For example, we may take U to be the sum of the squares of the deviations of the theoretical probabilities p_i from the corresponding observed frequencies $\hat{p}_i$ with certain coefficients of proportionality, or, alternatively, the maximal deviation of the empirical distribution function from the theoretical, and so on. Obviously in all cases U is a random variable whose distribution law depends on the distribution law of the random variable θ, the distribution of which is completely specified by the hypothesis, and also on the amount of statistical data.

If the proposed hypothesis is true, then the distribution law of the variable U is determined by the distribution law of the variable θ [as a function of $Q(t)$]. We assume that under the proposed hypothesis the distribution law of the variable U is known to us. Suppose that in processing the results we have found that the chosen measure of deviation, U, has taken on some value u. We want to establish whether we can explain the obtained value u solely by random causes, so that our hypothesis does not contradict the experimental data, or whether this quantity is too large and indicates the presence of an essential difference between the hypothesized and empirical distributions (so that our hypothesis is unlikely).

If we assume that our hypothesis is true and on this assumption compute the probability that the measure of deviation U turns out to be not smaller than the value u observed by us in the experiment (and this probability can be computed, since we have assumed that under the hypothesis the distribution law of U is known to us), then if this probability is small, our hypothesis is unlikely, while if it is large, the experimental data do not contradict our hypothesis.

For goodness-of-fit tests in statistics, one uses measures of deviation U whose distribution law has very simple properties and such that for sufficiently large amounts of experimental data this distribution is nearly independent of the theoretical distribution function $Q(t)$.

9.2.1   The Pearson χ^2 test

Suppose we know the n durations $\theta_1, \theta_2, \ldots, \theta_n$ of operating times to failure of some structure. We divide the entire range of observed values into $k < n$ intervals, whose boundaries are denoted by $t^{(1)}, t^{(2)}; t^{(2)}, t^{(3)}; t^{(3)}, t^{(4)}; \ldots, t^{(i)}, t^{(i+1)}; \ldots; t^{(k)}, t^{(k+1)}$. Computing the number of values m_i in the ith interval and dividing this number by the total number of operating times n, we find the observed frequency $\hat{p}_i = m_i/n$ corresponding to the given interval. Next we construct an empirical frequency table:

$t^{(1)}, t^{(2)}$	$t^{(2)}, t^{(3)}$	$t^{(3)}, t^{(4)}$	$\cdots$	$t^{(i)}, t^{(i+1)}$	$\cdots$	$t^{(k)}, t^{(k+1)}$
$\hat{p}_1$	$\hat{p}_2$	$\hat{p}_3$	$\cdots$	$\hat{p}_i$	$\cdots$	$\hat{p}_k$

Under the assumption that the obtained data are subject to a certain theoretical distribution law, we can also find the theoretical probability that the random variable of operating time to failure falls in each of these intervals:

$$p_1, p_2, p_3, \ldots, p_i, \ldots, p_k.$$

Here we should take into account the fact that, although we assume the distribution law known (e.g., Poisson, normal), we don't know, as a rule, the parameters of this law (for example, λ for the Poisson, a and σ for the normal, and so on), and in computing the theoretical probabilities we often have to set the unknown values of the parameters of the theoretical law equal to the statistical estimates obtained from the experiment. (Generally the effect of such modification of known procedures is not too simple to determine, even though it is done quite often in practice. Hence caution is recommended.)

To test the goodness-of-fit between theoretical and empirical distributions, we can start from the deviation between the theoretical probabilities p_i and the observed frequencies $\hat{p}_i$. As was shown by K. Pearson, if we take as the variable U

$$U = \chi^2 = \min \sum_{i=1}^{k} \frac{n}{p_i} (\hat{p}_i - p_i)^2 = \min \sum_{i=1}^{k} \frac{(m_i - np_i)^2}{np_i} \quad (9.2.1)$$

(the minimum is taken over the values of all unknown parameters of the theoretical distribution law), then, under reasonable conditions on the nature of the class of distributions specified by the parameters just referred to, for large n, the distribution law of the variable U (the variable U given in this form is usually denoted by χ^2) depends only on the number of intervals, k, and, as n increases, it approaches the known (tabled) χ^2 distribution. [A paper treating the problem of the χ^2 test with unknown parameters being estimated from the sample is the one by J. Neyman, "Contribution to the theory of the χ^2 test," in the "First Berkeley Symposium of Mathematical Statistics and Probability, U. of California Press, Berkeley, California, 1949.] The distribution depends on the parameter r, called the number of "degrees of freedom" of the distribution, which is equal to the number of intervals, k, minus the number of parameters, minus one. The loss of the final "degree of freedom" corresponds to the imposed condition

$$\sum_{i=1}^{k} \hat{p}_i = 1.$$

In order not to have to seek the minimum of the sum given by Formula (9.2.1) (which entails difficult computations, especially when the number of parameters is greater than one), we can use the approach proposed by

E. S. Ventsel' [1]. In place of the χ^2 statistic given by Formula (9.2.1), for a small number of intervals we can consider the approximation given by the quantity

$$\chi^2 \approx \sum_{i=1}^{k} \frac{(m_i - np_i)^2}{np_i},$$ (9.2.2)

where, in computing the theoretical probabilities, the unknown parameters of the theoretical distribution are replaced by statistical estimates. For example,

$$\hat{m}_t = \sum_{i=1}^{k} \frac{t^{(i+1)} + t^{(i)}}{2} \hat{p}_i = m_t,$$

if we assume that the theoretical mean value m_t and the estimated mean value $\hat{m}_t$ coincide;

$$\hat{D}_t = \sum_{i=1}^{k} \left(\frac{t^{(i+1)} + t^{(i)}}{2} - \hat{m}_t\right)^2 \hat{p}_i = D_t,$$

if we assume, moreover, that the theoretical and empirically determined variances coincide, and so on.

We have compiled special tables (Table 9.2.1) for the χ^2 distribution. Using these tables, we can find the probability p that a random variable U with a χ^2 distribution exceeds any obtained value of U and number of degrees of freedom r.

Thus estimating the goodness-of-fit of theoretical and empirical distributions by the χ^2 criterion reduces to the following:

1. Determine the measure of deviation:

$$\chi^2 \approx \sum_{i=1}^{k} \frac{(m_i - np_i)^2}{np_i}.$$

2. Determine the number of degrees of freedom r as the number of intervals minus the number of imposed constraints s:

$$r = k - s,$$

where s is the number of parameters plus one.

3. From Table 9.2.1 determine the probability that a quantity having χ^2 distribution with r degrees of freedom exceeds the given value of χ^2. If this probability is small, the hypothesis is rejected as unlikely. If this

Table 9.2.1   Values of χ^2 Depending on r and p

r \ p	0.001	0.01	0.02	0.05	0.10	0.20	0.30	0.50	0.70	0.80	0.90	0.95	0.98	0.99
1	10.83	6.64	5.41	3.84	2.71	1.642	1.074	0.455	0.148	0.064	0.016	0.004	0.001	0.000
2	13.82	9.21	7.82	5.99	4.60	3.22	2.41	1.386	0.713	0.446	0.211	0.103	0.040	0.020
3	16.27	11.34	9.84	7.82	6.25	4.64	3.66	2.37	1.424	1.005	0.584	0.352	0.185	0.115
4	18.46	13.28	11.67	9.49	7.78	5.99	4.88	3.36	2.20	1.649	1.064	0.711	0.429	0.297
5	20.5	15.09	13.39	11.07	9.24	7.29	6.06	4.35	3.00	2.34	1.610	1.145	0.752	0.554
6	22.5	16.81	15.03	12.59	10.64	8.56	7.23	5.35	3.83	3.07	2.20	1.635	1.134	0.872
7	24.3	18.48	16.62	14.07	12.02	9.80	8.38	6.35	4.67	3.82	2.83	2.17	1.564	1.239
8	26.1	20.1	18.17	15.51	13.36	11.03	9.52	7.34	5.53	4.59	3.49	2.73	2.03	1.646
9	27.9	21.7	19.68	16.92	14.68	12.24	10.66	8.34	6.39	5.38	4.17	3.32	2.53	2.09
10	29.6	23.2	21.2	18.31	15.99	13.44	11.78	9.34	7.27	6.18	4.86	3.94	3.06	2.56
11	31.3	24.7	22.6	19.68	17.28	14.63	12.90	10.34	8.15	6.99	5.58	4.58	3.61	3.05
12	32.9	26.2	24.1	21.0	18.55	15.81	14.01	11.34	9.03	7.81	6.30	5.23	4.18	3.57
13	34.6	27.7	25.5	22.4	19.81	16.98	15.12	12.34	9.93	8.63	7.04	5.89	4.76	4.11
14	36.1	29.1	26.9	23.7	21.1	18.15	16.22	13.34	10.82	9.47	7.79	6.57	5.37	4.66
15	37.7	30.6	28.3	25.0	22.3	19.31	17.32	14.34	11.72	10.31	8.55	7.26	5.98	5.23
16	39.3	32.0	29.6	26.3	23.5	20.5	18.42	15.34	12.62	11.15	9.31	7.96	6.61	5.81
17	40.8	33.4	31.0	27.6	24.8	21.6	19.51	16.34	13.53	12.00	10.08	8.67	7.26	6.41
18	42.3	34.8	32.3	28.9	26.0	22.8	20.6	17.34	14.44	12.86	10.86	9.39	7.91	7.02
19	43.9	36.2	33.7	30.1	27.2	23.9	21.7	18.34	15.35	13.72	11.65	10.11	8.57	7.63
20	45.3	37.6	35.0	31.4	28.4	25.0	22.8	19.34	16.27	14.58	12.44	10.85	9.24	8.26
21	46.8	38.9	36.3	32.7	29.6	26.2	23.9	20.3	17.18	15.44	13.24	11.59	9.92	8.90
22	48.3	40.3	37.7	33.9	30.8	27.3	24.9	21.3	18.10	16.31	14.04	12.34	10.60	9.54
23	49.7	41.6	39.0	35.2	32.0	28.4	26.0	22.3	19.02	17.19	14.85	13.09	11.29	10.20
24	51.2	43.0	40.3	36.4	33.2	29.6	27.1	23.3	19.94	18.06	15.66	13.85	11.99	10.86
25	52.6	44.3	41.7	37.7	34.4	30.7	28.2	24.3	20.9	18.94	16.47	14.61	12.70	11.52
26	54.1	45.6	42.9	38.9	35.6	31.8	29.2	25.3	21.8	19.82	17.29	15.38	13.41	12.20
27	55.5	47.0	44.1	40.1	36.7	32.9	30.3	26.3	22.7	20.7	18.11	16.15	14.12	12.88
28	56.9	48.3	45.4	41.3	37.9	34.0	31.4	27.3	23.6	21.6	18.94	16.93	14.85	13.56
29	58.3	49.6	46.6	42.6	39.1	35.1	32.5	28.3	24.6	22.5	19.77	17.71	15.57	14.26
30	59.7	50.9	48.0	43.8	40.3	36.2	33.5	29.3	25.5	23.4	20.6	18.49	16.31	14.95

probability is relatively high, the hypothesis can be acknowledged as not being contradicted by the experimental data.

Example 9.2.1 Suppose that from observations on some structure we have obtained $n = 42$ times to failure, and the empirical distribution of these quantities has the following form:

LIMITS FOR TIMES TO FAILURE	NUMBER OF CASES WHERE THE TIMES TO FAILURE FALLS IN THE GIVEN LIMITS	FREQUENCY OF THE EVENT $\hat{p}_i = m_i/n$
0–50	5	0.12
50–100	6	0.14
100–150	11	0.26
150–200	8	0.19
200–250	7	0.17
250–300	5	0.12

The statistical estimate of the mean production is

$$
\hat{m}_i = T = \sum_{i=1}^{6} \frac{t^{(i+1)} + t^{(i)}}{2} \hat{p}_i
$$
$$
= 25 \cdot 0.12 + 75 \cdot 0.14 + 125 \cdot 0.26 + 175 \cdot 0.19 + 225 \cdot 0.17 + 275 \cdot 0.12
$$
$$
= 150.5 \quad \text{hr}.
$$

We want to test the hypothesis that the operating time of the structure up to failure has an exponential distribution with parameter

$$
\lambda = \frac{1}{\hat{m}_t} = \frac{1}{T} = \frac{1}{150.0} \approx 0.0066 \ \frac{1}{\text{hr}}
$$

—that is,

$$
\mathcal{P}\{\theta \leq t\} = Q(t) = 1 - e^{-\lambda t}.
$$

Solution. Since the density of the exponential distribution is

$$
f(t) = \frac{d}{dt}(Q(t)) = \frac{d}{dt}(1 - e^{-\lambda t}) = \lambda e^{-\lambda t},
$$

then the probability that the random variable takes on only one of the values in the interval $t^{(i)}$, $t^{(i+1)}$ is

$$
p_i = \int_{t^{(i)}}^{t^{(i+1)}} \lambda e^{-\lambda t} \, dt = e^{-\lambda t^{(i)}} - e^{-\lambda t^{(i+1)}}.
$$

Then:

$$p_1 = e^{-\lambda t^{(1)}} - e^{-\lambda t^{(2)}} = e^{-0.0066 \cdot 0} - e^{-0.0066 \cdot 50} = 1 - 0.7189 \approx 0.28,$$
$$p_2 = e^{-\lambda t^{(2)}} - e^{-\lambda t^{(3)}} = e^{-0.0066 \cdot 50} - e^{-0.0066 \cdot 100} = 0.7189 - 0.5169 \approx 0.20,$$
$$p_3 = e^{-\lambda t^{(3)}} - e^{-\lambda t^{(4)}} = e^{-0.0066 \cdot 100} - e^{-0.0066 \cdot 150} = 0.5169 - 0.3716 \approx 0.15,$$
$$p_4 = e^{-\lambda t^{(4)}} - e^{-\lambda t^{(5)}} = e^{-0.0066 \cdot 150} - e^{-0.0066 \cdot 200} = 0.3716 - 0.2671 \approx 0.10,$$
$$p_5 = e^{-\lambda t^{(5)}} - e^{-\lambda t^{(6)}} = e^{-0.0066 \cdot 200} - e^{-0.0066 \cdot 250} = 0.2671 - 0.1920 \approx 0.08,$$
$$p_6 = e^{-\lambda t^{(6)}} - e^{-\lambda t^{(7)}} = e^{-0.0066 \cdot 250} - e^{-0.0066 \cdot 300} = 0.1920 - 0.1381 \approx 0.05.$$

By Formula (9.2.2) we find

$$\chi^2 \approx \sum_{i=1}^{k} \frac{n}{p_i} (\hat{p}_i - p_i)^2 = \sum_{i=1}^{6} \frac{42}{p_i} (\hat{p}_i - p_i)^2$$

$$= \frac{42}{0.12} (0.28 - 0.12)^2 + \frac{42}{0.14} (0.20 - 0.14)^2 + \frac{42}{0.26} (0.15 - 0.26)^2$$

$$+ \frac{42}{0.19} (0.10 - 0.19)^2 + \frac{42}{0.17} (0.08 - 0.17)^2 + \frac{42}{0.12} (0.05 - 0.12)^2$$

$$\approx 17.49.$$

The number of imposed constraints is $s = 2$, namely:

$$\sum_{i=1}^{k} \hat{p}_i = 1, \qquad m_t = \hat{m}_t.$$

Thus,

$$r = k - s = 6 - 2 = 4.$$

For $r = 4$ and $\chi^2 = 17.49$ we find from Table 9.2.1 that

$$p < 0.01,$$

—that is, we reject as unlikely the hypothesis that the operating time of the given structure has an exponential distribution.

Example 9.2.2   Under the conditions of Example 9.2.1, test the hypothesis of a truncated normal distribution law for the operating time to failure.

Solution.   The normal law depends on two parameters, m_t and D_t. Let

$$m_t = \hat{m}_t = 150.5 \text{ hr} \qquad \text{(see Example 9.2.1)},$$
$$D_t = \hat{D}_t.$$

We compute the quantity $\hat{D}_t$. By definition,

$$
\begin{aligned}
\hat{D}_t &= \sum_{i=1}^{k} \left(\frac{t^{(i+1)} + t^{(i)}}{2} - \hat{m}_t \right)^2 \hat{p}_i \\
&= [(25.0 - 150.5)^2 \cdot 0.12 + (75.0 - 150.5)^2 \cdot 0.14 \\
&\quad + (125.0 - 150.5)^2 \cdot 0.26 + (175.0 - 150.5)^2 \cdot 0.19 \\
&\quad + (225.0 - 150.5)^2 \cdot 0.17 + (275.0 - 150.5)^2 \cdot 0.12] \\
&\approx 5775 \quad \text{hr}^2,
\end{aligned}
$$

whence

$$
\widehat{\sigma}_t = \sqrt{\hat{D}_t} = \sqrt{5775} \approx 76.0 \quad \text{hr.}
$$

Since we have imposed the conditions

$$
m_t = \hat{m}_t, \qquad D_t = \hat{D}_t,
$$

the density of the truncated normal law is given by the expression

$$
f(t) = \frac{c}{\widehat{\sigma}_t \sqrt{2\pi}} \, e^{-(t-m_t)^2 / (2\widehat{\sigma_t^2})},
$$

where

$$
c = \frac{1}{\Phi\left(\dfrac{\hat{m}_t}{\widehat{\sigma}_t} \right)} = \frac{1}{\Phi\left(\dfrac{150.5}{76.0} \right)} = \frac{1}{\Phi(1.98)} .
$$

Using Appendix 3.1, we obtain

$$
c = \frac{1}{\Phi(1.98)} = \frac{1}{0.976} \approx 1.02.
$$

Since the probability that a random variable with truncated normal distribution falls in the interval from $t^{(i)}$ to $t^{(i+1)}$ is

$$
p_i = c \left[\Phi\left(\frac{\hat{m}_t - t^{(i)}}{\widehat{\sigma}_t} \right) - \Phi\left(\frac{\hat{m}_t - t^{(i+1)}}{\widehat{\sigma}_t} \right) \right]
$$

$\left(\text{where, as before, } \Phi(x) = \dfrac{1}{\sqrt{2\pi}} \displaystyle\int_{-\infty}^{x} e^{-u^2/2} \, dy \right)$, then, substituting the value c and using Appendix 3.1, we find

$$p_1 = 1.02 \left[\Phi \left(\frac{150.5 - 0.0}{76.0} \right) - \Phi \left(\frac{150.5 - 50.0}{76.0} \right) \right]$$
$$= 1.02(0.976 - 0.907) \approx 0.07,$$

$$p_2 = 1.02 \left[\Phi \left(\frac{150.5 - 50.0}{76.0} \right) - \Phi \left(\frac{150.5 - 100.0}{76.0} \right) \right]$$
$$= 1.02(0.907 - 0.745) \approx 0.16,$$

$$p_3 = 1.02 \left[\Phi \left(\frac{150.5 - 100.0}{76.0} \right) - \Phi \left(\frac{150.5 - 150.0}{76.0} \right) \right]$$
$$= 1.02(0.745 - 0.528) \approx 0.22,$$

$$p_4 = 1.02 \left[\Phi \left(\frac{150.5 - 150.0}{76.0} \right) - \Phi \left(\frac{150.5 - 200.0}{76.0} \right) \right]$$
$$= 1.02(0.528 - 0.258) \approx 0.28,$$

$$p_5 = 1.02 \left[\Phi \left(\frac{150.5 - 200}{76.0} \right) - \Phi \left(\frac{150.5 - 250.0}{76.0} \right) \right]$$
$$= 1.02(0.258 - 0.095) \approx 0.17,$$

$$p_6 = 1.02 \left[\Phi \left(\frac{150.5 - 250.0}{76.0} \right) - \Phi \left(\frac{150.5 - 300.0}{76.0} \right) \right]$$
$$= 1.02(0.095 - 0.026) \approx 0.07.$$

Now we can compute the quantity χ^2:

$$\chi^2 \approx \sum_{i=1}^{k} \frac{n}{p_i} (\hat{p}_i - p_i)^2 = \sum_{i=1}^{6} \frac{42}{p_i} (\hat{p}_i - p_i)^2$$
$$= \frac{42}{0.07} (0.12 - 0.07)^2 + \frac{42}{0.16} (0.14 - 0.16)^2 + \frac{42}{0.22} (0.26 - 0.22)^2$$
$$+ \frac{42}{0.28} (0.19 - 0.28)^2 + \frac{42}{0.17} (0.17 - 0.17)^2 + \frac{42}{0.07} (0.12 - 0.07)^2$$
$$\approx 4.64.$$

Since the number of imposed constraints is $s = 3$, namely

$$\sum_{i=1}^{k} \hat{p}_i = 1, \qquad m_t = \hat{m}_t, \qquad D_t = \hat{D}_t,$$

the number of degrees of freedom is

$$r = k - s = 6 - 3 = 3.$$

For $r = 3$ and $\chi^2 = 4.64$ we find from Table 9.2.1 that

$$p \sim 0.2$$

—that is, we reject as unlikely the hypothesis that the operating time of the given structure has a truncated normal distribution.

9.2.2 Kolmogorov's test

As a measure of deviation U between theoretical and empirical distributions we now consider the maximal value of the modulus of difference between the empirical distribution function $\hat{Q}(t)$ and the corresponding theoretical distribution function:

$$U = D = \max |\hat{Q}(t) - Q(t)|.$$

A. N. Kolmogorov has proved that for any distribution function $Q(t)$ of a continuous random variable θ, when the number of independent observations (number of trials, runs) increases unboundedly, the probability of the inequality

$$D\sqrt{n} \geq y$$

approaches the limit

$$\mathscr{P}\{D\sqrt{n} \geq y\} = P(y) = 1 - K(y),$$

where $K(y)$ is the tabled function whose values are given in Table 9.2.2.

Determining the goodness-of-fit between the theoretical and empirical distributions by Kolmogorov's criterion reduces to the following sequence of steps:

1. construct the empirical distribution function $\hat{Q}(t)$ and the proposed theoretical distribution function $Q(t)$;

2. determine the maximum, D, of the modulus of difference between them;

3. determine the quantity

$$y = D\sqrt{n}$$

and from Table 9.2.2 find the probability $P(y)$ that [if the variable θ actually has distribution law $Q(t)$] the maximal deviation of $\hat{Q}(t)$ from $Q(t)$ is not less than the actual observed deviation. If the probability $P(y)$ is very small, the hypothesis is rejected as being unlikely. If it is relatively large, the hypothesis can be considered compatible with the experimental data.

Example 9.2.3 Suppose we know that under the conditions of Example 9.2.1 the maximum of the modulus of the difference between the

Table 9.2.2   Values of Kolmogorov's Function

y	$K(y)$	y	$K(y)$	y	$K(y)$
0.30	0.0^3009	0.90	0.607	1.50	0.978
0.35	0.0^3303	0.95	0.672	1.60	0.988
0.40	0.0^2281	1.00	0.730	1.70	0.994
0.45	0.013	1.05	0.780	1.80	0.997
0.50	0.036	1.10	0.822	1.90	0.9^2854
0.55	0.077	1.15	0.858	2.00	0.9^3329
0.60	0.136	1.20	0.888	2.10	0.9^3705
0.65	0.208	1.25	0.912	2.20	0.9^3874
0.70	0.289	1.30	0.932	2.30	0.9^3949
0.75	0.373	1.35	0.948	2.40	0.9^3980
0.80	0.456	1.40	0.960	2.50	0.9^4925
0.85	0.535	1.45	0.970	3.00	0.9^5997

empirical distribution and the uniform distribution with density

$$
f(t) = \begin{cases} \dfrac{1}{300}\dfrac{1}{\text{hr}}, & 0 \le t \le 300 \quad \text{hr}, \\[2mm] 0, & t > 300 \quad \text{hr}, \end{cases}
$$

is $D = 0.07$.

Use Kolmogorov's test and Pearson's χ^2 test to test the goodness-of-fit of the experimental data with the uniform distribution.

Solution.   1.   Kolmogorov's test. Since $n = 42$, the quantity y is

$$
y = D \sqrt{n} = 0.07 \sqrt{42} \approx 0.45.
$$

From Table 9.2.2 we determine the probability $P(y)$ to be

$$
P(y) = P(0.45) = 1 - K(0.45) = 1 - 0.013 = 0.987
$$

—that is, the proposed uniform distribution hypothesis can be considered compatible with the experimental data.

2.   Pearson's χ^2 test. The theoretical probabilities are obviously

$$
p_1 = p_2 = \cdots = p_i = \cdots = p_6 = \tfrac{1}{6}.
$$

Using the conditions of Example 9.2.1, we obtain

$$
\chi^2 \approx \sum_{i=1}^{k} \frac{n}{p_i} (\hat{p}_i - p_i)^2 = \sum_{i=1}^{6} \frac{42}{\frac{1}{6}} (\hat{p}_i - \tfrac{1}{6})^2
$$

$$
\begin{aligned}
&= 252[(0.12 - 0.17)^2 + (0.14 - 0.17)^2 + (0.26 - 0.17)^2 \\
&\quad + (0.19 - 0.17)^2 + (0.17 - 0.17)^2 + (0.12 - 0.17)^2] \\
&\approx 3.63.
\end{aligned}
$$

Thus, in this case, the number of degrees of freedom is

$$r = k - 1 = 6 - 1 = 5,$$

and from Table 9.2.1 with $r = 5$ and $\chi^2 = 3.63$ we find that

$$p \sim 0.6$$

—that is, the proposed hypothesis can be considered compatible with the experimental data.

Remark. In parts 1 and 2 of this example we obtained qualitative agreement of the results, but quantitatively they are essentially different. It should be kept in mind that in those cases where the parameters of the theoretical distribution are estimated from the statistical data (which we did tacitly), the use of Kolmogorov's test leads to larger values of the probability $P(y)$.

9.3 ESTIMATING CONFIDENCE BOUNDS FOR THE PARAMETER OF THE BINOMIAL DISTRIBUTION

Often the testing for reliability of a group of N structures is carried out so that the outcome of tests on one structure is in no way related to the outcomes for the remaining structures. Suppose that as a result of testing for a time t we observe d structure failures. We have the problem of furnishing confidence bounds for the unknown probability P of failure-free operation of the structure for time t on the basis of the known values of N and d.

A two-sided confidence interval $(\underline{P}, \bar{P})$ with confidence coefficient $1 - \alpha$, given the observed value d, can be found from the equations

$$\sum_{i=0}^{d} C_N^i \bar{P}^{N-i}(1 - \bar{P})^i = \frac{\alpha}{2},$$

$$\sum_{i=d}^{N} C_N^i \underline{P}^{N-i}(1 - \underline{P})^i = 1 - \frac{\alpha}{2}.$$

The values $(\underline{P}, \bar{P})$ are tables for various values of N and d. Table 9.3.1 gives bounds for $\underline{P}$ and $\bar{P}$ for the confidence coefficient 0.95. We illustrate the way to use Table 9.3.1 in an example.

Example 9.3.1 We have tested 20 structures for t hours each. Two structures failed during the testing time.

We want to determine upper and lower bounds corresponding to the confidence coefficient 0.95 for the unknown value P.

Table 9.3.1   Ninety-Five Per Cent Confidence Limits for the Parameter of the Binomial Distribution*

Each cell gives the upper / lower confidence limit.

$N - n$

n	0	1	2	3	4	5	6	7	8	9	10	11	12	13	14	15	16	17	18	19	20
0	— / —	0.975 / 0.000	0.842 / 0.000	0.708 / 0.000	0.602 / 0.000	0.522 / 0.000	0.459 / 0.000	0.410 / 0.000	0.369 / 0.000	0.336 / 0.000	0.308 / 0.000	0.285 / 0.000	0.265 / 0.000	0.247 / 0.000	0.232 / 0.000	0.218 / 0.000	0.206 / 0.000	0.195 / 0.000	0.185 / 0.000	0.176 / 0.000	0.168 / 0.000
1	1.000 / 0.025	0.987 / 0.013	0.906 / 0.008	0.806 / 0.006	0.716 / 0.005	0.641 / 0.004	0.579 / 0.004	0.527 / 0.003	0.483 / 0.003	0.445 / 0.003	0.413 / 0.002	0.385 / 0.002	0.360 / 0.002	0.339 / 0.002	0.319 / 0.002	0.302 / 0.002	0.287 / 0.001	0.273 / 0.001	0.260 / 0.001	0.249 / 0.001	0.238 / 0.001
2	1.000 / 0.158	0.992 / 0.094	0.932 / 0.068	0.853 / 0.053	0.777 / 0.043	0.710 / 0.037	0.651 / 0.032	0.600 / 0.028	0.556 / 0.025	0.518 / 0.023	0.484 / 0.021	0.454 / 0.019	0.428 / 0.018	0.405 / 0.017	0.383 / 0.016	0.364 / 0.015	0.347 / 0.014	0.331 / 0.013	0.317 / 0.012	0.304 / 0.012	0.292 / 0.011
3	1.000 / 0.292	0.994 / 0.194	0.947 / 0.147	0.882 / 0.118	0.816 / 0.099	0.755 / 0.085	0.701 / 0.075	0.652 / 0.067	0.610 / 0.060	0.572 / 0.055	0.538 / 0.050	0.508 / 0.047	0.481 / 0.043	0.456 / 0.040	0.434 / 0.038	0.414 / 0.036	0.396 / 0.034	0.379 / 0.032	0.363 / 0.030	0.349 / 0.029	0.336 / 0.028
4	1.000 / 0.398	0.995 / 0.284	0.957 / 0.223	0.901 / 0.184	0.843 / 0.157	0.788 / 0.137	0.738 / 0.122	0.692 / 0.109	0.651 / 0.099	0.614 / 0.091	0.581 / 0.084	0.551 / 0.078	0.524 / 0.073	0.499 / 0.068	0.476 / 0.064	0.456 / 0.061	0.437 / 0.057	0.419 / 0.054	0.403 / 0.052	0.388 / 0.050	0.374 / 0.047
5	1.000 / 0.478	0.996 / 0.359	0.963 / 0.290	0.915 / 0.245	0.863 / 0.212	0.813 / 0.187	0.766 / 0.167	0.723 / 0.151	0.684 / 0.139	0.646 / 0.128	0.616 / 0.118	0.587 / 0.110	0.560 / 0.103	0.535 / 0.097	0.512 / 0.091	0.491 / 0.087	0.471 / 0.082	0.453 / 0.078	0.436 / 0.075	0.421 / 0.071	0.407 / 0.068
6	1.000 / 0.541	0.996 / 0.421	0.968 / 0.349	0.925 / 0.299	0.878 / 0.262	0.833 / 0.234	0.789 / 0.211	0.749 / 0.192	0.711 / 0.177	0.677 / 0.163	0.646 / 0.152	0.617 / 0.142	0.590 / 0.133	0.565 / 0.126	0.543 / 0.119	0.522 / 0.113	0.502 / 0.107	0.484 / 0.102	0.467 / 0.098	0.451 / 0.094	0.436 / 0.090
7	1.000 / 0.590	0.997 / 0.473	0.972 / 0.400	0.933 / 0.348	0.891 / 0.308	0.849 / 0.277	0.808 / 0.251	0.770 / 0.230	0.734 / 0.213	0.701 / 0.198	0.671 / 0.184	0.643 / 0.173	0.616 / 0.163	0.592 / 0.154	0.570 / 0.146	0.549 / 0.139	0.529 / 0.132	0.512 / 0.126	0.494 / 0.121	0.478 / 0.116	0.463 / 0.111
8	1.000 / 0.631	0.997 / 0.517	0.975 / 0.444	0.940 / 0.390	0.901 / 0.349	0.861 / 0.316	0.823 / 0.289	0.787 / 0.266	0.753 / 0.247	0.722 / 0.230	0.692 / 0.215	0.665 / 0.203	0.639 / 0.191	0.616 / 0.181	0.593 / 0.172	0.573 / 0.164	0.553 / 0.156	0.535 / 0.149	0.518 / 0.143	0.502 / 0.138	0.487 / 0.132
9	1.000 / 0.664	0.997 / 0.555	0.977 / 0.482	0.945 / 0.428	0.909 / 0.386	0.872 / 0.351	0.837 / 0.323	0.802 / 0.299	0.770 / 0.278	0.740 / 0.260	0.711 / 0.244	0.685 / 0.231	0.660 / 0.218	0.636 / 0.207	0.615 / 0.197	0.594 / 0.188	0.575 / 0.180	0.557 / 0.172	0.540 / 0.165	0.524 / 0.159	0.508 / 0.153
10	1.000 / 0.692	0.998 / 0.587	0.979 / 0.516	0.950 / 0.462	0.916 / 0.419	0.882 / 0.384	0.848 / 0.354	0.816 / 0.329	0.785 / 0.308	0.756 / 0.289	0.728 / 0.272	0.702 / 0.257	0.678 / 0.244	0.655 / 0.232	0.634 / 0.221	0.614 / 0.211	0.595 / 0.202	0.577 / 0.194	0.560 / 0.186	0.544 / 0.179	0.528 / 0.173

* Experiments indicate that for many parts (in particular, mechanical) whose failure is due to deterioration or aging, the time of failure-free operation has a truncated normal distribution.

Solution. In Table 9.3.1 we find the values $(1 - \underline{P}, 1 - \bar{P})$ corresponding to the values $N = 20$ and $d = 2$ at the intersection of row number 2 and the column numbered $20 - 2 = 18$.

The desired confidence bounds are $1 - \underline{P} = 0.317$ and $1 - \underline{P} = 0.012$ —that is, $\underline{P} = 0.683$ and $\bar{P} = 0.988$.

Remark. If we are interested in a one-sided confidence interval, then we should consider that, using the table with confidence coefficient $1 - \alpha$, we will obtain results corresponding to confidence coefficient $1 - \alpha/2$.

For example, if we are interested in only a lower confidence bound in Example 9.3.1, then the value $\underline{P} = 0.683$ corresponds to the confidence coefficient 0.975 and not 0.950.

9.4 OBTAINING CONFIDENCE BOUNDS FOR PARAMETERS OF THE NORMAL DISTRIBUTION

9.4.1 Confidence bounds for the mean value

If as the result of d trials we have obtained a collection of results $\theta_1, \theta_2, \ldots, \theta_d$ and somehow we know that the random variable θ has a normal distribution,[1] then we can determine two-sided confidence bounds for the unknown mean value T corresponding to a confidence coefficient $1 - \alpha$ by the formulas

$$\underline{T} = \hat{T}_0 - t_{\alpha/2}(r) \frac{s}{\sqrt{d-1}},$$

$$\bar{T} = \hat{T}_0 + t_{\alpha/2}(r) \frac{s}{\sqrt{d-1}},$$

where $\hat{T}_0$ is the arithmetic average of the θ_i variables:

$$\hat{T}_0 = \frac{1}{d} \sum_{i=1}^{d} \theta_i,$$

s is the square root of an unbiased estimate for the variance of θ,

$$s = \sqrt{\frac{1}{d-1} \sum_{i=1}^{d} (\theta_i - \hat{T}_0)^2};$$

$t_\alpha(r)$ are the quantities determined by Table 9.4.1, with $r = d - 1$.

---

[1] Experiments indicate that for many parts (in particular, mechanical) whose failure is due to deterioration or aging, the time of failure-free operation has a tuncated normal distribution.

Example 9.4.1   In testing certain structures whose operating time up to failure is normally distributed, we obtained ten realizations of the operating time up to failure (in hours): 120, 110, 80, 130, 120, 140, 80, 150, 130, and 140.

We wish to find confidence bounds for the unknown value T with confidence coefficient 0.99.

Solution.   First we determine the quantities $\hat{T}_0$ and s:

$$\hat{T}_0 = \tfrac{1}{10}(120 + 110 + 80 + 130 + 120 + 140 + 80 + 150 + 130 + 140)$$
$$= 120 \quad \text{hr.}$$

$$s = \sqrt{\frac{1}{10 - 1} \cdot (0 + 10^2 + 40^2 + 10^2 + 0 + 20^2 + 40^2 + 30^2 + 10^2 + 20^2)}$$
$$= \tfrac{1}{3} \sqrt{5100} = 24 \quad \text{hr.}$$

Further, from Table 9.4.1, in the column in which $\alpha/2 = 0.005$ at the ninth row (since the number of degrees of freedom of the distribution is $r = d - 1 = 10 - 1 = 9$), we find the value $t_{0.005} \approx 3.69$.

Table 9.4.1   Critical Values of the t Distribution

r	α					
	0.50	0.25	0.10	0.05	0.01	0.005
1	1.000	2.414	6.314	12.706	63.657	127.32
2	0.817	1.604	2.920	4.303	9.925	14.089
3	0.765	1.423	2.353	3.183	5.841	7.453
4	0.741	1.344	2.132	2.776	4.604	5.598
5	0.727	1.301	2.015	2.571	4.032	4.773
6	0.718	1.273	1.943	2.447	3.707	4.317
7	0.711	1.254	1.895	2.365	3.500	4.029
8	0.706	1.240	1.860	2.306	3.355	3.832
9	0.703	1.230	1.833	2.262	3.250	3.690
10	0.700	1.221	1.813	2.228	3.169	3.581
15	0.691	1.197	1.753	2.132	2.947	3.286
20	0.687	1.185	1.725	2.086	2.8453	3.153
25	0.684	1.178	1.708	2.060	2.787	3.078
30	0.683	1.173	1.697	2.042	2.750	3.030
40	0.681	1.167	1.689	2.021	2.705	2.971
60	0.679	1.162	1.671	2.000	2.660	2.915
120	0.677	1.156	1.658	1.980	2.617	2.860
∞	0.674	1.150	1.645	1.960	2.576	2.807

Now we determine the lower and upper confidence bounds to be

$$\underline{T} = 120 - 3.69 \cdot \tfrac{24}{3} = 90.4 \quad \text{hr},$$
$$\bar{T} = 120 + 3.69 \cdot \tfrac{24}{3} = 149.6 \quad \text{hr}.$$

Remark. If we are interested in a one-sided confidence interval with confidence coefficient $1 - \alpha$, then we should seek the value t_p directly in the column labeled with the value $p = \alpha$ rather than $p = \alpha/2$.

Thus if in Example 9.4.1 we are interested in only a lower confidence bound with confidence coefficient 0.99, then we look up the value $t_{0.01}(9)$ in the column in which $p = 0.01$ again in the ninth row of Table 9.4.1. In this case $t_{0.01}(9) = 3.25$ and

$$\underline{T} = 120 - 3.25 \cdot \tfrac{24}{3} = 94 \quad \text{hr}.$$

9.4.2 Confidence bounds for the variance

If as the result of tests we have obtained a set of results $\theta_1, \theta_2, \ldots, \theta_n$ and we are willing to assume that the random variable θ has a normal distribution, then we can determine confidence bounds (with confidence coefficient $1 - \alpha$) for the unknown value of the variance by the formulas

$$\underline{D} = \frac{ds^2}{\chi^2_{\alpha/2}(d-1)}, \qquad \bar{D} = \frac{ds^2}{\chi^2_{1-(\alpha/2)}(d-1)},$$

where s^2 is an unbiased estimate for the variance of the θ's,

$$s^2 = \frac{1}{d-1} \sum_{i=1}^{d} (\theta_i - \hat{T}_0)^2,$$

and

$$\hat{T}_0 = \frac{1}{d} \sum_{i=1}^{d} \theta_i,$$

$\chi^2_{\alpha/2}, \chi^2_{1-(\alpha/2)}$ are quantities determined from Table 9.2.1.

Example 9.4.2   Under the conditions of Example 9.4.1, determine two-sided confidence bounds for the unknown value σ^2 with confidence coefficient 0.90.

From Table 9.2.1 we find the value $\chi^2_{0.05}(9)$ in the column in which $p = 0.05$ in the ninth row to be $\chi^2_{0.05}(9) = 16.92$ and the value $\chi^2_{0.95}(9)$ in the column in which $p = 0.95$ again in the ninth row to be

$$\chi^2_{0.95}(9) = 3.32.$$

Using the value $s = 24$ hours computed in Example 9.4.1, we find

$$\underline{D} = \frac{10 \cdot 576}{16.92} = 341 \quad \text{hr}^2,$$

$$\bar{D} = \frac{10 \cdot 576}{3.32} = 1740 \quad \text{hr}^2.$$

Confidence bounds for the standard deviation are

$$\underline{\sigma} = \sqrt{\underline{D}} = 18.5 \quad \text{hr},$$

$$\bar{\sigma} = \sqrt{\bar{D}} = 41.6 \quad \text{hr}.$$

9.5   ESTIMATING THE RELIABILITY MEASURES OF A STRUCTURE FROM EXPERIMENTAL DATA FOR THE CASE OF AN EXPONENTIALLY DISTRIBUTED OPERATING TIME TO FAILURE

The theoretical material of this section was borrowed with very insignificant changes from B. V. Gnedenko, Yu. K. Belyaev, and A. D. Solov'ev [3, chap. 3].

9.5.1   Possible testing schemes

Consider the following scheme for testing reliability. At the initial moment of time $t = 0$ we begin testing N structures (under electrical load, effect of mechanical stress, and so on). The conditions under which we test these structures are said to determine the testing scheme.

The letter R will denote schemes where each failed structure is replaced in a negligibly small amount of time by a new, identical structure; this is testing *with replacement*.

In testing under a scheme of type R the same number of structures N must always be under load. Obviously such tests can be carried out in practice if we have operative repair of failed structures. Under the assumption of exponentially distributed operating time to failure, all the results given below also apply to the case where in processing the data we discard those intervals of time for which the number of structures being tested is less than N (the rest are being repaired) and the corresponding failures arising during this time. We assume that observations to detect failed structures are made continuously and, as a result, failures are eliminated immediately.

The letter W stands for those schemes in which structures that fail during the testing period are not replaced by new ones; this is testing *without replacement*.

The moment of termination of testing can be assigned in various ways. Some examples are as follows:

T schemes where testing is done during a time interval of pre-assigned duration T,

r schemes where testing is done up to the moment of occurrence of the rth failure,

(r, T) schemes where testing is done either up to the moment t_r of occurrence of the rth failure if $t_r < T$ or up to the moment T if $t_r \geq T$.

It is easy to see that there are six possible types of schemes based on these rules of termination and on the replacement scheme:

1. $[N, R, T]$, the scheme where N structures are put under test; failed structures are replaced (R); testing is done up to a previously assigned moment of time T.

2. $[N, R, r]$, the scheme where N structures are put under test; failed structures are replaced (R); testing is done up to the moment of occurrence of the rth failure.

3. $[N, R, (r, T)]$, the scheme where N structures are put under test; failed structures are replaced (R); testing is done either up to the moment t_r of occurrence of the rth failure if $t_r < T$ or up to a previously assigned moment of time T if $t_r \geq T$.

4. $[N, W, T]$, the scheme where N structures are put under test; failed structures are not replaced (W); testing is done up to a previously assigned moment T.

5. $[N, W, r]$, the scheme where N structures are put under test; failed structures are not replaced (W); testing is done up to the moment of occurrence of the rth failure.

6. $[N, W, (r, T)]$, the scheme where N structures are put under test; failed structures are not replaced (W); testing is done either up to the moment t_r of occurrence of the rth failure if $t_r < T$ or up to the moment T if $t_r \geq T$.

We let $d(t)$ denote the number of failures occurring up to moment t. The function $d(t)$ takes on the successive values $0, 1, 2, \ldots$. The points of increase of the function $d(t)$ correspond to the random moments of time t_i, $i = 1, 2, \ldots$, where t_i is the moment of the ith failure.

Figure 9.5.1 illustrates the behavior of the function $d(t)$ for the various testing schemes.

Besides the six schemes listed, testing can be designed on the basis of assigning a total running time to the tested structures. If $N(t)$ denotes the number of structures operating failure-free up to the moment t, that is, $N(t) = N - d(t)$, then the value $S(t)$ of the total running time at time t

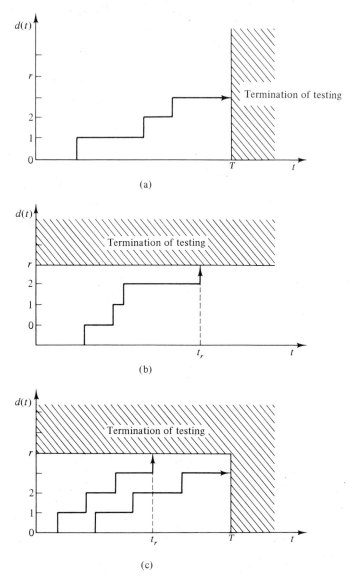

Figure 9.5.1   Nature of the Behavior of the Function $d(t)$ for Various Testing Schemes: (a) schemes $[N, R, T]$, $[N, W, T]$, (b) schemes $[N, R, r]$, $[N, W, r]$, (c) schemes $[N, R, (r, T)]$, $[N, W, (r, T)]$.

is defined as the sum of the times during which the structures being tested have operated failure-free. Thus

$$S(t) = \sum_{i=1}^{d(t)} t_i + N(t)t, \tag{9.5.1}$$

where t_i is the moment of failure of the ith unit, $i = 1, 2, \ldots, d(t)$.

Since for schemes of type R we always have

$$S(t) = Nt,$$

then we do not obtain any fundamentally new schemes of type R by basing them on the total running time.

For schemes of type W, however, there are two more schemes to add to the above six:

7. $[N, W, HS_0]$, the scheme in which N structures are tested; failed structures are not replaced (W); testing is done up to a moment t^* such that $S(t^*) = S_0$, where S_0 is a previously assigned total running time for the structures; if $S(t_N) \leq S_0$, then $t^* = T_N$, where T_N is the moment of failure of the last (Nth) structure.

8. $[N, W, (r, HS_0)]$, the scheme where N structures are tested; failed structures are not replaced (W); testing is done up to the moment t^*, which is defined as the moment when one of the following events first occurs: either $S(t^*) = S_0$ and $d(t^*) < r$, or $t^* = t_r$ but $S(t^*) < S_0$, where t_r is, as before, the moment of occurrence of the rth failure.

9.5.2 Estimating the parameter of the exponential law

This subsection investigates estimation of the failure intensity of structures [the parameter λ of the exponential law $Q(t, \lambda) = 1 - e^{-\lambda t}$] for various schemes.

It is known that no matter what desirable properties are possessed by estimators $\hat{\lambda}$ of the parameter λ, for example unbiasedness and efficiency, in many cases of great practical interest it turns out to be undesirable to characterize the reliability of structures with only point estimators. Very often the time T of testing is limited and a small value of λ means that the mean number of failures observed during testing is also small. (It may turn out that no failures are observed during testing.) In this case the natural measure of scattering of the values of the random estimate, the coefficient of variation, which is the ratio of the standard deviation to the mathematical expectation, is high (greater than one), and therefore the values of the estimator vary sharply from test to test and cannot serve as a stable characteristic of reliability for the structures.

Table 9.5.1 Estimates of λ

TESTING SCHEME	SUFFICIENT STATISTIC	ESTIMATE	VARIANCE OF THE ESTIMATE	CONFIDENCE INTERVAL WITH CONFIDENCE COEFFICIENT NOT LESS THAN $1 - (\epsilon_1 + \epsilon_2)$
$[N, R, T]$	$d(T)$ is the number of failures in time T	$\hat{\lambda} = \dfrac{d(T)}{NT}$ (unbiased)	$D[\hat{\lambda}] = \dfrac{\lambda}{NT}$	$\underline{\lambda} = \dfrac{\Delta_{1-\epsilon_2}(d-1)}{NT}$, $\bar{\lambda} = \dfrac{\Delta_{\epsilon_1}(d)}{NT}$, $d = d(T)$
$[N, R, r]$	t_r is the moment of the rth failure	$\hat{\lambda} = \dfrac{r-1}{Nt_r}$, $\left(\begin{array}{c}\text{unbiased,}\\ \text{if } r > 1\end{array}\right)$	$D[\hat{\lambda}] = \dfrac{\lambda^2}{r-2}$, $r > 2$	$\underline{\lambda} = \dfrac{\Delta_{1-\epsilon_2}(r-1)}{Nt_r}$, $\bar{\lambda} = \dfrac{\Delta_{\epsilon_1}(r-1)}{Nt_r}$
$[N, R, (r, T)]$	$\begin{cases} t_r, & \text{if } t_r \le T; \\ d(T), & \text{if } t_r > T \end{cases}$	$\hat{\lambda} = \begin{cases} \dfrac{d(T)}{NT}, & t_r > T; \\ \dfrac{r-1}{N(t_r)}, & t_r \le T \end{cases}$ $\left(\begin{array}{c}\text{unbiased,}\\ \text{if } r > 1\end{array}\right)$	$D[\hat{\lambda}] = \dfrac{\lambda}{NT} L_{r-2}(\lambda NT)$ $+ \dfrac{\lambda^2}{r-2}[1 - L_{r-2}(\lambda NT)]$	If $t_r \le T$, we use the plan $[N, R, r]$ for $\underline{\lambda}$ and $\bar{\lambda}$; If $t_r > T$, we use the formulas of plan $[N, R, T]$
$[N, W, T]$	$d(T)$ and $S_W(T)$	$\hat{\lambda} = \dfrac{d(T)}{S_W(T)}$, or $\hat{\lambda} = \dfrac{d(T)}{\left[N - \dfrac{d(T)}{2}\right]T}$, if $\dfrac{d(T)}{N}$ small (biased)	—	$\underline{\lambda} = \dfrac{\Delta'(d)}{T}$, $\bar{\lambda} = \dfrac{\bar{\Delta}'(d)}{T}$

$[N, W, r]$	$S_W(t_r)$—total operation at time t_r	$\hat\lambda = \dfrac{r-1}{S_W(t_r)}$ (unbiased)	$D[\lambda] = \dfrac{\lambda^2}{r-2}$, $r > 2$	$\underline\lambda = \dfrac{\Delta_{1-\epsilon_2}(r-1)}{S_W(t_r)}$, $\bar\lambda = \dfrac{\Delta_{\epsilon_1}(r-1)}{S_W(t_r)}$
$[N, W, (r, T)]$	$\begin{cases} d(T), S_W(T), \text{if } t_r > T; \\ S_W(t_r), \text{if } t_r \le T \end{cases}$	$\hat\lambda = \begin{cases} \dfrac{d(T)}{S_W(T)}, & t_r > T \\ \dfrac{r-1}{S_W(t_r)}, & t_r \le T \end{cases}$ (biased)	—	If $t_r > T$, then we use the formula for $\underline\lambda, \bar\lambda$ of plan $[N, W, T]$, but if $t_r \le T$, then $\underline\lambda$ and $\bar\lambda$ are found as solutions of the equations $$\sum_{i=0}^{d} C_N^i (1 - e^{-\underline\lambda t_0})^i e^{-(N-i)\underline\lambda t_0} = \epsilon_1,$$ $$\sum_{i=d}^{N} C_N^i (1 - e^{-\bar\lambda t_0})^i e^{-(N-i)\bar\lambda t_0} = \epsilon_2$$
$[N, W, (r, HS_0)]$	$\begin{cases} S_W(t_r), \text{if } t_r < T; \\ d(t_\theta), \text{if } S_0 \\ = S_W(t_\theta) < S_W(t_r) \end{cases}$ where t_θ is the end of testing	$\hat\lambda = \begin{cases} \dfrac{d(t_\theta)}{S_0}, & S_0 < S_W(t_r); \\ \dfrac{r-1}{S_W(t_r)}, & S_0 \ge S_W(t_r) \end{cases}$ (unbiased)	$D(\hat\lambda) = \dfrac{\lambda}{S_0} L_{r-2}(\lambda S_0) + \dfrac{\lambda^2}{r-2}[1 - L_{r-2}(\lambda S_0)]$	$\underline\lambda = \dfrac{\Delta_{1-\epsilon_2}(d-1)}{S_0}$. $\bar\lambda = \dfrac{\Delta_{\epsilon_1}(d)}{S_0}$, if $S_0 < S_W(t_r)$; $\underline\lambda = \dfrac{\Delta_{1-\epsilon_2}(r-1)}{S_W(t_r)}$, $\bar\lambda = \dfrac{\Delta_{\epsilon_1}(r-1)}{S_W(t_r)}$, if $S_0 \ge S_W(t_r)$

Here it is convenient to use the method of confidence intervals for estimating the parameter λ of the exponential law.

We recall the following three definitions:

1.   A two-sided confidence interval for the parameter λ with confidence coefficient not less than $1 - \alpha$ is a *random* interval $(\underline{\lambda}(X), \bar{\lambda}(X))$ whose end points (confidence bounds) [or confidence limits] $\underline{\lambda}(X) < \bar{\lambda}(X)$ depend only on the outcomes X and for any $\lambda > 0$

$$\mathcal{P}\{\underline{\lambda}(X) < \lambda < \bar{\lambda}(X)\} \geq 1 - \alpha.$$

2.   An upper confidence limit $\bar{\lambda}(X)$ and a lower confidence limit $\underline{\lambda}(X)$ are end points of random intervals $(0, \bar{\lambda}(X))$ and $(\underline{\lambda}(X), \infty)$, respectively, such that for any $\lambda > 0$, respectively,

$$\mathcal{P}\{0 < \lambda < \bar{\lambda}(X)\} \geq 1 - \alpha, \qquad \mathcal{P}\{\underline{\lambda}(X) < \lambda\} \geq 1 - \alpha.$$

3.   In Table 9.5.1 we give formulas for estimators and confidence intervals for the parameter λ of the exponential law, with the following notation:

$$\epsilon_1 + \epsilon_2 = \alpha \qquad \text{(generally } \epsilon_1 = \epsilon_2),$$
$$\Delta = \lambda N T,$$
$$L_d(\Delta) = \sum_{k=0}^{d} \frac{\Delta^k}{k!} e^{-\Delta}, \tag{9.5.2}$$

the quantity $\Delta_\alpha(d)$ is defined by the functional relation

$$L_d(\Delta_\alpha(d)) = \alpha. \tag{9.5.3}$$

Tables 9.5.2 and 9.5.3 are designed for making numerical computations according to the formulas of Table 9.5.1, and we give the quantiles $\Delta_{1-\alpha}(d)$ of the Poisson distribution and the values of the quantities $\underline{\Delta}'_{1-\alpha}(d)$ and $\bar{\Delta}'_{1-\alpha}(d)$ for $1 - \alpha = 0.95$, which are necessary for the construction of confidence intervals in the case of an $[N, W, T]$ scheme.

Example 9.5.1   Scheme $[N, R, T]$. Suppose that during the time of testing by the scheme $[N = 500, R, T = 100$ hours], $d(T) = 5$ structures have failed.

We wish to find: (1) an upper confidence bound with confidence coefficient $1 - \alpha = 0.9$, (2) a two-sided confidence bound corresponding to the value $1 - \alpha = 0.9$ with $\epsilon_1 = \epsilon_2 = 0.05$; (3) an estimate $\hat{\lambda}$ of the parameter λ.

Solution.   1.   From Table 9.5.1 it follows that

$$\bar{\lambda} = \frac{\Delta_{\epsilon_1}(d)}{NT} = \frac{\Delta_\alpha(d)}{NT}.$$

Table 9.5.2 Quantiles $\Delta_{1-\alpha}(d)$ of the Poisson Distribution,
$L_d[\Delta_{1-\alpha}(d)] = 1 - \alpha$

d \ $1-\alpha$	0.99993	0.9999	0.9993	0.999	0.993	0.99	0.95
0	0.000070	0.000100	0.000700	0.00100	0.00702	0.01005	0.05129
1	0.01188	0.01421	0.03789	0.04540	0.12326	0.14855	0.35536
2	0.07633	0.08618	0.16824	0.19053	0.38209	0.43604	0.81769
3	0.21115	0.23180	0.38894	0.42855	0.74108	0.82325	1.36632
4	0.41162	0.44446	0.68204	0.73937	1.17032	1.27911	1.97015
5	0.66825	0.71375	1.03236	1.10710	1.65152	1.78528	2.61301
6	0.97222	1.03040	1.42874	1.52034	2.17293	2.33021	3.28532
7	1.31628	1.38697	1.86297	1.97081	2.72659	2.90611	3.98082
8	1.69465	1.77758	2.32894	2.45242	3.30682	3.50746	4.69523
9	2.10271	2.19758	2.82197	2.96052	3.90942	4.13020	5.42541
10	2.53672	2.64323	3.33840	3.49148	4.53118	4.77125	6.16901
11	2.99367	3.11150	3.87531	4.04244	5.16960	5.42818	6.92421
12	3.47103	3.59988	4.43033	4.61106	5.82265	6.09907	7.68958
13	3.96672	4.10632	5.00152	5.19544	6.48871	6.78235	8.46394
14	4.47896	4.62904	5.58725	5.79398	7.16642	7.47673	9.24633
15	5.00626	5.16657	6.18615	6.40533	7.85464	8.18111	10.03596
16	5.54732	5.71762	6.79705	7.02835	8.55241	8.89457	10.83214

d \ $1-\alpha$	0.93	0.90	0.80	0.70	0.60	0.50	0.40
0	0.07257	0.10536	0.22314	0.35667	0.51082	0.69315	0.91629
1	0.43081	0.53181	0.82439	1.09735	1.37642	1.67835	2.02231
2	0.94230	1.10206	1.53504	1.91378	2.28508	2.67406	3.10538
3	1.53414	1.74477	2.29679	2.76371	3.21132	3.67206	4.17526
4	2.17670	2.43259	3.08954	3.63361	4.14774	4.67091	5.23662
5	2.85488	3.15190	3.90366	4.51714	5.09098	5.67016	6.29192
6	3.55984	3.89477	4.73366	5.41074	6.03924	6.66964	7.34265
7	4.28584	4.65612	5.57606	6.31217	6.99137	7.66925	8.38977
8	5.02895	5.43247	6.42848	7.21993	7.94661	8.66895	9.43395
9	5.78633	6.22130	7.28922	8.13293	8.90441	9.66871	10.47568
10	6.55583	7.02075	8.15702	9.05036	9.86440	10.66852	11.51533
11	7.33581	7.82934	9.03090	9.97161	10.82624	11.66836	12.55317
12	8.12496	8.64594	9.91010	10.89620	11.78972	12.66823	13.58944
13	8.92222	9.46962	10.79398	11.82373	12.75462	13.66811	14.62431
14	9.72672	10.29962	11.68206	12.75388	13.72081	14.66802	15.65793
15	10.53773	11.13530	12.57389	13.68639	14.68814	15.66793	16.69043
16	11.35465	11.97613	13.46913	14.62103	15.65651	16.66785	17.72191

d \ $1-\alpha$	0.30	0.20	0.10	0.05	0.025	0.01	0.005
0	1.20397	1.60944	2.30258	2.99573	3.68888	4.60517	5.29832
1	2.43922	2.99431	3.88972	4.74386	5.57164	6.63835	7.43013
2	3.61557	4.27903	5.32232	6.29579	7.22469	8.40595	9.27379
3	4.76223	5.51504	6.68078	7.75366	8.76727	10.04512	10.97748
4	5.89036	6.72098	7.99359	9.15352	10.24159	11.60462	12.59409
5	7.00555	7.90599	9.27467	10.51303	11.66833	13.10848	14.14976
6	8.11105	9.07538	10.53207	11.84240	13.05947	14.57062	15.65968
7	9.20895	10.23254	11.77091	13.14811	14.42268	15.99996	17.13359
8	10.30068	11.37977	12.99471	14.43465	15.76319	17.40265	18.57822
9	11.38727	12.51875	14.20599	15.70522	17.08480	18.78312	19.99842
10	12.46951	13.65073	15.40664	16.96222	18.39036	20.14468	21.39783
11	13.54798	14.77666	16.59812	18.20751	19.68204	21.48991	22.77926
12	14.62316	15.89731	17.78158	19.44257	20.96158	22.82084	24.14494
13	15.69544	17.01328	18.95796	20.66857	22.23040	24.13912	25.49669
14	16.78512	18.12509	20.12801	21.88648	23.48962	25.44609	26.83598
15	17.83246	19.23316	21.29237	23.09713	24.74022	26.74289	28.16406
16	18.89769	20.33782	22.45158	24.30118	25.98300	28.03045	29.48196

Table 9.5.3 Values $\Delta'_{1-\alpha}(d)$ and $\overline{\Delta}'_{1-\alpha}(d)$, $1 - \alpha = 0.95$

c \ N	50	60	80	100	150	200
0	0.07378 0.00000	0.06148 0.00000	0.04611 0.00000	0.03689 0.00000	0.02459 0.00000	0.01844 0.00000
1	0.11257 0.00051	0.09365 0.00042	0.07009 0.00032	0.05600 0.00025	0.03727 0.00017	0.02793 0.00013
2	0.14750 0.00489	0.12248 0.00407	0.09146 0.00305	0.07298 0.00243	0.04849 0.00162	0.03631 0.00121
3	0.18090 0.01263	0.14994 0.01049	0.11172 0.00783	0.08902 0.00625	0.05904 0.00415	0.04417 0.00311
4	0.21362 0.02248	0.17672 0.01863	0.13136 0.01388	0.10454 0.01106	0.06921 0.00734	0.05173 0.00549
5	0.24607 0.03384	0.20317 0.02800	0.15066 0.02082	0.11972 0.01657	0.07912 0.01097	0.05909 0.00820
6	0.27853 0.04640	0.22949 0.03832	0.16975 0.02842	0.13471 0.02259	0.08886 0.01493	0.06630 0.01115
7	0.31115 0.05995	0.25582 0.04942	0.18875 0.03657	0.14957 0.02902	0.09848 0.01915	0.07341 0.01429
8	0.34408 0.07440	0.28226 0.06120	0.20772 0.04518	0.16435 0.03580	0.10801 0.02358	0.08044 0.01758
9	0.37742 0.08966	0.30889 0.07360	0.22671 0.05419	0.17910 0.04289	0.11748 0.02819	0.08741 0.02100
10	0.41126 0.10570	0.33578 0.08657	0.24576 0.06358	0.19385 0.05025	0.12691 0.03297	0.09434 0.02453
11	0.44568 0.12247	0.36297 0.10008	0.26490 0.07331	0.20863 0.05785	0.13631 0.03788	0.10123 0.02816
12	0.48077 0.13996	0.39053 0.11411	0.28417 0.08336	0.22344 0.06568	0.14569 0.04293	0.10810 0.03189
13	0.51659 0.15818	0.41848 0.12865	0.30359 0.09372	0.23831 0.07372	0.15507 0.04809	0.11495 0.03569
14	0.55322 0.17711	0.44688 0.14368	0.32318 0.10437	0.25325 0.08198	0.16445 0.05337	0.12178 0.03957
15	0.59074 0.19677	0.47577 0.15921	0.34295 0.11531	0.26828 0.09042	0.17384 0.05875	0.12861 0.04352
16	0.62922 0.21717	0.50519 0.17524	0.36294 0.12654	0.28341 0.09906	0.18325 0.06123	0.13544 0.04753
17	0.66875 0.23833	0.53518 0.19177	0.38314 0.13804	0.29864 0.10788	0.19267 0.06980	0.14226 0.05160
18	0.70942 0.26027	0.56578 0.20881	0.40860 0.14982	0.31399 0.11688	0.20213 0.07547	0.14909 0.05574
19	0.75131 0.28303	0.59704 0.22637	0.42431 0.16188	0.32947 0.12607	0.21161 0.08123	0.15592 0.05993

From Table 9.5.2 we find that

$$\Delta_\alpha(d) = \Delta_{1-0.9}(5) = \Delta_{0.1}(5) = 9.27467.$$

Then

$$\bar{\lambda} = \frac{\Delta_\alpha(d)}{NT} = \frac{9.27467}{500 \cdot 100} \approx 18.55 \cdot 10^{-5} \; \frac{1}{hr},$$

—that is, with confidence coefficient 0.9 the true value λ does not exceed the obtained value $\bar{\lambda}$.

2.   From Table 9.5.1 it follows that

$$\underline{\lambda} = \frac{\Delta_{1-\epsilon_2}(d-1)}{NT}, \qquad \bar{\lambda} = \frac{\Delta_{\epsilon_1}(d)}{NT}.$$

In our case we find from Table 9.5.2 that

$$\Delta_{1-\epsilon_2}(d-1) = \Delta_{1-0.05}(5-1) = \Delta_{0.95}(4) = 1.97015,$$
$$\Delta_{\epsilon_1}(d) = \Delta_{0.05}(5) = 10.51303.$$

Then

$$\underline{\lambda} = \frac{\Delta_{1-\epsilon_2}(d-1)}{NT} = \frac{1.97015}{500 \cdot 100} \approx 3.94 \cdot 10^{-5} \; \frac{1}{hr},$$

$$\bar{\lambda} = \frac{\Delta_{\epsilon_1}(d)}{NT} = \frac{10.51303}{500 \cdot 100} \approx 21.03 \cdot 10^{-5} \; \frac{1}{hr}$$

—that is, with confidence coefficient 0.9 the true value λ lies in the interval with the obtained end points $\underline{\lambda}$, $\bar{\lambda}$.

3.   An estimate $\hat{\lambda}$ of the parameter λ is

$$\hat{\lambda} = \frac{d(T)}{NT} = \frac{5}{500 \cdot 100} = 10 \cdot 10^{-5} \; \frac{1}{hr}.$$

Example 9.5.2 Scheme $[N, R, r]$. Suppose that during the time of testing by the scheme $[N = 500, R, r = 15]$ we have obtained $t_{15} = 1211$ hours.

We want to find an upper confidence bound with confidence coefficient $1 - \alpha = 0.99$.

Solution.   From Table 9.5.1 it follows that

$$\bar{\lambda} = \frac{\Delta_{\epsilon_1}(r-1)}{Nt_r}.$$

In our case we find from Table 9.5.2 that

$$\Delta_{\epsilon_1}(r-1) = \Delta_\alpha(r-1) = \Delta_{1-0.99}(15-1)$$
$$= \Delta_{0.01}(14) = 25.44609.$$

Then

$$\bar{\lambda} = \frac{\Delta_{\epsilon_1}(r-1)}{Nt_r} = \frac{25.44609}{500 \cdot 1211} \approx 4.24 \cdot 10^{-5} \quad \frac{1}{hr}.$$

Example 9.5.3   Scheme $[N, W, T]$. Suppose that during the time of testing according to the scheme $[N = 150, W, T = 100$ hours$]$, $d(T) = 5$ structures have failed.

We wish to find an estimate and a two-sided confidence interval with confidence coefficient $1 - \alpha = 0.95$.

Solution.   1.   From Table 9.5.1 we have

$$\hat{\lambda} = \frac{d(T)}{\left[N - \dfrac{d(T)}{2}\right]T}, \quad \text{since} \quad \frac{d(T)}{N} = \frac{5}{150} \ll 1.$$

Substituting numerical values, we obtain

$$\hat{\lambda} = \frac{5}{[150 - \frac{5}{2}] \cdot 100} = 33.9 \cdot 10^{-5} \quad \frac{1}{hr}.$$

2.   From Tables 9.5.1 and 9.5.3 it follows that

$$\lambda = \frac{\Delta'(d)}{T} = \frac{0.01097}{100} = 10.97 \cdot 10^{-5} \quad \frac{1}{hr},$$

$$\bar{\lambda} = \frac{\bar{\Delta}'(d)}{T} = \frac{0.07912}{100} = 79.12 \cdot 10^{-5} \quad \frac{1}{hr}.$$

Example 9.5.4   Scheme $[N, W, r]$. Suppose that during the time of testing structures by the scheme $[N = 150, W, r = 5]$ we have found that the total running time of all units at the moment t_5 is $S_W(t_5) = 13,202$ hr.

We want to find an estimate and a two-sided confidence interval for λ with confidence coefficient $1 - \alpha = 0.90$ with $\epsilon_1 = \epsilon_2 = 0.05$.

Solution.   From Table 9.5.1 we have

$$\hat{\lambda} = \frac{r-1}{S_W(t_r)} = \frac{5-1}{S_W(t_5)} = \frac{4}{13,202} = 30.3 \cdot 10^{-5} \quad \frac{1}{hr},$$

$$\lambda = \frac{\Delta_{1-\epsilon_2}(r-1)}{S_W(t_r)} = \frac{\Delta_{0.95}(4)}{13,202} \quad \frac{1}{hr},$$

$$\bar{\lambda} = \frac{\Delta_{\epsilon_1}(r-1)}{S_W(t_r)} = \frac{\Delta_{0.05}(4)}{13,202} \quad \frac{1}{hr}.$$

Using Table 9.5.2, we obtain

$$\Delta_{0.95}(4) = 1.97015, \qquad \Delta_{0.05}(4) = 9.15352.$$

Then

$$\underline{\lambda} = \frac{\Delta_{0.95}(4)}{13{,}202} = \frac{1.97015}{13{,}202} \approx 14.9 \cdot 10^{-5} \; \frac{1}{\text{hr}},$$

$$\bar{\lambda} = \frac{\Delta_{0.05}(4)}{13{,}202} = \frac{9.15352}{13{,}202} \approx 69.3 \cdot 10^{-5} \; \frac{1}{\text{hr}}.$$

Thus, the interval $(\underline{\lambda}, \bar{\lambda})$ covers the value of the unknown parameter λ with confidence coefficient 0.9.

Example 9.5.5 Scheme $[N, W, (r, HS_0)]$. Suppose testing of structures is done according to the plan $[N = 500, W, (r = 20, HS_0 = 20{,}000 \text{ hours})]$. At the moment t_θ of termination of testing, when the total running time is $HS_0 = 20{,}000$ hours, 16 structures have failed.

Find an upper confidence bound with confidence coefficient $1 - \alpha = 0.99$ and an estimate.

Solution. 1. Since the value $S(t_{20}) > S_0$; that is, the total running time of the structures has exceeded the level S_0 assigned at the beginning of testing before failure of the twentieth structure, an upper confidence bound according to Table 9.5.1 is found by the formula

$$\bar{\lambda} = \frac{\Delta_{\epsilon_1}(d)}{S_0} = \frac{\Delta_{0.01}(16)}{20{,}000} \; \frac{1}{\text{hr}}.$$

By Table 9.5.2 we have

$$\Delta_{0.01}(16) = 28.03045.$$

Then

$$\bar{\lambda} = \frac{28.03045}{20{,}000} \approx 140 \cdot 10^{-5} \; \frac{1}{\text{hr}}$$

—that is, with confidence coefficient 0.99 the true value of the parameter λ being estimated does not exceed the quantity $140 \cdot 10^{-5} \; \frac{1}{\text{hr}}$.

2. From Table 9.5.1 it follows that

$$\hat{\lambda} = \frac{d(t_\theta)}{S_0} = \frac{16}{20{,}000} = 80 \cdot 10^{-5} \; \frac{1}{\text{hr}}.$$

9.5.3 Estimating the probability of failure-free operation

In the case of the exponential law one of the most important reliability measures, the probability of failure-free operation of a structure during a given time $[0, t_0]$, is defined as

$$P(t_0) = e^{-\lambda t_0}.$$

Using the values given in Table 9.5.1 for estimates $\hat{\lambda}$ of the parameter λ, for sufficiently large N (the number of structures being tested) as an estimate for $P(t_0)$ we may use the quantity

$$\hat{P}(t_0) = e^{-\hat{\lambda} t_0}. \qquad (9.5.4)$$

However, for small values of N this estimate is biased. This bias may be quite significant in a case of great practical importance: when $P(t_0)$ is close to one.

In Table 9.5.4 we give unbiased estimates for two schemes: $[N, R, T]$, $[N, W, r]$.

Example 9.5.6 Scheme $[N, R, T]$. Suppose that during the testing time for the scheme $[N = 2, R, T = 1000$ hours$]$, $d(T) = 9$ failures have been obtained.

We want to find an estimate for the probability of failure-free operation of the structure for a period of $t_0 = 24$ hours.

Solution. From Table 9.5.4 we have

$$\hat{P}(t_0) = \left(1 - \frac{t_0}{NT}\right)^{d(T)} = \left(1 - \frac{24}{2 \cdot 1000}\right)^9 \approx 0.898.$$

Example 9.5.7 Scheme $[N, W, r]$. Suppose that during the time of testing of structures according to the scheme $[N = 8, W, r = 2]$, the second failure was observed at the moment of time t_2, when the total running time of the structures was $S_W(t_2) = 7452$ hours.

We want to find an estimate for the probability of failure-free operation of the structure for $t_0 = 24$ hours.

Solution. Using Table 9.5.4, we obtain

$$\hat{P}(t_0) = \left(1 - \frac{t_0}{S_W(t_r)}\right)^{r-1} = 1 - \frac{24}{7452} \approx 0.997.$$

9.5.4 Lower confidence bound for the probability of failure-free operation of a system of m structures connected in series

We consider a system consisting of m structures of l types connected in series. The number of structures of ith type is m_i; that is, $\sum_{i=1}^{l} m_i = m$.

Since the probability of failure-free operation of a structure of ith type is $e^{-\lambda_i t_0}$, the probability of failure-free operation of the system

Table 9.5.4 Unbiased Estimates for Probability of Failure-Free Operation

TESTING SCHEME	SUFFICIENT STATISTIC	ESTIMATE	VARIANCE
$[N, R, T]$	$d(T)$	$\left(1 - \dfrac{t_0}{NT}\right)^{d(T)}$ (unbiased) for $t_0 < NT$	$\exp\left\{\lambda NT\left[\left(1 - \dfrac{t_0}{NT}\right)^2 - 1\right]\right\} - \exp(-2\lambda t_0)$
$[N, W, r]$	$S_W(t_r)$	$\begin{cases}\left(1 - \dfrac{t_0}{S_W(t_r)}\right)^{r-1}, & S_W(t_r) > t_0; \\ 0, & S_W(t_r) \leq t_0\end{cases}$ (unbiased)	—

during time t_0 is

$$P(t_0) = e^{-\sum\limits_{i=1}^{l} m_i \lambda_i t_0}.$$

We assume that the estimates $\hat{\lambda}_i$ of the unknown values of the parameters λ_i were obtained separately for each type of structure by the testing scheme $[N_i, W, r_i]$ (see Table 9.5.1). Suppose the obtained values of total runs (being sufficient statistics for these schemes) are, respectively,

$$S_W(t_{r_i}) = s_i, \qquad i = 1, 2, \ldots, l.$$

In this case, with confidence coefficient not less than $1 - \alpha$, a lower confidence bound for the probability of failure-free operation of the system can be determined from the formula

$$\underline{P}(t_0) = \exp\left\{-\left[\sum_{i=1}^{l} m_i \frac{r_i - 1}{S_i} + u_{1-\alpha}\sqrt{\sum_{i=1}^{l} m_i^2 \frac{r_i - 1}{S_i^2}}\right]t_0\right\}, \quad (9.5.5)$$

where $u_{1-\alpha}$ is the $(1 - \alpha)$th quantile of the normal distribution (see Table A3.1 of the Appendix).

Thus with confidence coefficient not less than $1 - \alpha$ the true value of the probability of failure-free operation of the system exceeds the quantity $\underline{P}(t_0)$.

Example 9.5.8   Suppose we have a system of $m = 6$ structures of three types. The number of structures of the first type is $m_1 = 2$, of the second $m_2 = 3$, and of the third $m_3 = 1$. We have tested structures of each type by the scheme $[N_i, W, r_i]$, with the following results:

1.   for structures of the first type after the planned third failure ($r_1 = 3$) the total run of the structures being tested was

$$S_W(t_{r_i}) = S_1 = 2500 \quad \text{hr};$$

2.   for structures of the second type $S_2 = 1200$ hours after $r_2 = 2$;
3.   for structures of the third type $S_3 = 6000$ hours after $r_3 = 4$.

From the experimental data on the individual structures we want to determine a lower confidence bound with confidence coefficient $1 - \alpha = 0.90$ for the probability of failure-free operation of the system for time $t_0 = 24$ hours.

Solution.   We use Formula (9.5.5):

$$\underline{P}(t_0) = \exp\left[-\left(\sum_{i=1}^{l} m_i \frac{r_i - 1}{S_i} + u_{1-\alpha}\sqrt{\sum_{i=1}^{l} m_i^2 \frac{r_i - 1}{S_i^2}}\right)t_0\right].$$

First of all we find the quantity

$$\sum_{i=1}^{l} m_i \frac{r_i - 1}{S_i} = m_1 \frac{r_1 - 1}{S_1} + m_2 \frac{r_2 - 1}{S_2} + m_3 \frac{r_3 - 1}{S_3}$$

$$= 2 \cdot \frac{3 - 1}{2500} + 3 \cdot \frac{2 - 1}{1200} + 1 \cdot \frac{4 - 1}{6000} = 0.0046 \; \frac{1}{hr},$$

$$\sum_{i=1}^{l} m_i^2 \frac{r_i - 1}{S_i^2} = m_1^2 \frac{r_1 - 1}{S_1^2} + m_2^2 \frac{r_2 - 1}{S_2^2} + m_3^2 \frac{r_3 - 1}{S_3^2}$$

$$= 4 \cdot \frac{2}{(2500)^2} + 9 \cdot \frac{1}{(1200)^2} + \frac{3}{(6000)^2} \approx 761 \cdot 10^{-8} \; \frac{1}{hr^2},$$

whence

$$\sqrt{\sum_{i=1}^{l} m_i^2 \frac{r_i - 1}{S_i^2}} = \sqrt{761 \cdot 10^{-8}} = 0.0028 \; \frac{1}{hr} \cdot$$

From Table A3.1 we find

$$u_{1-\alpha} = u_{0.9} = 1.28.$$

Then

$$P(t_0) = P(t_0 = 24 \; hr)$$

$$= \exp\left[-\left(\sum_{i=1}^{l} m_i \frac{r_i - 1}{S_i} + u_{1-\alpha} \sqrt{\sum_{i=1}^{l} m_i^2 \frac{r_i - 1}{S_i^2}} \right) t_0 \right].$$

$$= \exp\left[-(0.0046 + 1.28 \cdot 0.0028)24 \right] \approx e^{-0.197} \approx 0.82.$$

REFERENCES

1. VENTSEL', E. S., *Probability Theory*, chaps. 7, 14. Moscow: Fizmatgiz, 1962.
2. GNEDENKO, B. V., BELYAEV, Y. K., AND SOLOL'EV, A. D., *Mathematical Methods in Reliability Theory*, chaps. 3, 4. Moscow: Nauka, 1965.
3. DUNIN-BARKOVSKII, I. V., AND SMIRNOV, N. V., *Probability Theory and Mathematical Statistics in Engineering* (general part), chaps. 5, 6. Moscow: GITTL, 1955.
4. LLOYD, D., AND LIPOW, M., *Reliability: Management, Methods, and Mathematics*. Englewood Cliffs, N.J.: Prentice-Hall, Inc., 1962.
5. SHOR, Y. B., *Statistical Methods of Analysis and Control of Quality and Reliability*, chaps. 7, 20, 24, 25, 26. Moscow: Soviet Radio, 1962.

APPENDIX 1

A1.1 BASIC CONCEPTS AND BRIEF SUMMARY OF PROBABILITY THEORY

Random events are events that may or may not occur as a result of a performance of the random experiment (this latter is referred to as a trial). We denote a random event by the symbol A.

The sure event is a random event that must occur without fail. We denote a sure event by the symbol E.

The impossible event is the random event that a fortiori can not occur. We denote the impossible event by the symbol U.

Nondisjoint (disjoint) events are events such that the occurrence of one does not exclude (excludes) the occurrence of another.

Dependent (independent) events are events such that the occurrence of one influences (does not influence) the probability of occurrence of another.

The complementary event (relative to some chosen event A) is the event that consists of the nonoccurrence of this chosen event A. We denote the complementary event by $\bar{A}$.

An exhaustive collection of events is a collection of events such that at least one event from this collection must occur as the result of a trial (of the random experiment).

Remark. The events A and $\bar{A}$ constitute an exhaustive collection of events, since as the result of a trial only two situations are possible: either the event A occurs, or the event A does not occur (that is, the event $\bar{A}$ occurs).

The union of events $A_1, A_2, \ldots, A_n$ is an event A whose occurrence in a

356

trial is equivalent to the occurrence of at least one of the events A_1, A_2, ..., A_n in the same trial. We denote the union of events by

$$A = A_1 + A_2 + \cdots + A_n = \sum_{i=1}^{n} A_i.$$

The intersection of events A_1, A_2, ..., A_n is an event A such that its occurrence in a trial is equivalent to the occurrence of all the events A_1, A_2, ..., A_n simultaneously in the same trial. We denote the intersection of events by

$$A = A_1 \wedge A_2 \wedge \cdots \wedge A_n = \prod_{i=1}^{n} A_i. \tag{1}$$

A random variable is a variable quantity that as the result of a trial may take on some numerical value.

A discrete random variable is a random variable that may take on only a finite or countable set of values.

A continuous random variable is a random variable that may take on any value in a certain (possibly infinite) interval.

A random process is a collection of random variables depending on a parameter (for example, on time).

The probability of a random event (classical definition) is defined as follows: Consider the equiprobable events A_1, A_2, ..., A_n: events such that there exists no objective reason for more frequent occurrence of any one of them. (It is assumed that all these events are disjoint and form an exhaustive group of events.) Such events are called elementary. Suppose an event of interest, A, can be divided into some number k of elementary events A_1, A_2, ..., A_k; that is, the occurrence of any one of the elementary events is equivalent to the occurrence of the event A. The probability of the event A is defined as the ratio of the number k of favorable[1] outcomes (elementary events) to the total number m of all elementary outcomes:

$$P(A) = \frac{k}{m}. \tag{2}$$

Basic Properties of Probability.   Probabilities of random events possess the following basic properties:

$$\begin{aligned} P(U) &= 0, \\ P(E) &= 1, \\ 0 = P(U) \leq P(A) &\leq P(E) = 1. \\ P(A) + P(\bar{A}) &= 1. \end{aligned} \tag{3}$$

[1] By favorable outcome we mean an outcome of a trial where the event A is observed.

Table A1.1 Continuous Distributions

DISTRIBUTION	REGION OF VALUES	DISTRIBUTION DENSITY	MATHEMATICAL EXPECTATION	VARIANCE	MODE
Uniform	(a, b)	$\dfrac{1}{b-a}$	$\dfrac{a+b}{2}$	$\dfrac{(b-a)^2}{12}$	—
Normal	$(-\infty, \infty)$	$\dfrac{1}{\sigma\sqrt{2\pi}}e^{-(x-a)^2/2\sigma^2}$	a	σ^2	a
Logarithmic normal	$(0, \infty)$	$\dfrac{1}{x\sigma\sqrt{2\pi}}e^{-(\ln x - a)^2/2\sigma^2}$	$e^{a+(\sigma^2/2)}$	$e^{2a+\sigma^2}(e^{\sigma^2}-1)$	$e^{a-\sigma^2}$
Weibull	$(0, \infty)$	$\alpha c x^{\alpha-1}e^{-cx^\alpha}$	$\dfrac{\Gamma\left(1+\dfrac{1}{\alpha}\right)}{c^{(1/\alpha)}}$	$\dfrac{\Gamma\left(1+\dfrac{2}{\alpha}\right)-\Gamma^2\left(1+\dfrac{1}{\alpha}\right)}{c^{2/\alpha}}$	$(\text{for } \alpha > 1)$ $\sqrt[\alpha]{\dfrac{\alpha-1}{c\alpha}}$
Gamma distribution	$(0, \infty)$	$\dfrac{\beta^\alpha}{\Gamma(\alpha)}x^{\alpha-1}e^{-\beta x}$	$\dfrac{\alpha}{\beta}$	$\dfrac{\alpha}{\beta^2}$	$\begin{array}{l}-\ (\alpha \le 1)\\[4pt] \dfrac{\alpha-1}{\beta}\ (\alpha > 1)\end{array}$

Exponential	$(0, \infty)$	$\lambda e^{-\lambda x}$	$\dfrac{1}{\lambda}$	$\dfrac{1}{\lambda^2}$	—
χ^2	$(0, \infty)$	$\dfrac{2^{(k/2)-1}}{\Gamma\left(\dfrac{k}{2}\right)2^{k/2}} e^{-(x/2)}$	k	$2k$	$k - 2$
Beta	$(0, 1)$	$x^{a-1}(1-x)^{b-1}$	$\dfrac{a}{a+b}$	$\dfrac{ab}{(a+b)^2(a+b+1)}$	$\dfrac{a-1}{a+b-2}$
Student's	$(-\infty, \infty)$	$\left[2^{(n-1)/2}\Gamma\left(\dfrac{n}{2}\right)\sqrt{\pi n}\right]^{-1}\left(1+\dfrac{x^2}{2}\right)^{-(n+1)/2}$	0	$\dfrac{1}{\dfrac{n}{2}-2}$	0
Fisher	$(0, \infty)$	$\dfrac{\Gamma\left(\dfrac{n_1+n_2}{2}\right)}{\Gamma\left(\dfrac{n_1}{2}\right)\Gamma\left(\dfrac{n_2}{2}\right)}\left(\dfrac{n_1}{n_2}\right)^{n_1/2}$ $\times\, x^{(n_1/2)-1}\left(1+\dfrac{n_1}{n_2}x\right)^{-(n_1+n_2)/2}$	$\dfrac{n_2}{n_2-2}$	$\dfrac{2n_2^2(n_1+n_2-2)}{n_1(n_2-2)^2(n_2-4)}$	$\dfrac{(n_1-2)n_2}{2n_1^2+n_2}$

Table A1.2 Discrete Distributions

DISTRIBUTION	POSSIBLE VALUES	PROBABILITY	MATHEMATICAL EXPECTATION	VARIANCE
Binomial	$0, 1, 2, \ldots, n$	$P_n(m) = C_n^m p^m q^{n-m}$	np	npq
Hypergeometric	$0, 1, \ldots, \min(M, n)$	$P_m = \dfrac{C_M^m C_{N-M}^{n-m}}{C_N^n}$	$n\dfrac{M}{N}$	$\dfrac{M(N-M)n(N-n)}{N^2(N-1)}$
Poisson	$0, 1, 2, \ldots$	$P_m = \dfrac{\lambda^m}{m!} e^{-\lambda}$	λ	λ
Geometric	$0, 1, 2, \ldots$	$P_m = pq^{m-1}$	$\dfrac{1}{p}$	$\dfrac{q}{p^2}$
Negative binomial	$r, r+1, \ldots$	$P_m = C_{m-1}^{r-1} p^r q^{m-r}$	$\dfrac{r}{p}$	$\dfrac{qr}{p^2}$

Theorem for Addition of Probabilities. If A_1, A_2, ..., A_n are disjoint events and A is the union of these events, then the probability of the event A is equal to the sum of the probabilities of the events A_1, A_2, ..., A_n; that is,

$$P(A) = P\left(\sum_{i=1}^{n} A_i\right) = P(A_1) + P(A_2) + \cdots + P(A_n) = \sum_{i=1}^{n} P(A_i). \quad (4)$$

Corollary 1. If the disjoint random events A_1, A_2, ..., A_n form an exhaustive group of events, then

$$P\left(\sum_{i=1}^{n} A_i\right) = P(E) = 1. \quad (5)$$

Corollary 2. For any random events A_1 and A_2 we have

$$P(A_1 \wedge A_2) + P(A_1 \wedge \bar{A}_2) = P(A_1). \quad (6)$$

Indeed

$$\begin{aligned} P(A_1 \wedge A_2) + P(A_1 \wedge \bar{A}_2) &= P(A_1 \wedge A_2 + A_1 \wedge \bar{A}_2) \\ &= P(A_1 \wedge (A_2 + \bar{A}_2)) = P(A_1 \wedge E) \\ &= P(A_1)P(E) = P(A_1). \end{aligned}$$

The conditional probability of the event A_1 given the occurrence of the event A_2 is the probability of the event A_1 computed under the assumption that the event A_2 has occurred. We denote this conditional probability by $P(A_1 \mid A_2)$:

$$P(A_1 \mid A_2) = \frac{P(A_1 \wedge A_2)}{P(A)_2}. \quad (7)$$

Corollary 3. It follows from the definition that for independent events A_1 and A_2

$$P(A_1 \mid A_2) = P(A_1) \quad \text{and} \quad P(A_2 \mid A_1) = P(A_2).$$

Theorem for Multiplication of Probabilities. The probability of simultaneous occurrence of two events A_1 and A_2 on a given experimental trial is equal to the probability of one of them multiplied by the conditional probability of the other computed under the assumption that the first event has occurred—that is,

$$P(A_1 \wedge A_2) = P(A_1 \mid A_2)P(A_2). \quad (8)$$

In the general case the multiplication theorem can be written in the form

$$P(A_1 \wedge A_2 \wedge \cdots \wedge A_n) = P(A_1 \mid A_2 \wedge \cdots \wedge A_n)$$
$$\times P(A_2 \mid A_3 \wedge \cdots \wedge A_n) \cdots P(A_{n-1} \mid A_n)P(A_n). \quad (9)$$

Corollary 4.   Interchanging A_1 and A_2 in Formula (8), we obtain

$$P(A_1 \wedge A_2) = P(A_2 \mid A_1)P(A_1),$$

—that is,

$$P(A_1)P(A_2 \mid A_1) = P(A_2)P(A_1 \mid A_2). \quad (10)$$

Theorem for Multiplication of Probabilities for Independent Events.   Using Corollary 3, we can rewrite the theorem for multiplication of probabilities for independent events as

$$P(A_1A_2 \cdots A_n) = P\left(\prod_{i=1}^{n} A_i\right)$$

$$= P(A_1)P(A_2) \cdots P(A_n) = \prod_{i=1}^{n} P(A_i). \quad (11)$$

Corollary 5.   For disjoint random events A_1 and A_2

$$P(A_1 \wedge A_2) = P(U) = 0. \quad (12)$$

Corollary 6.   The independent random events A_1 and A_2 are always nondisjoint if $P(A_1) > 0$ and $P(A_2) > 0$, since

$$P(A_1 \wedge A_2) > P(A_1)P(A_2) > 0. \quad (13)$$

Theorem for Addition for Nondisjoint Random Events.   Consider two nondisjoint events A_1 and A_2. For them we may write

$$P(A_1 + A_2) = P(A_1) + P(A_2) - P(A_1 \wedge A_2). \quad (14)$$

For n nondisjoint random events $A_1, A_2, \ldots, A_n$ the formula for addition of probabilities (14) takes the form

$$P\left(\sum_{i=1}^{n} A_i\right) = \sum_{i=1}^{n} P(A_i) - \sum_{i=1}^{n} \sum_{j>i}^{n} P(A_i \wedge A_j)$$

$$+ \sum_{i=1}^{n} \sum_{j>i}^{n} \sum_{k>j}^{n} P(A_i \wedge A_j \wedge A_k) + \cdots + (-1)^{n+1}P\left(\prod_{i=1}^{n} A_i\right). \quad (15)$$

The formula of total probability is a generalization of the formulas for multiplication and addition of probabilities. If the event A_0 may occur only under the condition that some event A_i, of the disjoint events A_1,

$A_2, \ldots, A_n$ with known probabilities, has occurred, and if we know the conditional probabilities $P(A_0 \mid A_i)$ (for all $i = 1, 2, \ldots, n$), then the probability of the event $P(A_0)$ can be computed from the formula for total probability

$$P(A_0) = \sum_{i=1}^{n} P(A_i)P(A_0 \mid A_i). \qquad (16)$$

Indeed

$$A_0 = \sum_{i=1}^{n} (A_0 \wedge A_i).$$

From the rule for addition of probabilities we have

$$P(A_0) = P\left(\sum_{i=1}^{n} (A_0 \wedge A_i)\right) = \sum_{i=1}^{n} P(A_0 \wedge A_i)$$

and, further, from the rule for multiplication of probabilities, each of the terms can be represented in the form

$$P(A_0 \wedge A_i) = P(A_i)P(A_0 \mid A_i),$$

whence we finally obtain

$$P(A) = \sum_{i=1}^{n} P(A_i)P(A_0 \mid A_i).$$

Usually the events $A_1, A_2, \ldots, A_n$, when only the event A_0 can be observed, are called hypotheses relative to A_0.

Formula for Probabilities of Hypotheses (Bayes Formula). Let $A_1, A_2, \ldots, A_n$ be disjoint hypotheses relative to the event A_0. The conditional probability of the hypothesis A_i computed under the assumption that A_0 has occurred is determined by the formula

$$P(A_i \mid A_0) = \frac{P(A_i)P(A_0 \mid A_i)}{\sum_{i=1}^{n} P(A_i)P(A_0 \mid A_i)} . \qquad (17)$$

The distribution function $F(x)$ (cumulative distribution function of a random variable ζ)[2] is the probability of the event $(\zeta \leq x)$, where x is a real number:

$$F(x) = \mathcal{P}(\zeta \leq x). \qquad (18)$$

---

[2] The basic distribution functions used in reliability theory are given in Tables A1.1 and A1.2.

Corollary 7. From the definition of distribution function it follows that

$$F(-\infty) = 0, \qquad F(+\infty) = 1, \qquad 0 \leq F(x) \leq 1. \tag{19}$$

Corollary 8. From the definition of distribution function it follows that

$$P\{x_1 < \zeta \leq x_2\} = F(x_2) - F(x_1). \tag{20}$$

Corollary 9. If a discrete random variable takes on the value x_i from a finite number n of possible values with probability p_i, then the distribution function may be written in the form

$$F(x) = \begin{cases} 0, & x < x_1, \\ p_1, & x_1 \leq x < x_2, \\ p_1 + p_2, & x_2 \leq x < x_3, \\ p_1 + p_2 + p_3, & x_3 \leq x < x_1, \\ \cdots\cdots\cdots\cdots\cdots\cdots\cdots\cdots\cdots\cdots\cdots \\ p_1 + p_2 + \cdots + p_{n-2}, & x_{n-2} \leq x < x_{n-1}, \\ 1 - p_n, & x_{n-1} \leq x < x_n, \\ 1, & x > x_n. \end{cases} \tag{21}$$

Remark. If the number of possible values is $n = 1$; that is, the random variable can take on one and only one value c, then this random variable is said to be degenerate. The distribution function of such a random variable can be written as

$$F(x) = \begin{cases} 0, & x < c, \\ 1, & x \geq c. \end{cases} \tag{22}$$

The probability density of a random variable is the limit of the ratio of the probability that the random variable ζ on a trial takes on a value in the interval $(x, x + \Delta x)$ to the length of the interval Δx as $\Delta x \to 0$:

$$f(x) = \lim_{\Delta x \to 0} \frac{F(x + \Delta x) - F(x)}{\Delta x}. \tag{23}$$

In other words, the probability density is the first derivative of the cumulative distribution function

$$f(x) = \frac{d}{dx} F(x) = F'(x). \tag{24}$$

Corollary 10. From the definition of probability density (21) and Corollary 8, it follows[3] that

$$\mathcal{P}\{a \leq x \leq b\} = \int_a^b f(x)\, dx. \tag{25}$$

---

[3] Strictly speaking, (25) is actually the definition of probability density.

The conditional probability density of a random variable ζ at the point x is the probability density of ζ computed under the condition that the random variable is greater than x.

$$\varphi(x) = \frac{f(x)}{1 - F(x)}. \tag{26}$$

Remark. Clearly conditional densities can also be defined for other conditions; however, this conditional density is actually the most important in reliability theory, since it is the failure intensity of a system when ζ is the time to failure of the system.

The mathematical expectation (mean value) of a random variable is defined as follows:

(a)   for discrete random variables:

$$M(x) = \sum_i x_i \mathcal{P}\{x = x_i\}, \tag{27}$$

(b)   for continuous random variables:

$$M(x) = \int_{-\infty}^{\infty} xf(x)\, dx, \tag{28}$$

(c)   for nonnegative random variables:

$$M(x) = \int_{0}^{\infty} [1 - F(x)]\, dx, \tag{29}$$

if the integrals (28) and (29) exist.

Remark. The mathematical expectation is often called the moment of first order about the origin of the distribution $F(x)$. The moment of nth order about the origin is defined as the quantity

$$M^{(n)}(x) = \int_{-\infty}^{\infty} x^n f(x)\, dx.$$

Corollary 11.   From Formulas (22) and (29) we have

$$M(c) = c. \tag{30}$$

Corollary 12.   From the rule for taking a constant from behind an integral sign it follows that

$$M(cx) = cM(x). \tag{31}$$

Corollary 13.   From the rule for evaluating the integral of a sum of functions the expected value of a sum of any random variables is

$$M\left(\sum_{i=1}^{n} x_i\right) = \sum_{i=1}^{n} M(x_i). \tag{32}$$

Corollary 14. From Corollaries 11 and 13 it follows that

$$M(c + x) = c + M(x). \tag{33}$$

Corollary 15. For the product of independent random variables, it follows from the factorization of the joint distribution function that

$$M\left(\prod_{i=1}^{n} x_i\right) = \prod_{i=1}^{n} M(x_i). \tag{34}$$

The variance of a random variable is defined by the (equivalent) formulas

$$D(x) = M[x - M(x)]^2 \tag{35}$$

or

$$D(x) = M(x^2) - [M(x)]^2. \tag{36}$$

For discrete random variables

$$D(x) = \sum_{i=1}^{n} \mathcal{P}\{x = x_i\}[x_i - M(x)]^2 \tag{37}$$

or

$$D(x) = \sum_{i=1}^{n} \mathcal{P}\{x = x_i\}x_i^2 - \left[\sum_{i=1}^{n} \mathcal{P}\{x = x_i\}x_i\right]^2. \tag{38}$$

For continuous random variables

$$D(x) = \int_{-\infty}^{\infty} f(x)[x - M(x)]^2 \, dx \tag{39}$$

$$D(x) = \int_{-\infty}^{\infty} x^2 f(x) \, dx - \left[\int_{-\infty}^{\infty} x f(x) \, dx\right]^2. \tag{40}$$

Remark. The variance is often called the central moment of second order of the distribution $F(x)$.

Corollary 16. From the definition of the variance and (30) it follows that

$$D(c) = 0. \tag{41}$$

Corollary 17. From (35) and (31) we have

$$D(cx) = c^2 D(x). \tag{42}$$

Corollary 18. From (37) and (41) it follows that

$$D(c + x) = D(x). \tag{43}$$

Corollary 19. For the sum of independent random variables we have

$$D\left(\sum_{i=1}^{n} x_i\right) = \sum_{i=1}^{n} D(x_i). \tag{44}$$

The standard deviation (root-mean-square deviation) is defined by the formula

$$\sigma_x = |\sqrt{D(x)}|. \tag{45}$$

A1.2 BASIC CONCEPTS AND A BRIEF SURVEY OF MATHEMATICAL STATISTICS

A test (or trial) is a realization in practice of some complex of conditions. (The crucial aspect of an experimental trial is its repeatability. The conditions of the trial must be so explicit that anyone with suitable training can perform it properly.)

A realization of a random event is an outcome obtained as the result of making a trial.

A realization of a random variable is a numerical quantity obtained as the result of making a trial.

The frequency (relative frequency) of a random event is defined as follows. If we repeat a trial N times in which it is possible for a certain event A to occur, and this event actually occurs n_N times, then the frequency of occurrence of this event is

$$W_N(A) = \frac{n_N}{N}. \tag{46}$$

The statistical definition of the probability of a random event is made as follows: In all cases of interest it is observed that as the number of trials N increases, the value of $W_N(A)$ becomes more and more stable and appears to approach some number $P(A)$. The probability of a random event can be defined as the limit of $W_N(A)$ as the number of trials N increases without limit:

$$\lim_{N \to \infty} W_N(A) = \lim_{N \to \infty} \frac{n_N}{N} = P(A). \tag{47}$$

Remark. It is said that the quantity W_N converges in probability to the quantity P if for all arbitrarily small positive numbers ϵ and δ we can choose an N such that the probability of satisfying the inequality $|W_N - P| < \epsilon$ becomes greater than $1 - \delta$.

The order statistics of n random variables is the collection of variables being investigated arranged in increasing (nondecreasing) order of magnitude of their realizations:

$$x_1 \leq x_2 \leq \cdots \leq x_n.$$

The empirical distribution function (for n realizations $x_1 \leq x_2 \leq \cdots \leq x_n$ of a given random variable) is the function defined by the equalities

$$F_n(x) = \begin{cases} 0 & \text{for } x \leq x_1, \\ \dfrac{k}{n} & \text{for } x_k \leq x \leq x_{k+1}, \\ 1 & \text{for } x \geq x_n. \end{cases} \tag{48}$$

Glivenko's Theorem. As the number of trials increases without limit, the maximal deviation between the empirical and theoretical distribution functions converges to zero with probability one:

$$\mathcal{P}\{ \lim_{n \to \infty} \max_x |F(x) - F_n(x)| = 0 \} = 1. \tag{49}$$

A point estimate of a parameter φ is a random variable or statistic $\hat{\varphi}$ depending only on the results of trials $(x_1, x_2, \ldots, x_n)$ and known quantities, but not on the unknown parameter.

An unbiased estimate $\hat{\varphi}$ of the parameter φ is an estimate whose mathematical expectation coincides with the estimated parameter.

The estimate $\hat{\varphi}$ is said to be consistent if, as the number of trials increases without limit, it converges to the estimated parameter φ in probability. [Actually consistency refers to a sequence of estimates, $\hat{\varphi}_1(x_1), \ldots, \hat{\varphi}_n(x_1, \ldots, x_n), \ldots$, as is clear from the above.]

An efficient estimate $\hat{\varphi}$ is one characterized as having minimum variance among all unbiased estimates of the parameter φ.

A sufficient estimate (sufficient statistic) is an estimate $\hat{\varphi}$ of the parameter φ such that the conditional distribution of the vector of results of trials $(x_1, x_2, \ldots, x_n)$ given a value of $\hat{\varphi}$ does not depend on the parameter φ. (In a sense, a sufficient statistic contains all of the information about the parameter that is available from the data.)

Remark. Every unbiased sufficient estimate is also at the same time efficient.

A confidence interval is an interval covering the unknown value of the parameter φ being estimated with probability not less than some preassigned value called the confidence coefficient.

Confidence bounds (upper and lower) are end points of the confidence interval.

A two-sided confidence interval for a positive parameter φ with confidence coefficient $1 - \alpha$ is a random interval $[\underline{\varphi}(x), \bar{\varphi}(x)]$ whose end points

$$\underline{\varphi}(x) \leq \varphi \leq \bar{\varphi}(x)$$

depend only on outcomes of trials x and for any $\varphi > 0$

$$\mathcal{P}\{\underline{\varphi}(x) \le \varphi \le \bar{\varphi}(x)\} \ge 1 - \alpha.$$

Upper $[0, \bar{\varphi}(x)]$ and lower $[\underline{\varphi}(x), +\infty]$ one-sided confidence intervals are random intervals for which for any $\varphi > 0$ we have, respectively,

$$\mathcal{P}\{0 < \varphi < \bar{\varphi}(x)\} \ge 1 - \alpha,$$
$$\mathcal{P}\{\underline{\varphi}(x) < \varphi\} \ge 1 - \alpha.$$

The confidence level (confidence probability) is the probability β that the confidence interval does not cover the true parameter value φ. It follows that

$$\beta = \alpha,$$

where $1 - \alpha$ is the confidence coefficient.

The sample mean (arithmetic average) based on n realizations of the random variable is defined as

$$\bar{x} = \frac{1}{n} \sum_{i=1}^{n} x_i.$$

The sample variance based on n realizations of a random variable is the quantity found by the formula

$$s^2 = \frac{1}{n-1} \sum_{i=1}^{n} (x_i - \bar{x})^2.$$

A histogram is a function defined by the equality

$$f_n(x) = \frac{k_x}{n},$$

where k_x is the number of realizations of the random variable lying in the interval $x^{(k)} \le x \le x^{(k+1)}$ and $[x^{(')}, x^{('')}]$, $[x^{(''')}, x^{('''')}]$, ... are appropriately chosen intervals of values.

APPENDIX 2

Some mathematical definitions, formulas, and probability distributions frequently encountered in reliability problems.

A2.1 CONSTANTS

$$e = 2.718282\ldots$$
$$e^{-1} = 0.367879\ldots$$
$$\pi = 3.141593\ldots$$
$$\sqrt{\pi} = 1.772454\ldots$$

A2.2 ELEMENTARY COMBINATORIAL FORMULAS

1. $n! = 1 \cdot 2 \cdot \ldots \cdot n.$
2. $0! = 1.$
3. $(-n)! = 0.$
4. $n! = \left(\dfrac{n}{e}\right)^n \sqrt{2\pi n}\left(1 + \dfrac{1}{12n} + \dfrac{1}{288n^2} + \cdots\right).$ (STIRLING'S FORMULA)
5. $\ln(n!) \approx (n + \tfrac{1}{2})\ln n - n + \ln\sqrt{2\pi}.$
6. $C_n^m = \dfrac{n!}{m!(n-m)!}.$
7. $C_n^1 = n.$
8. $C_n^n = C_n^0 = 1.$

9. $C_n^{n+m} = 0 \ (m \geq 1)$.

10. $C_n^m = C_n^{m-n}$.

11. $\displaystyle\sum_{k=0}^{n} C_n^k = 2^n$.

12. $C_{n+m}^{s} = \displaystyle\sum_{k=0}^{s} C_n^k C_m^{s-k}$.

A2.3 SERIES

1. Arithmetic progression:

$$\sum_{k=0}^{n} (a + kb) = (n + 1)a + c_{n+1}^2 b$$

$$\sum_{k=1}^{n} (a + kb) = na + c_{n+1}^2 b$$

2. Geometric progression:

$$\sum_{k=0}^{n} aq^k = \frac{a(q^{n+1} - 1)}{q - 1}.$$

$$\sum_{k=0}^{\infty} aq^k = \frac{a}{1 - q}.$$

3. Binomial formula:

$$(a + b)^n = \sum_{k=0}^{n} C_n^k a^{n-k} b^k.$$

4. Taylor series (for a function of one variable):

$$f(x) = f(a) + \frac{x - a}{1!} f'(a) + \frac{(x - a)^2}{2!} f''(a) + \cdots$$

$$+ \frac{(x - a)^n}{n!} f^{(n)}(a) + \cdots$$

or

$$f(a + h) = f(a) + \frac{h}{1!} f'(a) + \frac{h^2}{2!} f''(a) + \cdots + \frac{h^n}{n!} f^{(n)}(a) + \cdots.$$

5. $(1 + x)^a = 1 + ax + \dfrac{a(a - 1)}{2!} x^2 + \cdots$

$$+ \dfrac{a(a - 1) \cdots (a - k + 1)}{k!} x^k + \cdots,$$

where a is any number.

6. $\dfrac{1}{(1 + x)^n} = \displaystyle\sum_{k=0}^{\infty} (-1)^k C_n^k x^k.$

7. $\dfrac{1}{(1 - x)^n} = \displaystyle\sum_{k=0}^{\infty} C_n^k x^k.$

8. $\dfrac{1}{1 + x} = 1 - x + x^2 - x^3 + \cdots = \displaystyle\sum_{k=0}^{\infty} (-1)^k x^k.$

9. $\dfrac{1}{1 - x} = 1 + x + x^2 + x^3 + \cdots = \displaystyle\sum_{k=0}^{\infty} x^k.$

10. $e^x = 1 + \dfrac{x}{1!} + \dfrac{x^2}{2!} + \dfrac{x^3}{3!} + \cdots = \displaystyle\sum_{k=0}^{\infty} \dfrac{x^k}{k!}.$

11. $e^{-x} = 1 - \dfrac{x}{1!} + \dfrac{x^2}{2!} - \dfrac{x^3}{3!} + \cdots = \displaystyle\sum_{k=0}^{\infty} (-1)^k \dfrac{x^k}{k!}.$

12. $e^{-x^2} = \displaystyle\sum_{k=0}^{\infty} (-1)^k \dfrac{x^{2k}}{k!}.$

13. $\ln (1 + x) = x - \tfrac{1}{2}x^2 + \tfrac{1}{3}x^3 - \tfrac{1}{4}x^4 + \cdots = \displaystyle\sum_{k=0}^{\infty} (-1)^k \dfrac{x^{k+1}}{k + 1}.$

14. $\ln x = (x - 1) - \tfrac{1}{2}(x - 1)^2 + \tfrac{1}{3}(x - 1)^3 - \cdots$

$$= \displaystyle\sum_{k=0}^{\infty} (-1)^k \dfrac{(x - 1)^{k+1}}{k + 1}.$$

A2.4   INTEGRALS

(a) Indefinite integrals

1. $\displaystyle\int e^{ax} \, dx = \dfrac{1}{a} e^{ax}.$

2. $\displaystyle\int xe^{ax}\,dx = e^{ax}\left(\frac{x}{a} - \frac{1}{a^2}\right).$

3. $\displaystyle\int x^2 e^{ax}\,dx = e^{ax}\left(\frac{x^2}{a} - \frac{x2}{a^2} + \frac{2}{a^3}\right).$

4. $\displaystyle\int x^n e^{ax}\,dx = e^{ax}\left[\frac{x^n}{a} + \sum_{k=1}^{n}(-1)^k \frac{1}{a^{k+1}} \frac{n!}{(n-k)!}\, x^{n-k}\right].$

(b) Definite integrals

5. $\displaystyle\int_{t_1}^{t_2} e^{-ax}\,dx = \frac{1}{a}e^{-at_1}(1 - e^{-a(t_2-t_1)}).$

6. $\displaystyle\int_{0}^{t} e^{-ax}\,dx = \frac{1}{a}(1 - e^{-at}).$

7. $\displaystyle\int_{0}^{\infty} e^{-ax}\,dx = \frac{1}{a}.$

8. $\displaystyle\int_{t_1}^{t_2} xe^{-ax}\,dx = \frac{e^{-at_1}}{a^2}[1 + at_1 - e^{-a(t_2-t_1)}(1 + at_2)].$

9. $\displaystyle\int_{0}^{t} xe^{-ax}\,dx = \frac{1}{a^2}[1 - e^{-at}(1 + at)].$

10. $\displaystyle\int_{0}^{\infty} xe^{-ax}\,dx = \frac{1}{a^2}.$

11. $\displaystyle\int_{0}^{t} x^n e^{-ax}\,dx = \frac{n!}{a^{n+1}} - e^{-at}\sum_{k=0}^{n}\frac{n!}{k!}\cdot\frac{t^k}{a^{n-k+1}}.$

12. $\displaystyle\int_{0}^{\infty} x^n e^{-ax}\,dx = n!a^{-(n+1)}.$

A2.5 LAPLACE TRANSFORMS

The Laplace transform of a function $p(t)$ is a function $a(s)$ defined by the equation

$$a(s) = \int_{0}^{\infty} p(t)e^{-st}\,dt, \qquad t \geq 0.$$

The function $p(t)$ is called the original and $a(s)$ the transform.

Below is a short table of Laplace transforms of basic functions used in reliability computations.

Table A2.1 Laplace Transforms

$p(t)$	$a(s)$
1	$\dfrac{1}{s}$
t	$\dfrac{1}{s^2}$
$bt + c$	$\dfrac{b + cs}{s^2}$
$\frac{1}{2}t^2$	$\dfrac{1}{s^2}$
e^{bt}	$\dfrac{1}{s - b}$
$\dfrac{1}{b} e^{-(t/b)}$	$\dfrac{1}{1 + bs}$
$\dfrac{1}{b} (e^{bt} - 1)$	$\dfrac{1}{s(s - b)}$
$1 - e^{-(t/b)}$	$\dfrac{1}{s(1 + bs)}$
te^{bt}	$\dfrac{1}{(s - b)^2}$
$\dfrac{1}{b^2} te^{-(t/b)}$	$\dfrac{1}{(1 + bs)^2}$
$\dfrac{e^{bt} - e^{ct}}{b - c}$	$\dfrac{1}{(s - b)(s - c)}$
$p'(t)$	$-p(0) + sa(s)$

APPENDIX 3

A3.1 NORMAL DISTRIBUTION FUNCTION

$$\Phi(u) = \frac{1}{\sqrt{2\pi}} \int_{-\infty}^{u} e^{-x^2/2}\, dx.$$

μ	$\Phi(\mu)$	μ	$\Phi(\mu)$	μ	$\Phi(\mu)$	μ	$\Phi(\mu)$	μ	$\Phi(\mu)$
0.00	0.500	1.20	0.885	2.40	0.9^2180	3.60	0.9^3841	4.80	0.9^6207
0.05	0.520	1.25	0.894	2.45	0.9^2286	3.65	0.9^3869	4.85	0.9^6383
0.10	0.540	1.30	0.903	2.50	0.9^2379	3.70	0.9^3892	4.90	0.9^6521
0.15	0.560	1.35	0.911	2.55	0.9^2461	3.75	0.9^4116	4.95	0.9^6629
0.20	0.579	1.40	0.919	2.60	0.9^2534	3.80	0.9^4277	5.00	0.9^6713
0.25	0.599	1.45	0.926	2.65	0.9^2598	3.85	0.9^4409	5.05	0.9^6779
0.30	0.618	1.50	0.933	2.70	0.9^2653	3.90	0.9^4519	5.10	0.9^6830
0.35	0.637	1.55	0.939	2.75	0.9^2702	3.95	0.9^4609	5.15	0.9^6870
0.40	0.655	1.60	0.945	2.80	0.9^2745	4.00	0.9^4683	5.20	0.9^7004
0.45	0.674	1.65	0.951	2.85	0.9^2781	4.05	0.9^4744	5.25	0.9^7240
0.50	0.691	1.70	0.955	2.90	0.9^2813	4.10	0.9^4793	5.30	0.9^7421
0.55	0.709	1.75	0.960	2.95	0.9^2841	4.15	0.9^4834	5.35	0.9^7560
0.60	0.726	1.80	0.964	3.00	0.9^2865	4.20	0.9^4867	5.40	0.9^7667
0.65	0.742	1.85	0.968	3.05	0.9^2886	4.25	0.9^4893	5.45	0.9^7748
0.70	0.758	1.90	0.971	3.10	0.9^3032	4.30	0.9^5146	5.50	0.9^7810
0.75	0.773	1.95	0.974	3.15	0.9^3184	4.35	0.9^5319	5.55	0.9^7857
0.80	0.788	2.00	0.977	3.20	0.9^3313	4.40	0.9^5459	5.60	0.9^7893
0.85	0.802	2.05	0.980	3.25	0.9^3423	4.45	0.9^5501	5.65	0.9^7920
0.90	0.816	2.10	0.982	3.30	0.9^3517	4.50	0.9^5660	5.70	0.9^7940
0.95	0.829	2.15	0.984	3.35	0.9^3596	4.55	0.9^5732	5.75	0.9^7955
1.00	0.841	2.20	0.986	3.40	0.9^3663	4.60	0.9^5789	5.80	0.9^7967
1.05	0.853	2.25	0.988	3.45	0.9^3720	4.65	0.9^5834	5.85	0.9^7975
1.10	0.864	2.30	0.989	3.50	0.9^3767	4.70	0.9^5870	5.90	0.9^7982
1.15	0.875	2.35	0.9^2061	3.55	0.9^3807	4.75	0.9^5898	5.95	0.9^7987

Remarks. 1. For a more complete table see Ya. Yanko, *Mathematical-Statistical Tables* (Gosstatizdat, 1961), p. 109.

2. Exponents indicate the number of nines.

3. For a negative argument u the value $\Phi(u)$ is defined by: $\Phi(-u) = 1 - \Phi(u)$.

A3.2 ORDINATES OF THE NORMAL DISTRIBUTION

$$\varphi(u) = \frac{1}{\sqrt{2\pi}} e^{-u^2/2}.$$

u	$\varphi(u)$	u	$\varphi(u)$	u	$\varphi(u)$	u	$\varphi(u)$	u	$\varphi(u)$
0.00	0.399	0.80	0.290	1.60	0.111	2.40	0.022	3.20	0.0²24
0.05	0.398	0.85	0.278	1.65	0.102	2.45	0.020	3.25	0.0²20
0.10	0.397	0.90	0.266	1.70	0.094	2.50	0.018	3.30	0.0²17
0.15	0.395	0.95	0.254	1.75	0.086	2.55	0.015	3.35	0.0²15
0.20	0.391	1.00	0.242	1.80	0.079	2.60	0.014	3.40	0.0²12
0.25	0.387	1.05	0.230	1.85	0.072	2.65	0.012	3.45	0.0²10
0.30	0.381	1.10	0.218	1.90	0.066	2.70	0.010	3.50	0.0³9
0.35	0.375	1.15	0.206	1.95	0.060	2.75	0.009	3.55	0.0³7
0.40	0.368	1.20	0.194	2.00	0.054	2.80	0.008	3.60	0.0³6
0.45	0.361	1.25	0.183	2.05	0.049	2.85	0.007	3.65	0.0³5
0.50	0.352	1.30	0.171	2.10	0.044	2.90	0.006	3.70	0.0³4
0.55	0.343	1.35	0.160	2.15	0.040	2.95	0.005	3.75	0.0³4
0.60	0.333	1.40	0.150	2.20	0.036	3.00	0.0²44	3.80	0.0³3
0.65	0.323	1.45	0.139	2.25	0.032	3.05	0.0²38	3.85	0.0³2
0.70	0.312	1.50	0.130	2.30	0.028	3.10	0.0²33	3.90	0.0³2
0.75	0.301	1.55	0.120	2.35	0.025	3.15	0.0²28	3.95	0.0³2

Remarks. 1. For a more complete table see Ya. Yanko, *Mathematical-Statistical Tables* (Gosstatizdat, 1961), p. 112.

2. Exponents indicate the number of zeros.

A3.3 EXPONENTIAL DISTRIBUTION FUNCTION

x	e^{-x}	x	e^{-x}	x	e^{-x}	x	e^{-x}	x	e^{-x}	x	e^{-x}
0.001	0.9990	0.029	0.9714	0.057	0.9446	0.085	0.9185	0.23	0.7945	0.60	0.5488
0.002	0.9980	0.030	0.9704	0.058	0.9437	0.086	0.9176	0.24	0.7866	0.65	0.5220
0.003	0.9970	0.031	0.9695	0.059	0.9427	0.087	0.9167	0.25	0.7788	0.70	0.4966
0.004	0.9960	0.032	0.9685	0.060	0.9418	0.088	0.9158	0.26	0.7711	0.75	0.4724
0.005	0.9950	0.033	0.9675	0.061	0.9409	0.089	0.9149	0.27	0.7634	0.80	0.4493
0.006	0.9940	0.034	0.9666	0.062	0.9399	0.090	0.9139	0.29	0.7483	0.85	0.4274
0.007	0.9930	0.035	0.9656	0.063	0.9390	0.091	0.9130	0.30	0.7408	0.90	0.4066
0.008	0.9920	0.036	0.9646	0.064	0.9380	0.092	0.9121	0.31	0.7334	0.95	0.3867
0.009	0.9910	0.037	0.9637	0.065	0.9371	0.093	0.9112	0.32	0.7261	1.00	0.3679
0.010	0.9900	0.038	0.9627	0.066	0.9362	0.094	0.9103	0.33	0.7189	1.05	0.3499
0.011	0.9891	0.039	0.9618	0.067	0.9352	0.095	0.9094	0.34	0.7118	1.15	0.3166
0.012	0.9881	0.040	0.9608	0.068	0.9343	0.096	0.9085	0.35	0.7047	1.20	0.3012
0.013	0.9871	0.041	0.9598	0.069	0.9334	0.097	0.9075	0.36	0.6977	1.25	0.2865
0.014	0.9861	0.042	0.9589	0.070	0.9324	0.098	0.9066	0.37	0.6907	1.30	0.2725
0.015	0.9851	0.043	0.9579	0.071	0.9315	0.099	0.9057	0.38	0.6839	1.35	0.2592
0.016	0.9841	0.044	0.9570	0.072	0.9306	0.100	0.9048	0.39	0.6771	1.40	0.2466
0.017	0.9831	0.045	0.9560	0.073	0.9297	0.11	0.8958	0.40	0.6703	1.45	0.2346
0.018	0.9822	0.046	0.9551	0.074	0.9287	0.12	0.8869	0.41	0.6637	1.50	0.2231
0.019	0.9812	0.047	0.9541	0.075	0.9278	0.13	0.8781	0.42	0.6570	2.00	0.1353
0.020	0.9802	0.048	0.9532	0.076	0.9269	0.14	0.8694	0.43	0.6505	3.00	0.0498
0.021	0.9792	0.049	0.9522	0.077	0.9260	0.15	0.8607	0.44	0.6440	4.00	0.0183
0.022	0.9782	0.050	0.9512	0.078	0.9250	0.16	0.8521	0.45	0.6378	5.00	$0.0^{2}674$
0.023	0.9773	0.051	0.9503	0.079	0.9241	0.17	0.8437	0.46	0.6313	6.00	$0.0^{2}248$
0.024	0.9763	0.052	0.9494	0.080	0.9231	0.18	0.8353	0.47	0.6250	7.00	$0.0^{3}912$
0.025	0.9753	0.053	0.9484	0.081	0.9222	0.19	0.8270	0.48	0.6188	8.00	$0.0^{3}335$
0.026	0.9743	0.054	0.9475	0.082	0.9213	0.20	0.8187	0.49	0.6126	9.00	$0.0^{3}123$
0.027	0.9734	0.055	0.9465	0.083	0.9203	0.21	0.8106	0.50	0.6065		
0.028	0.9724	0.056	0.9456	0.084	0.9194	0.22	0.8025	0.55	0.5769		

Remarks. 1. For a more complete table see I. N. Bronshtein and K. A. Semendyaev, *Mathematics Handbook* (Gostekhizdat, 1953), pp. 52–68. 2. Exponents indicate the number of zeros.

REFERENCES

1. BRONSHTEIN, I. N. AND SEMENDYAEV, K. A., *Mathematics Handbook*, 3d ed. Moscow: GITTL, 1953.
2. GRADSHTEIN, I. S. AND RYZHIK, I. M., *Tables of Integrals, Sums, Series, and Products*. Moscow: Fizmatgiz, 1962.
3. DWIGHT, G. T., *Tables of Integrals and Other Mathematical Formulas*. Moscow: Nauka, 1964.
4. DYOTCH, G., *Guide to Practical Use of the Laplace Transform*. Moscow: Fizmatgiz, 1958.
5. KANTOROVICH, M. I., *Operational Calculus and Nonstationary Phenomena in Electrical Circuits*. Moscow: GITTL, 1953.
6. YANKO, E. AND EMDE, F., *Tables of Functions with Formulas and Curves*. Moscow: GIFML, 1959.

BIBLIOGRAPHY

Listed here are books on reliability theory, probability theory, and mathematical statistics.

BASIC LITERATURE ON RELIABILITY

1. GNEDENKO, B. V., BELYAEV, Y. K., SOLOV'EV, A. D., *Mathematical Methods in Reliability Theory.* New York: Academic Press, Inc., 1968.
2. LLOYD, D. K. AND LIPOW, M., *Reliability: Management, Methods, and Mathematics.* Englewood Cliffs, N.J.: Prentice-Hall, Inc., 1962.
3. POLOVKO, A. M., *Foundations of Reliability Theory.* Moscow: Nauka, 1965.
4. SHISHONOK, N. A., REPKIN, V. F., AND BARVINSKII, L. L. *Foundations of Reliability Theory and Use of Radioelectronic Equipment.* Moscow: Soviet Radio, 1964.
5. SHOR, Y. B., *Statistical Methods for Analyzing Control of Quality and Reliability.* Moscow: Soviet Radio, 1962.

ADDITIONAL LITERATURE ON RELIABILITY

1. ASTAF'EV, A. V., *Environment and Reliability of Radiotechnical Equipment.* Moscow: Energiya, 1965.
2. BAZOVSKY, I., *Reliability: Theory and Practice.* Englewood Cliffs, N.J.: Prentice-Hall, Inc., 1961.
3. BELOV, F. I. AND SOLOVEICHIK, F. S., *Problems in Reliability of Radioelectronic Equipment.* Moscow: Gosenergoizdat, 1961.
4. VASIL'EV, B. V., KOZLOV, B. A., AND TKACHENKO, L. G., *Reliability and Efficiency of Radioelectronic Structures.* Moscow: Soviet Radio, 1964.

379

5. *Problems of Reliability of Radioelectronic Equipment.* Translated from English by I. I. Morozov. Moscow: Soviet Radio, 1959.

6. *Problems of Accuracy and Reliability in Machine Construction.* Collection edited by N. G. Bruevicha. Moscow: AN SSSR, 1962.

7. DRUZHININ, G. V., *Reliability of Automatic Structures.* Moscow: Energiya, 1964.

8. EPIFANOV, A. D., *Reliability of Automatic Systems.* Moscow: Mashinostroenie, 1964.

9. ZIZEMSKII, E. I., *Reliability of Radioelectronic Equipment.* Moscow: Sudpromgiz, 1963.

10. *Cybernetics in the Service of Communism,* vol. 2. Collection edited by I. Berg, N. G. Bruevich, and B. V. Gnedenko. Moscow: Energiya, 1964.

11. KSENZ, S. P., *Search for Defects in Radioelectronic Systems by the Method of Functional Trials.* Moscow: Soviet Radio, 1965.

12. LEBEDEV, A. V., *Reliability and Durability.* Moscow: Profizdat, 1961.

13. LEVIN, B. R., *Theory of Random Processes and Its Application to Radiotechnology,* 2d ed. Moscow: Soviet Radio, 1960.

14. LEONT'EV, L. P., *Introduction to the Theory of Reliability for Radioelectronic Equipment.* Moscow: AN Latv. SSR, 1963.

15. LUTZKII, V. A., *Computing Reliability and Efficiency of Radio-Electronic Equipment.* Moscow: AN SSSR, 1963.

16. MALIKOV, I. M., POLOVKO, A. A., ROMANOV, N. A., AND CHUKREEV, P. A., *Basic Theory and Computation of Reliability,* 2d ed. Moscow: Sudpromgiz, 1960.

17. *Small-size Radio Equipment: Problems of Construction, Production, and Use.* Collection translated from English. Edited by V. I. Siforov. Moscow: Foreign Literature, 1954.

18. MESYTSEV, P. P., *Application of Probability Theory and Mathematical Statistics to Construction and Manufacture of Radio Equipment.* Moscow: Oborongiz, 1958.

19. *Reliability of Surface Radioelectronic Equipment.* Translated from English. Edited by N. M. Shuleikin. Moscow: Soviet Radio,

20. *Reliability of Semiconducting Structures.* Translated from English. Edited by A. A. Maslov. Moscow: Foreign Literature, 1963.

21. *Reliability of Radioelectronic Equipment.* Collected papers. Moscow: Soviet Radio, 1958.

22. *Reliability of Radioelectronic Equipment.* Collected papers. Moscow: Soviet Radio, 1960.

23. NECHIPORENKO, V. I., *Functionally Reliable Electronic Circuits.* Moscow: Gostekhizdat, 1963.

24. PAROL', N. V., *Reliability of Receiving-Amplifying Tubes.* Moscow: Soviet Radio, 1964.

25. *Problems of Reliability of Radioelectronic Equipment.* Translated from English. Edited by B. E. Berdichevskii. Moscow: Oborongiz, 1960.

26. *Repairability of Radioelectronic Equipment.* Collection translated from English. Edited by O. F. Poslavskii. Moscow: Soviet Radio, 1964.

27. SAPOZHNIKOV, R. A., BESSONOV, A. A., AND SHOLOMITZKII, A. G., *Reliability of Automatic Controlling Systems*. Moscow: Vysshaya shkola, 1964.
28. SORIN, Y. M. *Reliability of Radioelectronic Equipment*. Moscow: Gorsenergoizdat, 1961.
29. SORIN, Y. M., AND LEBEDEV, A. V., *Lectures on Reliability*. Moscow: Znanie, 1964.
30. KHINNEI, K., AND WALSH, K., *Radio Components and Their Reliability Problems*. Translated from English. Edited by V. M. Traeev. Moscow: Soviet Radio, 1960.
31. SHCHUKIN, A. M., *Probability Theory and Experimental Determination of Characteristics of Complicated Objects*. Moscow: Gosenergoizdat, 1959.

BASIC LITERATURE ON PROBABILITY THEORY

1. VENTSEL', E. S., *Probability Theory*, 3rd ed. Moscow: Nauka, 1964.
2. GNEDENKO, B. V., *Course in Probability Theory*, 4th ed. Moscow: Nauka, 1965.
3. GNEDENKO, B. V., BELYAEV, Y. K., and SOLOV'EV, A. D., *Mathematical Methods in Reliability Theory*. New York: Academic Press, Inc., 1969.
4. GENDENKO, B. V., AND KHINCHIN, A. Y., *Elementary Introduction to Probability Theory*, 6th ed. Moscow: Nauka, 1964.
5. DUNIN-BARKOVSKII, I. V., AND SMIRNOV, N. V., *Probability Theory and Mathematical Statistics* (general part). Moscow: GTTI, 1955.
6. MESHALKIN, L. D., *Collection of Problems in Probability Theory*. Moscow: MGU, 1964.
7. *Guide for Engineers in Solving Problems in Probability Theory*. Edited by A. A. Sveshnikov. Moscow: Sudpromgiz, 1962.
8. *Collection of Problems in Probability Theory, Mathematical Statistics and the Theory of Random Functions*. Edited by A. A. Sveshnikov. Moscow: Nauka, 1965.
9. FELLER, W., *Introduction to Probability Theory and Its Applications*, 3rd ed., vol. 1. New York: John Wiley & Sons, Inc., 1968.
10. YAGLOM. A. M., AND YAGLOM. I. M., *Probability and Information*. Moscow.

ADDITIONAL LITERATURE ON PROBABILITY THEORY

1. ANGO, A., *Mathematics for Electro- and Radioengineers*. Translated from French. Edited by K. S. Shifrin. Moscow: Nauka, 1964.
2. BARTLETT, M. S., *Introduction to the Theory of Random Processes*. New York: Cambridge University Press, 1955.
3. BOEV, G. P., *Probability Theory*. Moscow: Gostekhizdat, 1956.
4. GILENKO, N. D., *Problem Book on Probability Theory*. Moscow: Uchpedgiz, 1943.

5. GIKHMAN, I. I., AND SKOROKHOD, A. V., *Introduction to the Theory of Random Processes*. Moscow: Nauka, 1965.
6. GLIVENKO, V. I., *Course in Probability Theory*. Moscow: GONTI, 1939.
7. GNEDENKO, B. V., AND KOVALENKO, I. N., *Lectures on Queueing Theory*. Kiev, 1963.
8. KEMENY, J., SNELL, J., AND THOMPSON, J., *Introduction to Finite Mathematics*. Englewood Cliffs, N.J.: Prentice-Hall, Inc., 1957.
9. KORDONSKII, K. B., *Applications of Probability Theory in Engineering*. Moscow: Fizmatgiz, 1963.
10. KOFMAN, A., AND KROON, R., *Queueing* (theory and applications). Translated from French. Edited by I. N. Novalenko. Moscow: Mir, 1965.
11. LEVIN, B. R., *Theory of Random Processes and Its Application to Radio Technology*, 2d ed. Moscow: Soviet Radio, 1960.
12. LEVIN, B. R., *Theoretical Foundations of Statistical Radiotechnology*. Moscow: Soviet Radio, 1966.
13. LANING, J. H., AND BATTIN, R. H., *Random Processes in Automatic Control*. New York: McGraw-Hill, Inc., 1956.
14. MERILL, G., GOLDBERG, H., AND HELMHOLTZ, R., *Operations Research* (armament launching). Princeton, N.J.: D. Van Nostrand Company, Inc., 1956.
15. MORSE, F. M., AND KIMBALL, G. E., *Methods of Operations Research*. New York: John Wiley & Sons, Inc., 1951.
16. MIDDLETON, D., *An Introduction to Statistical Communication Theory*. New York: McGraw-Hill, Inc., 1960.
17. POLETAEV, I. A., *Signal*. Moscow: Soviet Radio, 1958.
18. PUGACHEV, V. S., *Theory of Random Functions and Its Application to Automatic Control Problems*. Moscow: Fizmatgiz, 1960.
19. ROZANOV, Y. A., *Stationary Random Processes*. Moscow: Fizmatgiz, 1963.
20. ROMANOVSKII, V. I., *Discrete Markov Chains*. Moscow: Gostekhizdat, 1949.
21. RUMSHISKII, L. Z., *Elements of Probability Theory*. Moscow: Fizmatgiz, 1960.
22. SAATI, T., *Mathematical Methods of Operations Research*. New York: McGraw-Hill, Inc., 1959.
23. SAATI, T., *Elements of Queueing Theory and Its Applications*. New York: McGraw-Hill, Inc., 1961.
24. SARYMSAKOV, T. A., *Foundations of the Theory of Markov Processes*. Moscow: Gostekhizdat, 1954.
25. SVESHNIKOV, A. A., *Applied Methods of the Theory of Random Functions*. Moscow: Sudpromgiz, 1961.
26. SEDYAKIN, N. M., *Elements of the Theory of Random Impulse Flows*. Moscow: Soviet Radio, 1965.
27. SKOROKHOD, A. V., *Random Processes with Independent Increments*. Moscow: Nauka, 1964.
28. KHINCHIN, A. Y., *Asymptotic Laws of Probability Theory*. Moscow: GTTI, 1936.
29. KHINCHIN, A. Y., *Mathematical Methods of Queueing Theory*, Trudy Math Inst., V. A. Steklov, vol. 49. Moscow: AN SSSR, 1955.

30. KHINCHIN, A. Y., *Works on Mathematical Theory of Queueing*. Moscow: Fizmatgiz, 1963.
31. HOWARD, R. A., *Dynamic Programming and Markov Processes*. Cambridge, Mass.: M.I.T. Press, 1960.

BASIC LITERATURE ON MATHEMATICAL STATISTICS

1. VAN DER WAERDEN, B. L., *Mathematische Statistik*. Berlin: Springer-Verlag, 1957.
2. GNEDENKO, B. V., BELYAEV, Y. K., AND SOLOV'EV, A. D., *Mathematical Methods in Reliability Theory*. New York: Academic Press, Inc., 1968.
3. DUNIN-BARKOVSKII, I. V., AND SMIRNOV, N. V., *Probability Theory and Mathematical Statistics in Technology* (general part). Moscow: Gostekhizdat, 1955.
4. CRAMER, H., *Mathematical Methods of Statistics*. Princeton, N.J.: Princeton University Press, 1946.
5. SMIRNOV, N. V., AND DUNIN-BARKOVSKII, I. V., *Short Course in Mathematical Statistics for Technical Applications*. Moscow: Fizmatgiz, 1959.
6. HALD, A., *Statistical Theory with Engineering Applications*. New York: John Wiley & Sons, Inc., 1952.

ADDITIONAL LITERATURE ON MATHEMATICAL STATISTICS

1. BROWNLEE, K. A., *Statistical Theory and Methodology in Science and Engineering*. New York: John Wiley & Sons, Inc., 1965.
2. BUSLENKO, N. P., *Mathematical Models for Industrial Processes*. Moscow: Nauka, 1964.
3. BUSLENKO, N. P., AND SCHREIDER, Y. A., *Method of Statistical Tests* (Monte-Carlo) *and Its Realizations on Digital Computers*. Moscow: Fizmatgiz, 1961.
4. WALD, A., *Sequential Analysis*. New York: John Wiley & Sons, Inc., 1947.
5. GOLENKO, D. I., *Models and Statistical Analysis of Pseudo-random Numbers on Computers*. Moscow: Nauka, 1965.
6. GOL'DANSKII, V. I., KUTSENKO, A. V., AND PODGORETSKII, M. I., *Statistical Readings in Recording Nuclear Particles*. Moscow: Fizmatgiz, 1959.
7. GUTER, R. S., AND OVCHINSKII, B. V., *Elements of Numerical Analysis and Mathematical Processing of Test Results*. Moscow: Fizmatgiz, 1962.
8. DLIN, A. M., *Mathematical Statistics in Technology*. Moscow: Sovietskaya Nauka, 1958.
9. EZHOV, A. I., *Fitting and Computing of Series of Distributions*. Moscow: Gosstatizdat, 1961.
10. COWDEN, D., *Statistical Methods in Quality Control*. Englewood Cliffs, N.J.: Prentice-Hall, Inc., 1957.
11. KUTAI, A. K., AND KORDONSKII, K. B., *Analysis of Accuracy and Control of Quality in Machine Construction*. Moscow: Mashgiz, 1953.

12. LUKOMSKII, J. I., *Theory of Correlation and Its Application to Analysis of Production.* Moscow: Gosstatizdat, 1958.
13. MADELUNG, E., *Mathematical Tools of Physics.* Translated from German. Edited by V. I. Levin. Moscow: Fizmatgiz, 1960.
14. *Method of Statistical Testing* (Monte-Carlo method). Edited by Y. A. Shreider. Moscow: Fizmatgiz, 1962.
15. MITROPOL'SKII, A. K., *Technology of Statistical Computations.* Moscow: Fizmatgiz, 1961.
16. NALIMOV, V. V., *Applications of Mathematical Statistics in Analysis of Matter.* Moscow: Fizmatgiz, 1962.
17. ROMANOVSKII, V. I., *Application of Mathematical Statistics in Testing.* Moscow: Gostekhizdat, 1947.
18. FISHER, R. A., *Statistical Methods for Research Workers.* New York: Hafner Publishing Company, 1958.
19. CHUPROV, A. A., *Basic Problems of Correlation Theory.* Moscow: Gosstatizdat, 1960.
20. YULE, J. E., AND KENDALL, M. J., *An Introduction to the Theory of Statistics.* New York: Hafner Publishing Company, 1950.
21. YASTREMSKII, B. S., *Some Problems in Mathematical Statistics.* Moscow: Gosstatizdat, 1961.

TABLES

BOL'SHEV, L. N., AND SMIRNOV, N. V., *Tables of Mathematical Statistics.* Moscow: Nauka, 1965.
BRONSHTEIN, I. N., AND SEMENDYAEV, K. A., *Handbook on Mathematics.* Moscow: Fizmatgiz, 1959.
GRADSHTEIN, I. S., AND RYZHIK, I. M., *Tables of Integrals, Sums, Series, and Products.* Moscow: Gostekhizdat, 1948.
SEGAL, B. I., AND SEMENDYAEV, K. A., *Five-Place Mathematical Tables.* Moscow: Fizmatgiz, 1962.
SLUTSKII, E. E., *Tables for Computing the Incomplete γ-Function and Probabilities of χ^2.* Moscow: AN SSSR, 1950.
SMIRNOV, N. V., AND BOL'SHEV, L. N., *Tables for Computing the Two-Dimensional Normal Distribution Function.* Moscow: AN SSSR, 1962.
YANKE, E., AND EMDE, F., *Tables of Functions with Formulas and Curves.* Translated from German. Edited by I. N. Bronshtein. Moscow: Fizmatgiz, 1959.
YANKO, Y., *Mathematical-Statistical Tables.* Translated from Czech. Edited by A. M. Dlin. Moscow: Gosstatizdat, 1961.

INDEX

A

Addition, for nondisjoint random events, 362
of probabilities, 361
Additive efficiency measures, efficiency of systems with, 262

B

Bayes formula, 363
Binomial parameter, confidence bounds for, 335
Branching systems, efficiency for, 268
Bridge circuit, 234

C

Chi-square test, 325
Coefficient, of idleness, 2
of readiness, 3
of reliability, 3
Combinatorial formulas, 370
Component stand-by, 1
universal, 3

Confidence bound(s), for binomial parameter, 335
for normal parameter, 337
lower, for probability of failure-free operation, 352

D

Data, failure, processing of, 313
processing, 311
Density of failure, 1
Detection, of failed units, 309
of improper operation, 310
Discrete failure, 1
Duration of repair, 1

E

Efficiency, 1
of branching system, 268
of multifunctional systems, 265
of short-time action systems, 257
of sustained-action systems, 259
of systems, 256

007